Pittsburgh Series in Social and Labor History

City at the
P·O·I·N·T

*Essays on the
Social History of
Pittsburgh*

Samuel P. Hays

EDITOR

University of Pittsburgh Press

Published by the University of Pittsburgh Press, Pittsburgh, Pa., 15260
Copyright © 1989, University of Pittsburgh Press
All rights reserved
Baker & Taylor International, London
Manufactured in the United States of America
Paperback reprint 1991

Library of Congress Cataloging-in-Publication Data

City at the point : essays on the social history of
Pittsburgh / Samuel P. Hays, editor.
    p.   cm. — (Pittsburgh series in social and labor history)
    Bibliography: p.
    Includes index.
    ISBN 0-8229-3618-6—ISBN 0-8229-5447-8 (pbk.)
    1. Pittsburgh (Pa.)—History—Congresses. 2. Pittsburgh (Pa.)—
Social conditions—Congresses. I. Hays, Samuel P. II. Series.
F159.P657C58   1989
974.8'86—dc19   89-5434
                                                 CIP

▪ Contents

     Illustrations follow page 263

▪ *Foreword*

THIS VOLUME is the first in a series sponsored by the Pittsburgh Center for Social History. It flowed from papers commissioned by the center for a conference on the social history of Pittsburgh and Allegheny County held at the University of Pittsburgh in March 1987. Needless to say, the center was fortunate in being able to persuade Professor Samuel P. Hays to serve as the editor of this volume. He was the logical choice, given his long-standing interest in multidimensional local social history and Pittsburgh itself, and given the fact that four of the obvious candidates for this conference where his former students (Nora Faires, John N. Ingham, Paul Kleppner, and Linda K. Pritchard).

The Pittsburgh Center for Social History was formed in 1985 by the departments of history at the University of Pittsburgh and Carnegie Mellon University to facilitate substantive intellectual exchange among social historians in the Pittsburgh area: to provide a forum for efforts to evaluate and improve the teaching of social history; to define and address specific problems, issues, and concepts related to the field of social history; and to support the development and use of the region's archival resources. It is comprised of over seventy social historians and archivists, organized into four working groups concerned with State and Society, Work and Working-class Culture, Mentalities, and Comparative Economic Development.

The center is grateful to the dean of the Faculty of Arts and Sciences at the University of Pittsburgh and to the Allegheny County Bicentennial Commission for funds enabling us to bring these scholars together and to provide partial subvention of this volume.

Peter Karsten and Peter Stearns, Co-Directors

▪ *Introduction*

THIS BOOK marks a distinctive period in writing about the history of Pittsburgh. Since 1960 a considerable amount of research has been conducted into the city's history by graduate students at Carnegie Mellon and the University of Pittsburgh. As of 1988, twenty-two articles and seven books had appeared; two more books were in the process of being published. Twenty-four dissertations have been completed. But remaining in the files are over two hundred seminar papers which have never seen print. A smaller but significant number of articles and books have been produced by historians from other institutions.

The entire body of work represents a major effort to probe the evolution of a city which until 1960 had not received much scholarly attention. In this book we review what has been written and what has not, how the whole adds up to conclusions about the evolution of the city, and how it fits in the evolving body of the larger history of urbanization.

In the first instance, we focus on the research thus far completed, and especially highlight the work in unpublished seminar papers and doctoral dissertations. Much that is unpublished is publishable. Other papers, although not in themselves publishable, contain material of considerable significance that is worthy of use and synthesis by others. This research on the city of Pittsburgh has been a cooperative venture, though not formally organized as such, in which many have participated; through this volume we hope to provide all involved with a sense of joint contribution to a common effort.

We also seek an assessment of past research. What particular emphases have guided that research and how have those emphases shaped the peculiar ways in which we now view the history of the city? What has not been done and, from the perspective of that assessment, what directions should research now take? The project represented by this book was intended, therefore, to make students and faculty at the two institutions more sensitive to the limitations as well as the possibilities of past work, and to foster some ideas, not necessarily a formal agenda, about the directions research over the next quarter of a century might take.

To sharpen such new directions we have also sought to relate work on the city of Pittsburgh to studies of other cities. The larger field of urban history, it should be emphasized, is not our major objective; extended comparison of the history of Pittsburgh and the history of other cities was not intended. However, we do not wish the city's history to stand in isolation as purely a venture in local affairs; on the contrary we seek to contribute to ideas about urban history as a whole.

We have, therefore, brought some aspects of that larger context into the chapters where they serve as useful examples on which to draw for future research. We have also included two contributions specifically intended to foster larger ideas about urbanization as a whole. One relates the history of Pittsburgh to both the history of urbanization and the history of the entire nation; the other sharpens the meaning of the city's history by comparing it with similar cities in Europe.

This volume does not purport to be a comprehensive history of Pittsburgh. The authors have made little systematic attempt to undertake research independent of the seminar papers or dissertations already completed. Such a project awaits its own author. We have, however, sought to assist such a future task by pointing up the significance of work already done, identifying what yet needs to be done, and providing some ideas about its overall meaning for the process of the city's urbanization.

We write for a mixed audience, but primarily for the informed reader who is intrigued with both Pittsburgh and the history of urbanization in general. We wish to prompt those who are fascinated with the city's history to think beyond the celebration of its past to an understanding of its continuing processes of development. And we wish to provide to professional historians both a useful case study and some ideas about how cities evolve so as to contribute to the larger task of thinking about the general processes of urbanization.

The Authors and the Project

All of the authors of the following chapters have been directly associated with the history departments of the two institutions. Five were doctoral students: Nora Faires, John N. Ingham, Paul Kleppner, and Linda K. Pritchard at the University of Pittsburgh, and Michael P. Weber at Carnegie Mellon. Their chapters focus on the subjects of their dissertations, on which they have continued to specialize.

Nine are faculty members, or in one case a former faculty member, at the two institutions, all of whom have directed graduate work: Lawrence Glasco, Maurine Weiner Greenwald, Edward K.

Muller, Richard Oestreicher, Roy Lubove, and myself at the University of Pittsburgh, and Herrick Chapman, Joel A. Tarr, and Michael P. Weber (now at Duquesne University) at Carnegie Mellon. Their chapters reflect their own specializations in such fields as the history of blacks, women, labor or social welfare; political or regional history; the history of urban communities or city-building; or European urban history.

The entire project was conceived as a joint institutional venture, bringing together the work and talents of former graduate students and current and former faculty in a common effort to focus on both research on the history of Pittsburgh and the contributions of the two departments to that task. At the same time, the project reflects several developments, not immediately obvious, which have made this joint venture possible. One is the Archives of Industrial Society (AIS) at the Hillman Library at the University of Pittsburgh. This archive was established in 1962 as a repository for manuscript material pertaining to Pittsburgh and the surrounding area. Over the years, students have regularly consulted its files for their research: voting records of Allegheny County; business and industry manuscripts; records of voluntary organizations; collections of personal papers such as local representatives to the U.S. Congress; local publications; and the records, including student enrollments, of the city's public schools.

Two groups of manuscript records have been especially useful. One relates to labor, including records of the United Electrical Workers' national union, as well as a considerable number of local and regional organizations. These are housed in the Labor Archives division of the AIS. Another are the ethnic collections, gathered as a result of a major inventory conducted some ten years ago for twelve counties in southwestern Pennsylvania, which resulted in action to microfilm as well as to preserve. Both of these sets of records have played an important role in the research reported here.

Equally helpful has been the computerized data which makes available the U.S. census returns for the city of Pittsburgh for 1850 through 1880, a project conducted under the auspices of the Department of History at the University of Pittsburgh. Formerly in manuscript only, this information has been used by students for any project they might wish. Data about every individual in the city is included for those years. This record has provided a useful basis for training in the use of quantitative data, but, more important, has enabled students to make quick forays into the demography of their subject and then to augment the quantitative data with a wide range of qualitative sources. The theme in graduate instruction has been to combine both types of data and to make them useful within the context of significant conceptual problems.

A major emphasis in the research examined here has been to understand the city's history not just in terms of major events and leading figures, but of its people. There is an effort to bring into the context of history as wide a range of people as lived here, to understand ethnic and religious groups, workers as well as managers, the mass public as well as the elite, the communities as well as the city. We have extended a broad net for exploration and sought to integrate this vast array of parts into some relatively coherent whole. We have clearly done more of the first than of the second, but this volume will constitute a step in the latter direction as well. From either vantage point the tone of the research fostered at both institutions has been to understand Pittsburgh as a social order and not to focus on its more dramatic historical episodes.

The manuscript material collected by AIS has been crucial to the research on which the following chapters are based. But it has a limited time dimension, confined primarily to the years prior to 1920. This is one of the major reasons why the research papers themselves deal primarily with the years from 1850 to 1920. The title, Archives of Industrial Society, was explicitly chosen to focus on the post-1850 period of Pittsburgh history. This review of research about the city has made clear that one of the tasks of the future is to examine the years since 1920. Toward that end a significant expansion of the archival program to the more recent time period will be necessary. The close link between available manuscript resources and the success of research is one of the book's major implications.

The book should also make clear the importance for this project of the range of specialized topical skills represented by the participating faculty. Only a few of the writers are predominantly urban historians. Our historical geographer, Edward K. Muller, is the only one who teaches an urban history course. Yet each of us has specialized in some feature of either society or politics and has often used Pittsburgh as the place in which to work out a specific research problem. All this has provided a varied range of forays into the city's history and, at the same time, has enabled the authors to bring to bear on one city a perspective about the topic in hand that ranges far beyond the city itself. This gives the volume larger relevance than a concern with Pittsburgh alone.

The authors represent two broad emphases in urban history. One concerns the organizational context of the city, for which Paul

Kleppner, Roy Lubove, Edward K. Muller, and Joel A. Tarr provide specialized knowledge; and the other the demography of urban society, which is fundamental to the work of Nora Faires, Laurence Glasco, Maurine Weiner Greenwald, John N. Ingham, Richard Oestreicher, Linda K. Pritchard, and Michael P. Weber. These are two quite different ways of approaching urban history, and much of the insight that can be derived from the volume comes from the conjunction of these two sets of perspectives.

The Center for Social History, established as a joint venture of the history departments at Carnegie Mellon and the University of Pittsburgh in the spring of 1985, took up the project as one of its first major ventures. We selected topics which graduate students had researched most thoroughly and were central to the city's history, identified potential authors, and invited them to participate. All agreed to do so. The authors gathered for a two-day conference in late April 1987 to discuss their first drafts, which previously were circulated. In these extremely useful discussions each participant obtained a clearer sense of the range of individual contributions and their relationship to each other. The papers then were revised over the summer and fall of 1987 and put together for publication shortly thereafter.

It should be emphasized that this is the first book-length product of the Center for Social History. The center has also fostered several ongoing discussion groups that bring together faculty from the two universities in such subjects as work and working-class culture; popular symbols, myths, and ideas (known to historians as mentalities); society and the state; and comparative economic development. Four conferences are now planned for 1988–1990. Some of these , if not all, will lead to further publications. This volume, therefore, is the first of a larger, ongoing joint venture which will continue to coordinate specific projects between the two institutions.

An Assessment of the Research

Students and faculty have concentrated on some subjects rather than others; a retrospective look makes that obvious. There is, for example, the overwhelming emphasis on the years from 1850 to 1920, the iron and steel city, as we have already observed. The years before the 1850s and after 1920 remain relatively unexplored. There are some exceptions, such as Eugene Harper's dissertation on the economy and society of southwestern Pennsylvania in the 1790s or Bruce Stave's book on machine politics in Pittsburgh in the 1930s.

But these still are exceptions, and it is quite clear that a more adequate account of the city's evolution requires research in the years
before and after the iron and steel era.

Equally striking are the subjects about which little has been written. One especially noteworthy omission is the city's economic history. Few studies are available to chart the evolution of the iron and
steel industry from the vantage point of economic history, rather than
its social and political context. The seminar papers and dissertations
skirt this massive subject. Many of them explore the economic aspect
of their subjects, such as ethnicity or labor, but this is no substitute
for a direct analysis of economic development. From this beginning
many other neglected subjects emerge: art and culture, the law and
legal institutions, leisure and recreation, the family, the elderly, the
middle class, higher education, the suburbs and suburbanization,
health and medicine, the family and the history of consumption.

The reasons for several of these omissions might be suggested.
Among the many specialists in social history that the two departments have brought together, no one concentrates on the history of
the family. The predominant context of social analysis tends to be
the community, an emphasis Michael P. Weber's paper underscores.
With different faculty, family studies would have played a larger role
in our volume. Here is a clear case of the impact of faculty specialization on graduate student research.

Even more striking is the absence of work on the history of consumers and consumption. Historians do considerable research on
the "supply side," the history of occupations, industries, and production. But there is a "demand side" as well for which the story has yet
to be told for Pittsburgh as well as for the nation. This could put
more substance on the analysis of inequality emphasized in one of
the concluding chapters, for inequality in occupation and income
translates directly into inequality in the ability to consume. To develop a history of consumption in Pittsburgh is one of the challenges
that lies ahead.

The seminar papers, the dissertations, and the chapters of this
book reveal as even more serious limitation that is conceptual rather
than topical: the absence of a framework of analysis about the character and processes of long-run historical change. Almost all the
graduate student research has a limited time dimension. Some of
this, as we have already emphasized, comes from the time-bound
character of the evidence, such as the census data. But much of the
narrowed focus comes from the nature of graduate study itself, in
which time constraints for the research dictate time constraints in
the definition of the historical problem researched.

Because of this limitation, it falls upon others who would synthe-size the specific studies to provide that missing time dimension. Such is the task of the individual chapters to fellow. But the task is made more difficult because the research papers and dissertations were not designed to highlight change or the processes of change. Hence the chapters in this volume have considerable difficulty in focusing sharply on historical process. Social and political institu-tions are elaborated far more fully than are social processes, and especially those long-run processes of change which ought to be the hallmark of historical analysis.

Especially missing is a focus on the constant tension between things new and things old. America is a society of persistent change, as new facets of the economy, the society, and the political order arise repeatedly, decade after decade, to reshape the historical land-scape, to challenge the legitimacy of the old, and to create tension between the two. That central drama of history is not captured in this book, a limitation that is inherent in the short-term context of most of the research.

Greater success could be achieved on this score by changing the way in which research topics are formulated. One possibility is to place the topics in the context of a century of change. The New Deal, for example, could well be defined in terms of the role it played in the one hundred years that surrounded it, 1885 to 1985, rather than as a response to the more limited event of the Great Depres-sion. Such a perspective would build the context of long-run change into the research in the first place and hence into the minds of those undertaking later synthesis.

Another possibility is generational change. Value changes are implicit in birth and growing up, making a family and earning a living, retirement and growing old; the major breaks in human conti-nuity are those which occur between one generation and the next. Intergenerational family history—genealogy, some would call it—could provide a much needed time dimension to give historical meaning to social history. In this way one could work out the signifi-cance of a wide range of topics in terms of the changes between the grandparents of the families of any one decade and their grandchil-dren fifty years later.

However one might generate such a historical perspective, the chapters here that synthesize work on Pittsburgh call for a more deliberate attempt to organize historical inquiry around the basic problem of history—the processes of long-run change and the ways in which they transform a social order. This is the next large task of a satisfactory history of Pittsburgh and its region.

Toward Comparative Syntheses

The form of this study of urbanization differs from the more traditional mode of urban history. It reflects a cooperative, rather than an individual effort, and a synthesis coming from varied perspectives rather than a single vantage point. The result reflects an unusual amount of historical energy and a common product in which the whole is far greater than the sum of individual efforts. Because the writers place their contributions in the framework of their own larger ideas about their subject, the result constitutes not only an account of the city but a fertile source of ideas about many facets of urbanization.

This is not to say that this approach to the history of Pittsburgh is unique. One need only follow the research on Philadelphia which has been promoted, over roughly the same period of time, by historians at both Temple University and the University of Pennsylvania. Each of those ventures has had its own distinctive form, representing a variety of energies devoted to different ways of tackling the history of a city. The parallel but cooperative interinstitutional activity reflected in this book, however, suggests a model in which research about other cities could be brought together to evaluate the results and chart the future.

We are especially intrigued with the possibility of combining in a single forum work that compares research about one city with that about another. In the summary chapters in this book we outline the way in which Pittsburgh reflects aspects of urbanization which are, on the one hand unique, and on the other, common to large cities. At this time such ideas are little more than hypotheses amid the limited possibilities for structured comparative analyses. But historical work about urbanization in the United States has reached the point where such comparative thought would be highly productive. We look upon this book as a step in that direction.

Samuel P. Hays

City at the Point

Immigrants and Industry: Peopling the "Iron City"

NORA FAIRES

Folk hero Joe Magarac, the towering, brawny, immigrant steelworker who bent red-hot iron in his bare hands, seems a fitting cultural icon for the Pittsburgh region. Magarac symbolizes the working people who streamed from abroad to forge the metal upon which Pittsburgh's industrial fame rested. Like most of those immigrants, little is known of Joe Magarac's origins. Indeed, even his nativity is contested: not distinctly Serbian or Croatian, Polish or Hungarian, he seems instead an amalgam of nationalities, a composite image of the "Slavic" peasant who left his village in Southern or Eastern Europe to search "for bread," finding a home in the mighty factories of Pittsburgh during the industrial heyday of the "Steel City," the late nineteenth and early twentieth centuries.[1]

Coincidentally, just as the mythic Joe Magarac appeared in local folklore, a team of social investigators also published the results of their years-long study of the conditions of life in the Pittsburgh region's workshops, factories, and homes. The Pittsburgh Survey, a six-volume study completed in 1914, depicted the massive furnaces where men like Magarac made steel, the steam-filled laundries where women ironed sheets, and the fetid sweatshops where families fashioned stogies, offering glimpses into the lives of the region's working people. In *Homestead*, a volume of the Survey, dramatic pictures of the mills that lined the banks of the Monongahela River give way to even more compelling photographs of those who lived in their bleak shadows. One depicts a well-scrubbed native Pennsylvanian worker, but more common are the faces of immigrants—an aging, dignified German; a younger, determined "Slav"; and the hesitant, smiling faces of women and children, unidentified as to their nativity, but clearly new arrivals not only to Homestead but to the United States.[2]

None of these Homestead faces is triumphant; all show that

3

plenty and beauty are scarce in the new land. Joe Magarac, the immigrant worker who conquers the labor of the mill and transcends toil,
is myth. Rather than workers bending the job to their will, industry
typically dominated the lives of immigrants and their children. Yet
the powerful image of Joe Magarac endures, along with the portraits
of Homestead's residents. They underscore a central feature of the
history of the Pittsburgh region: the intimate tie between the
growth of industry and the onset of mass immigration.

The Pittsburgh Survey bequeathed a unique and impressive body
of material on the region and its immigrants to those who would
continue the exploration of "the conditions under which working
people live and labor in a great industrial city."[3] Unfortunately, few
studies appeared to build on this legacy. Not until the 1960s did the
region and its inhabitants—immigrant and native-born, nonwhite
and white, female and male—become once again the subject of intense research. This essay reviews some of the recent work written
on immigration to Pittsburgh, noting some of the major published
works but concentrating on unpublished material and publications
in regional journals, the bulk of which has been written by historians and other social scientists at the University of Pittsburgh and
Carnegie Mellon University. Taking stock of this research offers an
opportunity both to reflect on the saga of immigration to the Pittsburgh region and to consider this research in the light of broader
trends in the field of immigration and ethnic history.[4]

Early Years: Settlement Before 1830

From the perspective of the indigenous peoples who had for centuries made the headwaters of the Ohio River their home, the entry
of whites—first French, then British—into Western Pennsylvania
was a story of the ascent of foreign newcomers to dominance. But
from the perspective of the victors, immigration to Pittsburgh did
not begin until non-English settlers began trickling into the area
surrounding Fort Pitt. Like other trans-Appalachian settlements in
the new nation, Pittsburgh drew migrants from eastern states and
from abroad. Some of these foreign settlers were Welsh, others were
Irish or Scots-Irish, and still others were German, Swiss, and French.
Apart from studies of those settlers and their descendants who comprised the first Pittsburgh elite, we know little of the lives of early
immigrants and their offspring, although we do know that German
Protestants founded Pittsburgh's first church and that by 1808 the
growing numbers of Irish Roman Catholics in the area justified the
creation of a Catholic parish in the city.[5]

As Pittsburgh developed from a frontier outpost to a commercial center, the population of the town rose from several hundred in 1790 to over twelve thousand by 1830; in this same period, the settlement of Allegheny, situated across the Allegheny River from the Pittsburgh "Point," grew to nearly three thousand. During the early years a few industrial enterprises were also founded in the Pittsburgh region, but their limited success little hinted at the city's destiny as the "workshop of the nation." Pittsburgh's first iron foundry, for example, established by Alsatian native George Anshutz in 1793, failed when transportation costs for this heavy product proved too high; other industries were also hampered by poor access to markets. By the 1820s, however, improvements in transportation began to lower the costs of shipping at the same time that the application of steam power became more widespread. As Pittsburgh lost its lead in interregional trade to cities farther west like Cincinnati and St. Louis, fledgling industrialists, following in Anshutz's footsteps, set up iron mills to exploit the rich coal and ore resources of the region. These enterprises flourished, laying the foundation for a century of industrial growth.

Industry Takes Hold: Immigration to Pittsburgh, 1830–1880

The first flood of immigrant newcomers to Pittsburgh coincided with the city's transition from a trading to a manufacturing center. The torrent of foreigners who poured into the region after 1830 stemmed mostly from the lands represented among the city's pioneers. Yet their numbers were so much greater, their backgrounds so different, and, all too often from the point of view of many established residents, their beliefs and practices so at odds with prevailing customs, that their appearance drastically altered the society of the region. Along with a continued influx of native-born migrants to the city, the growing ranks of the foreign-born helped increase Pittsburgh's population to nearly forty-seven thousand in 1850 and to more than one hundred fifty-six thousand in 1880; neighboring Allegheny had a similarly rapid rate of growth, increasing from twenty-one thousand to seventy-eight thousand inhabitants in these thirty years.

Foremost among Pittsburgh's immigrants were the Irish. According to a detailed study by Victor Walsh, more than ten thousand Irish immigrants had settled in the "Iron City" by 1850, constituting 21.4 percent of the city's population.[6] Few of these newcomers resembled the impoverished, famine-fleeing peasants depicted in Oscar Hand-

lin's classic study of Boston;[7] instead, most came from areas of Ireland
that were economically more advanced than those ravaged by the
potato rot—farming areas of larger plots and less tenancy, and industrializing counties with burgeoning towns. Irish Catholics settled in
three enclaves in Pittsburgh: a diverse and relatively prosperous
neighborhood on the edge of the central business district and surrounding St. Paul's Cathedral, the seat of the Catholic diocese of
Pittsburgh; a poorer, heavily Celtic settlement on the Hill, filled with
single male day laborers; and a third enclave in the lower Strip district, where ironworks and glass factories crowded along the banks of
the Allegheny River, and where many Irish industrial workers lived
amidst English, Scottish, and Welsh newcomers and larger numbers
of Germans.[8] In all three neighborhoods the middle-class, or "lace-
curtain," Irish Catholics dominated the group's associational life,
acting as church trustees, serving in local government, and champion-
ing the cause of total abstinence—a cause led by the most prominent
Irishman of the city, Bishop Michael O'Connor. In contrast, not until
the 1870s were segments of the Irish working class drawn into activities beyond the parish. Contributing to a growing literature on Irish
nationalism, Walsh demonstrates that English-speaking Irish ironworkers, especially those in the heavily industrial Woods Run district
of Allegheny, more ardently supported the cause than did Gaelic
speakers, while the lace-curtain Irish channeled their efforts into the
land reform movement and away from more radical programs for
social change in their homeland.[9]

Complementing Walsh's research on the Irish of Woods Run,
John Bennett analyzes the process of class formation in this dense
settlement during the 1870s and 1880s.[10] In Woods Run small numbers of native-born, English, and German workers peppered the work
force of the iron mill, while Welsh immigrants, many of them Protestants, held most of the highly skilled jobs, and Irish Catholic
immigrants—the subjects of Walsh's study—filled most of the unskilled jobs. Residence reinforced lines of ethnic and religious cleavage in the mill: the Welsh clustered on one side of the main street,
the Germans on the other, and the Irish gathered in still another
section of the small district. Nevertheless, the skilled Welsh Protestants and unskilled Irish Catholics joined ranks to make Woods Run
a stronghold of labor solidarity for several decades. Significantly, the
atmosphere of class consciousness in this community also apparently nurtured Irish patriotism.[11]

The efforts of the skilled Welsh workers were vital in forging this
cohesive community. Regrettably, aside from Bennett's portrait of
the Welsh in Woods Run and in Johnstown (a mill town seventy

miles east of Pittsburgh), little is known about this small, but seemingly central, group of immigrants. Similarly, English and Scottish immigrants to the Pittsburgh region have attracted little scholarly attention, despite suggestions that, because of their key positions in manufacturing and their traditions of labor activism before migration, they played a crucial role in the development of the region's industry.[12]

Substantially more is known about German immigrants, the second largest segment of industrializing Pittsburgh's foreign-born population. As I detail elsewhere, these immigrants comprised between 10 and 15 percent of the residents of both Pittsburgh and Allegheny from 1850 to 1870.[13] The arrival of thousands of these distinctly foreign newcomers, few of whom spoke English, dramatized the growing ethnic diversity of both cities. In the eastern wards of Allegheny, the region's first recognizable ethnic enclave grew up. Known to non-Germans as Dutchtown, this neighborhood supported a collection of businesses, including bakeries, butchers' shops, and beer gardens, which catered to German customers. Dutchtown inaugurated a pattern that would be replicated in subsequent ethnic settlements in the city: the businesses established by middle-class immigrants—those outside of industry—became, in conjunction with the religious and secular organizations that dotted these commercial areas, the most visible symbols of immigrant settlements.

Withal Dutchtown did not house the majority of the cities' German immigrants. Germans also resided elsewhere in Allegheny and clustered in several Pittsburgh neighborhoods, notably the lower Strip district and Lawrenceville, both lying along the Allegheny River, and the Uptown district, which stretched along the banks of the Monongahela.[14] Moreover, most Germans did not own, or even work in, the stores which lined Dutchtown's streets. Behind the homogeneous, middle-class image that commercial Dutchtown conveyed lay great diversity among German immigrants. Perhaps more than any other nationality to settle in the cities, Germans were a fragmented immigrant group, divided by regional origin, class, religion, and political outlook.[15] A small number of Germans attained considerable wealth and some power,[16] but these well-to-do Germans, themselves splintered along religious lines, were far removed in property and standing from the German majority. At mid-century, few Germans in either city owned any property; most German men held manual jobs, with many employed at craft work and others toiling at strenuous, unskilled labor, while many German women worked as domestic servants or laundresses.[17] Economically, German immigrants generally resembled the Irish in 1850; twenty years later, in contrast to the Irish, more Germans had left unskilled labor and had acquired property.[18] Mirroring their counterparts in other American cities, Pittsburgh and Allegheny Germans concentrated heavily in the shoemaking, tailoring, butchering, and baking trades, and in a business commonly associated with the immigrant group, brewing lager beer.[19]

Pittsburgh Germans quickly founded an array of social organizations ranging from fraternal associations and reading societies to labor unions and "oom-pah" bands. In the spiritual realm as well, German institutions proliferated. German Christians founded churches which spanned the spectrum from the liturgical traditions of Catholicism and Lutheranism to the evangelical fervor of Methodism. Disagreeing not only on theology but on many leading social questions, such as the abolition of slavery and the value of public schooling, the members of these churches also differed in regional background and class.[20] As Paul Kleppner's work demonstrates, religious differences among Germans had political consequences.[21] In the 1850s, nativism and anti-Catholicism flourished in Pittsburgh, with the city experiencing both the national brand, in the form of Know-Nothingism, and a homegrown version, in the person of one-time mayor Joe Barker, elected as the "People's and Anti-Catholic" candidate.[22] German Protestants and Catholics responded differently to this assault and to the emergence of the Republican party. In the crucial election of 1860, German Protestants, identifying the Republican party with anti-Catholicism more than nativism, tended to support Lincoln's candidacy while German Catholics, with similar perceptions, clung to the Democratic party.[23] On another political issue, temperance, Pittsburgh's German Catholics and most German Protestants both denounced, albeit separately, antialcohol agitation.[24]

Like German Christians, the region's German Jews also established rival congregations.[25] The earliest were Rodef Shalom, which embraced a Reform religious orientation and whose members came mostly from southwestern German provinces, and Tree of Life, which defended the Orthodox tradition and drew its followers from eastern provinces.[26] In the 1840s newly arriving German Jews settled in Pittsburgh's thickly populated central business district and in Dutchtown, where they established societies and organizations, including the Concordia Club, the cornerstone of German Jewish social life. By 1880, a new area of settlement, on the west side of Allegheny in the Manchester district, sheltered many German Jewish immigrants. Twenty years later, most German Jews had abandoned Allegheny City altogether, probably in response to the growth of the iron mills in nearby Woods Run, and moved to more middle-class neighborhoods in Pittsburgh's eastern wards. True to the differences in their membership, the Reformed and Conservative congregations followed their adherents eastward, but established new quarters in different neighborhoods: in 1906 Rodef Shalom selected a site at the intersection of the three main areas of new settlement, Oakland, Shadyside, and Squirrel Hill, while Tree of Life, which had come to serve both German and Polish Jews, moved to a less central (and less affluent) location in lower Oakland the following year.[27]

In the decade preceding Rodef Shalom's removal to the East End, a group of women in the congregation had formed the Columbian Council, a society to promote education, practice philanthropy, and combat discrimination. Most of the council members were the daughters of German immigrants; the mothers of some had been active in the Hebrew Ladies' Aid Society, a congregational relief organization formed during the Civil War.[28] The goals and activities of the Columbian Council no doubt resembled those of many religious and secular societies established by immigrant and second-generation women in the Pittsburgh region. As yet, however, we know very little about such groups, even though it is apparent that women's organizations were central to the life of ethnic settlements.[29]

The dramatic increase in the migration of Jews from Southern and Eastern Europe to the Pittsburgh region prompted the organization of the Columbian Council. In founding a society to help these often impoverished arrivals, the council members bore witness to the tremendous reordering of Pittsburgh's economy and demography now underway. Organizing an anarchist circle in Allegheny, other German Jews came to conclusions different from those the women of Rodef Shalom had reached. They believed that the appropriate response to the world they saw around them was not philanthropy but the radical restructuring of society.[30] The most famous member of this group, Alexander Berkman, sought to dramatize the radicals' program by assassinating industrialist Henry Clay Frick; instead, his failed attempt on the steel magnate's life, coupled with Frick's defeat of the Homestead strikers in 1892, symbolized the triumph of a more vigorous and militant capitalism in the region.

The Heyday of Industry: Immigration to Pittsburgh, 1880–1930

The pace of industrialization accelerated in the Pittsburgh region in the late nineteenth century as a new phase of capitalist expansion began. Few cities, indeed, matched Pittsburgh's economic growth in this era, particularly during the boom years of the region's industry, the three decades preceding World War I. Between 1880 and 1930, the population of the city nearly tripled, soaring from two hundred thirty-five thousand to six hundred seventy thousand, with the greatest gains occurring in the 1890s and 1920s.[31] Immigrants accounted for much of this increase. During these five decades, foreign-born men and women streamed into the city and to the scores of mill towns that sprouted along the banks of the Ohio River and its tributaries. Of Pittsburgh's more than three hundred forty thousand inhabitants in 1890, nearly one hundred thousand were immigrants (28.9 percent). Adding those of "foreign stock" (American-born persons of foreign parentage) to this number indicates more clearly the impact of immigration on the city's growth: 65.9 percent of Pittsburgh's residents were either immigrants themselves or the sons or daughters of immigrants. Twenty years later much the same picture prevailed: 26.4 percent of Pittsburgh's five hundred thirty thousand residents were foreign-born and 35.8 percent were of foreign parentage; hence, 62.2 percent were either first- or second-generation Americans. By 1930, the immigrant share of the city's population had dropped to 19.4 percent, but those of foreign-born parentage comprised 34.7 percent of the city's residents. In all, immigrants and their children constituted between half and two-thirds of the city's residents between 1880 and 1930.[32]

The influx of these thousands of newcomers, most of whom came from Southern and Eastern Europe or from the Mediterranean, followed the unprecedented expansion of the city's industry, especially steel. This expansion created new unskilled and semiskilled jobs increasingly filled by immigrants. In turn, the massive infusion of foreign-born workers transformed the working class in Pittsburgh. More than at any time before or since, the city's working class became an ethnic mosaic.[33]

It is this Pittsburgh that caught the attention of the Russell Sage Foundation, prompting it to send a team of researchers to investigate conditions in the Steel City, and it is this Pittsburgh—and these immigrants—who were the subjects of the resulting Pittsburgh Survey. Not coincidentally, it is this era of Pittsburgh's immigration which has drawn most contemporary scholarly attention.

Two of the largest immigrant groups to settle in Pittsburgh during these years were Poles and Italians. Small numbers of each nationality had arrived before 1880, but their ranks swelled in the ensuing decades: by 1910 nearly twenty-one thousand Poles and more than fourteen thousand Italians resided in the city. In *Lives of Their Own*, John Bodnar, Roger Simon, and Michael P. Weber compare the experiences of these two immigrant groups with that of the city's blacks, who numbered more than twenty-six thousand in 1910.[34] These authors paint a rich portrait of the three groups, detailing their backgrounds and expectations upon arrival, tracing the community life they built, surveying the occupations they held, assessing their rates of residential persistence and levels of homeownership, and documenting the role that kin networks played in easing adjustment and securing jobs. According to this study, immigrants, both Polish and Italian, enjoyed a decisive advantage over black newcomers in obtaining work in Pittsburgh's expanding iron and steel industry. Moreover, once having gained jobs in industry, many Poles and Italians helped other co-ethnics secure work in the mills, while blacks, largely blocked from industrial employment, opened small businesses or, more often, took poorer paying service jobs. While blacks were more likely to send their children to schools than were either of the immigrant groups, the more economically successful Poles and Italians were better able to establish identifiable enclaves in the city and, within those enclaves, to buy homes. By the turn of the century, Italians had clustered in the Bloomfield and East Liberty sections of Pittsburgh, and Poles had settled in dense pockets in Lawrenceville, the South Side, and "Polish Hill," the bluff overlooking the Strip district. Whether living here or in the poorer areas of the city where most blacks dwelled, members of all three groups faced tremendous barriers: for Polish, Italian, and black newcomers life in the Steel City was hard. The ability to find and keep steady employment determined both individual and family welfare. The lives of immigrants and industry—or rather, the economy—were intertwined.[35]

The fate of ethnic institutions further demonstrates this connection, for both religious and secular organizations depended on their members' financial resources.[36] St. Stanislaus Kostka Church, for instance, organized in 1875 as the first Polish Catholic parish in Pittsburgh, grew dramatically as immigrants from Poland found work in the mills of the Strip district. In the next half century, the church established a school and numerous parish societies, adorned its structure with frescoes and elaborate furnishings, and erected a convent and a rectory. Then, during the 1930s, rocked by the economic crisis of the Depression and by a sharp decline in the area's Polish population, the church building itself was severely damaged—first by the disastrous flood that swept through Pittsburgh's low-lying districts on St. Patrick's Day, 1936; later that year, by the explosion of the nearby Pittsburgh Banana Company, which shook the foundations of the edifice.[37]

Prior to these depradations, St. Stanislaus Kostka served as the center of a vibrant ethnic enclave. Largely neglected by the Irish-dominated Roman Catholic hierarchy, St. Stanislaus Kostka, like other Pittsburgh Polish parishes, pursued its own course. Fighting off both secular and religious challenges to its primacy in the ethnic settlement, it mobilized against the rival Polish National Catholic church and sought to short-circuit the appeal of Polish nationalist organizations by demonstrating its own support for the homeland. During World War I, for example, it became a recruiting center for parishioners who wished to fight in France for the liberation of Poland. In this effort the church cooperated with a fraternal society which also played an important role in the life of the ethnic settlement, the Polish Falcons.[38]

The Polish Falcons were only one of a multitude of fraternal societies that flourished in Pittsburgh's ethnic enclaves. Finns, Czechs, Hungarians, Serbs, Croatians, and other immigrants established organizations to protect themselves and their families from the dire financial consequences of illness or industrial accident, both common occurrences in the unsanitary neighborhoods and dangerous factories where many immigrants lived and worked.[39] Over the years, many of these mutual aid organizations took on social functions, sponsoring dances, picnics, and other gatherings. A study by June Granatir Alexander demonstrates this evolution in Slovak fraternal societies.[40]

According to Alexander, Pittsburgh Slovaks organized their first fraternal lodge in 1889. As the number of Slovaks in the region grew from several hundred in 1890 to nearly three thousand in 1910, Slovak fraternals proliferated; by 1915 there were nearly three dozen lodges in Pittsburgh. This network of fraternal lodges helped to knit together the dispersed pockets of Slovak settlement in Frankstown, Woods Run, the Hill, the South Side, and other sections of the city. Moreover, several of these societies banded together to organize the first Slovak Catholic parish in Pittsburgh, while others supported the establishment of subsequent Catholic parishes. Another fraternal lodge, composed of Protestant Slovaks from the county of Liptov in northern Hungary, raised sufficient funds to found a Slovak Lutheran church. These churches, Alexander contends, in turn fostered a sense of "ethnic community" among Slovak immigrants and their children, promoting both a nostalgic connection to the homeland and adjustment to America.[41]

For many native-born Americans, the adjustment promoted by such immigrant churches seemed neither sufficiently rapid nor thorough. The reaction of some of these native-born Americans was to deprecate the customs of immigrant newcomers and to use the arm of the state to enforce social norms they believed more fitting. The *Presbyterian Banner*, for instance, a local publication, applauded the use of constables in breaking up what it regarded as unruly immigrant wedding celebrations and fought strenuously for the maintenance of Sabbatarian laws and the enactment of temperance legislation. At the same time, some Presbyterian churches took a different approach, launching missions to wean the foreign-born from their "alien" ideologies and bring them into the Presbyterian fold.[42] The planners of Vandergrift, a "model" industrial town established upstream from Pittsburgh on a tributary of the Allegheny River, adopted still another approach to the "problem" of the growing numbers of immigrants in the region: residential segregation. The many immigrants from Southern and Eastern Europe who labored in Vandergrift's industry were virtually excluded from living there; instead, they lived outside of town in neighborhoods dubbed "Hunkeyvilles."[43]

The isolation of the foreign- from the native-born was more extreme and certainly more conscious in Vandergrift than elsewhere. But throughout the region, few outside the ethnic enclave were familiar with the conditions of daily life common to most immigrant families. In particular, the lives of immigrant women remained beyond the purview of most observers. Oral histories of immigrant women and their daughters conducted in the 1970s provide some insight into these womens' lives, testifying to the contributions they made to their families' welfare.[44] Pittsburgh's Italian women typically helped their families by becoming thrifty and enterprising consumers, keeping boarders, or working in such family businesses as restaurants, barber shops, and landscaping firms. In addition to helping their families in these ways, Bulgarian, Polish, Serbian, and Slovak women often worked outside the home, supplementing the family's economy with wages from domestic service or factory labor.[45] The household budgets kept by Homestead women at the turn of the century show how crucial to a family's fortunes were women's contributions, both paid and unpaid.[46] In the case of taking in boarders, these contributions were made only through great exertion. It was not uncommon in heavily industrial neighborhoods for a woman to cook and clean for a half dozen single men in addition to her family. This is just the situation described in Thomas Bell's largely autobiographical novel, *Out of This Furnace*. Mary Dobrejcak, the daughter of a Slovak immigrant, keeps house for her husband, a Slovak steelworker, her three children, and six recent arrivals to Braddock. Her work day begins at four-thirty and ends at nine at night; no day off or holiday alters this

Sunday through Saturday routine. With no running water and plenty of soot in the air, the laundry days of women like Mary Dobrejcak were especially arduous.[47]

The hard work of men, of course, was typically done outside the home, in the mills and mines of the region's industry. Several studies of labor disputes document the climate of prejudice in which many immigrants labored and demonstrate a tactic which employers often used to control their labor force: pitting one group of workers against another. Herbert Gutman, for instance, analyzes a strike in 1874 among miners in Buena Vista, a small coal-mining town twenty-five miles southeast of Pittsburgh. In order to break the strike, the mine operators brought in some two hundred Italian immigrants to work as scabs. The result was tragic: the coal operators armed the strikebreakers, the strikers armed themselves, and, in an ensuing gunfight, three Italians were killed. Almost all the actors in the Buena Vista dispute were men, for mining was a male preserve. But one incident in this drama offers stirring testimony regarding both the role of Italian women in the immigrant group and the general separation between the sexes which prevailed in mining towns. Just after the third worker was shot two young Italian women strode into the open. Suddenly, the gunfire ceased. Moments before, the Italian workers had waved a white flag, but the crowd had continued to fire on them. The unexpected action of the two women brought the battle to an end.[48]

Immigrants played important roles in other equally dramatic and more well known labor disputes, including the major battles of 1892, 1919, and 1937 in the steel industry. The leaders of the momentous Homestead strike were native-born and second-generation Irish, German, and British workers. But the vitality of the strike depended on the support of Homestead's residents, most of them unskilled immigrants from Southern and Eastern Europe. The crushing of the workers by Carnegie and Frick resulted from their use of state power to overcome local resistance, not the severing of relations between skilled and unskilled, native- and foreign-born.[49] In contrast, immigrants from Southern and Eastern Europe were among the organizers of the effort to unionize the steel mills in 1919.[50] By 1937, the task of bringing unionism back to the region fell to these immigrants and their children. They made up the bulk of those organized into the newly founded industrial union, the Steel Workers Organizing Committee of the Congress of Industrial Organizations.[51] In the fictional *Out of This Furnace*, Johnny Dobrejcak, known to his workmates as Dobie, is drawn into the union fight; eventually this second-generation Slovak becomes a leader in the organizing drive.[52] The real-life counterparts of Dobie came from steel towns all over the region. Ewa Morawska provides the fullest exploration of the experience of immigrant and second-generation workers in one such city, Johnstown, during the half-century following 1890. She traces the "life-worlds" of eight east central European immigrant groups—Slovaks, Hungarians, Croatians, Serbs, Slovenes, Poles, Ukrainians, and Rusyns—in this citadel of employer's power from the years of labor's defeat to its era of triumph.[53]

Not all those immigrants involved in labor disputes were white; indeed, employers promoted racial as well as ethnic antagonisms to undermine workers' solidarity. The first Chinese to arrive in the area came as contract laborers to a cutlery works in Beaver Falls in 1872; regarded as a tractable labor force, they replaced native-born workers whom management saw as more unruly and strike-prone. By the end of the nineteenth century, several hundred Chinese had settled in the city of Pittsburgh, with few remaining in industrial employment. Most were natives of two counties in Kuangtung province, and virtually all were male. The profile of Pittsburgh's Chinese settlement mirrored that of other cities and resulted from the impact of the Chinese Exclusion Act of 1882, which barred further immigration of Chinese laborers and forbade immigration of wives of current Chinese settlers in America. Consequently, very few Chinese immigrants married, and new arrivals to the community were mostly the "sons"—either in fact or on paper—of those who had settled in the city by 1882. Discrimination against the group fostered the creation of a dense ethnic enclave, the "Chinatown" district near Second Avenue and Grant Street in downtown Pittsburgh, and severely restricted these immigrants' choice of employment. The group's concentration in a few occupations was extreme: the region housed only 435 Chinese residents in 1930 but boasted 185 Chinese laundries and restaurants.[54]

In the early years Greek immigration resembled that of the Chinese. Like the Chinese, many Greeks came to Pittsburgh as links in a chain of migration which stretched from a few villages in their homeland to the bustling Steel City thousands of miles away; most of the first Greek immigrants were men; and many of these men opened up shops, particularly restaurants. These similarities waned quickly. Prior to restrictions on immigration imposed in the 1920s, substantial numbers of Greek women, unlike their Chinese counterparts, could and did join men in the city's Greek settlement. Furthermore, the businesses that Greeks set up provided them with the possibility of upward mobility, an opportunity largely denied to the

Chinese. With both families and businesses established, Greek immigrants were able to organize a broad network of religious and
secular organizations in the city.[55]

Pittsburgh became the home of other migrants from the Mediterranean as well. The first immigrants from present-day Syria and
Lebanon settled in the Hill district in 1890; by 1920 Syrian and
Lebanese immigrants numbered over fourteen hundred. Belonging
to two competing religious bodies, a Maronite Catholic parish and a
Syrian Orthodox congregation, these immigrants hailed from a handful of villages. In Pittsburgh many took up the occupation commonly pursued by Lebanese and Syrian immigrants elsewhere in the
United States: peddling. Each day these immigrant men and women
traversed the hills and valleys of the region, selling the dry goods and
other wares they carried on their backs, returning at nightfall to the
small enclave on the Hill.[56]

Large numbers of Jewish immigrants lived near this settlement
of Syrians and Lebanese. Galician and Rumanian Jews established
synagogues in the Hill district in the 1890s, and, along with other
Eastern European Jews, founded an array of religious and fraternal
associations. By 1920 an enclave where Jews both lived and worked
stretched across all three wards of the Hill.[57] Some toiled in their
crowded apartments while others worked in the garment and stogy
factories or in bakeries, groceries, shoemaking shops, and tailoring
establishments. Often these enterprises were family businesses, and
typically women and children worked alongside their husbands, fathers, and brothers.[58]

Among the biggest employers on the Hill were the cigar-making
factories, where nearly two thousand adults and children labored
long hours for low pay in dimly lit, dirty rooms.[59] Two labor
organizations—the Knights of Labor and the Cigar Makers' International Union (affiliated with the American Federation of Labor)—
neglected the workers' attempts to organize but in 1913, as part of
the Industrial Workers of the World, the stogy-makers carried out a
successful strike. Ethnic solidarity helped these immigrant workers:
the enclave supported the strikers; feelings of obligation muted Jewish employers' hostility toward their co-ethnics; and immigrant cultural traditions, especially the legacy of Jewish socialism, buoyed
their cause. The IWW victory, however, was short-lived. After World
War I the locus of stogy manufacturing shifted, and, with the advent
of the cigarette, demand for cigars declined; by 1930, the cigar-
making industry had disappeared from the Hill.[60]

The Hill's garment industry suffered a similar fate. In 1917 several hundred Jewish tailors joined with a smaller number of Italians

to form a local of the Amalgamated Clothing Workers of America. In the 1920s and 1930s the garment trades declined, the number of Jewish tailors on the Hill decreased, and the local's membership became increasingly Italian. A more enduring stronghold of Jewish labor organization on the Hill was in an even smaller industry: the baking business. Immigrant workers in the Hebrew Bakers' Union were able to obtain comparatively good wages as early as 1910 and to maintain their rates in subsequent decades.[61] Like other Jewish labor organizations, this union relied on the support of the immigrant settlement. The entire baking industry, indeed, depended on the vitality of the ethnic enclave, for it was principally Jewish immigrants and their children who consumed the bakers' breads and pastries. Hence, the removal of this industry from the Hill after the 1920s signaled an important shift: the bakeries were following their clientele away from the crowded Hill district and into the thinly settled East End.[62]

The Syrians and Lebanese also began to abandon the Hill in the 1930s, leaving only a remnant of the immigrant group in the old ethnic enclave. Chinatown too declined in the 1930s, as Chinese immigration dwindled and immigrant businesses, like others in the city, succumbed to hard times. Unlike previous generations, a new wave of immigrant settlements did not succeed those disintegrating in the heart of the city. Instead, the 1930s ushered in an era of prolonged crisis in industry and sustained decline in immigration.

Industry Corrodes, Immigration Wanes: Pittsburgh After 1930

In the more than half-century since 1930, immigration to Pittsburgh declined precipitously. At the onset of the Depression, Pittsburgh's foreign-born numbered nearly one hundred ten thousand and comprised 16.2 percent of the city's residents; three decades later, fewer than forty-six thousand immigrants lived in Pittsburgh, accounting for only 7.4 percent of the population. During these decades, the newcomers settling in the city were typically migrants from the South rather than immigrants from abroad: Pittsburgh's black population swelled from fifty-five thousand in 1930 to one hundred thousand in 1960, doubling the group's share of the city's population from 8.2 percent to 16.7 percent.[63]

To some extent, Pittsburgh's experience from the 1930s to the 1960s mirrored the national trend, with immigration to the United States plummeting after the passage of restrictive legislation in the 1920s. Yet Pittsburgh's drop in immigration was more extreme and enduring. In the 1970s and 1980s a new wave of immigrants swept onto America's shores. Migrants, principally from the Western Hemisphere and Asia, crowded into neighborhoods in New York, Chicago, Los Angeles, Miami, San Antonio, and scores of smaller cities. Few came to Pittsburgh. In 1970 immigrants comprised only 4.4 percent of the residents of the Pittsburgh metropolitan area; ten years later the foreign-born made up only 1.3 percent of the region's residents. Even more telling is the nationality of these immigrants: fewer than four hundred Mexican immigrants lived in the Pittsburgh area in 1980, while the region's "Asian" immigrants—a designation including migrants from such countries as China, Iran, Israel, Lebanon, the Philippines, and Vietnam—totaled less than ten thousand. In 1980 the region's largest immigrant group by far was the nearly twenty thousand Italians, more than four-fifths of whom had left their homeland before 1960.[64]

In the contemporary era, as in former years, the rate of immigration to Pittsburgh reflects the fortunes of industry. Pittsburgh's factories, absorbing the labor of thousands of immigrants during its industrial heyday, no longer provide enough jobs for those born in the region. The roots of this industrial crisis are deep. In the 1920s Pittsburgh lost some of its competitive advantage in steelmaking to Midwestern cities; then the Depression hit the Steel City hard; and even during the "good times" of the postwar years Pittsburgh did not regain its industrial prominence. By the 1960s foreign competition and capital flight rendered Pittsburgh's industry a feeble shadow of its former mighty self.

With the corrosion of the region's industrial base, Pittsburgh began to evolve from a blue-collar steel center to a white-collar financial and service hub. But this evolution did not prompt an influx of migrants, either native- or foreign-born, to the city. The population of Pittsburgh declined from six hundred seventy thousand in both 1930 and 1940 to only six hundred thousand in 1960 and to fewer than four hundred twenty-five thousand in 1980. Suburban exodus accounted for only part of this decline, with the metropolitan region also losing population. In the 1980s Pittsburgh no longer ranked among the nation's ten largest metropolitan areas. Clearly, neither the rusting steel mills nor the postmodern office buildings of the new Pittsburgh could attract the immigrants of the late twentieth century.

Scholarly interest in the region's foreign-born and their descendants during the post-Depression era matched this decline in immigration. Few studies have traced the small ethnic settlements established since the 1920s, and research is also slim regarding the experiences of the second-, third-, and fourth-generation ethnics who streamed to Pittsburgh during the city's industrial expansion. Hence our knowledge of the ethnic experience in Pittsburgh after 1930 is more fragmentary than that of earlier years.

Part of the story is the continued movement away from older ethnic enclaves and the establishment, in some cases, of areas of "second settlement." In the expanding areas of Jewish settlement in the East End, for example, immigrant entrepreneurs established new retail centers. By 1940 the shops that lined Murray Avenue, the main street of Squirrel Hill, announced the arrival of a substantial Jewish population in that neighborhood, just as the stores of Dutchtown had proclaimed the influx of German newcomers in the mid-nineteenth century. Other groups resettled in suburban locations. By 1950, for instance, the remnant of the Syrian and Lebanese settlement on the Hill was gone—this neighborhood, like the rest of the Lower Hill, was leveled for construction of the Civic Arena, part of Pittsburgh's sweeping Renaissance redevelopment effort. Most of these immigrants had joined other Syrian and Lebanese in suburbs south and east of the city, notably Brookline, Dormont, and Homestead. The Maronites in the group built their new church in Brookline; the Syrian Orthodox, in contrast, erected their new edifice in Oakland, a location equidistant from the suburbs where their parishioners resided.[65] In contrast, no new center of Chinese settlement emerged after the demise of Chinatown. Instead the group became less residentially—and occupationally—segregated. With the easing of immigration restrictions against the Chinese after World War II, several hundred Chinese professionals, both men and women, settled in Pittsburgh. Their arrival established class barriers within the ethnic group. As late as 1982, the more recent immigrants had little contact with the prewar Chinese and supported a separate set of social institutions.[66] A study conducted in 1952 suggests a parallel development among Greeks, as postwar immigrants chose not to affiliate with the organizations frequented by earlier arrivals.[67]

Some of the city's ethnic enclaves persisted through the postwar years. Many second-generation Poles and Italians, enjoying a modicum of economic security, remained in the neighborhoods in which their immigrant parents had settled: Poles in the South Side and Polish Hill, and Italians in Bloomfield and East Liberty. Here sons and daughters of immigrants bought homes, staying near their families and using the kin network to gain jobs. But as old structures in the neighborhoods continued to age and few new structures were built, some second-generation Poles and Italians left the ethnic settlements. By 1960, all four neighborhoods became less ethnically homogeneous, but all save East Liberty retained strong identities as blue-collar, ethnic neighborhoods. In contrast, blacks, concentrated in the Hill district and, increasingly, in the Homewood-Brushton area of the East End, experienced less occupational success and had lower rates of homeownership.[68]

Despite the residential persistence of many second- and later generation Poles, by the mid-1970s most of Pittsburgh's Polish institutions faced a crisis in membership. The pillars of Polish immigrant life, churches and fraternal associations, suffered acute declines in numbers while ethnic associations that stressed Polish culture were more able to maintain their smaller and more elite memberships. A 1975 survey of these organizations suggests an overall erosion of ethnic ties among the city's Poles: a decline in the speaking of Polish, a diminished commitment to Polish nationalism, and the absence of a viable political organization. Yet, perhaps responding to the media's pronouncement of an ethnic revival, most Poles reported that they did not believe that the group was assimilating.[69] In contrast, most of the city's Greeks interviewed in a 1974 study saw decline in the group's identity, citing the decreased use of the Greek language among the young, the demise of ethnic institutions, and the dispersal of Greeks throughout the region.[70] Greeks in Aliquippa, a steel town twenty miles downstream from Pittsburgh on the Ohio River, may have retained more of their consciousness and cohesion, according to research conducted in 1976. Compared to the Greeks, many of whom owned small businesses, the town's Serbs, long clustered in mill work, sustained less ethnic identity.[71]

Whether identifying consciously as ethnic group members, Pittsburgh's Conservative Jews used ethnic traditions and kin connections to facilitate their upward mobility, according to a 1978 study.[72] Other Pittsburgh residents, including the descendants of the city's Irish immigrants, may have expressed their ethnic identity through their choice of recreation. An analysis of the programming offered on "The Irish Hour," a two-hour weekly radio broadcast, reveals that as late as 1977 contemporary and traditional Irish music commanded a following in the Pittsburgh region.[73] And all of the region's residents could express their appreciation for, if not identification with, ethnic culture by attending local festivals. In Aliquippa, for example, working-class and middle-class Italians joined together to sponsor the San Rocco festival, a celebration honoring the patron saint of Patrica, the village from which many of their ancestors migrated. First organized in 1925, the festival became in the postwar years a central part of the town's social life, encouraging both Italians and non-Italians to participate in what had once been an expression of loyalty to a single Italian village.[74]

As the history of the San Rocco celebration suggests, ethnic identity and ethnicity itself have been transformed in the modern era. Unfortunately, the nature of this transformation—in the Pittsburgh region and in American society generally—is little understood. Instead, most scholars of the Pittsburgh region have restricted their studies to the period from 1850 to 1930 and focused on the lives of immigrants rather than their descendants. In an exaggeration of a basic preoccupation in the field of ethnic and immigration history, scholars of Pittsburgh have gravitated to the dramatic story of industrial boom and immigrant influx, neglecting other aspects of the region's ethnic life. In a city still dominated by the legacy of heavy industry and ossified in its ethnic composition, this preoccupation is readily understandable. The consequence, however, is a narrow and skewed vision of the region's past.

Beyond "Immigrants and Industry": An Agenda for Research

Industrialization was only part of the capitalist transformation within the region. It is this overall transformation and the complicated roles that immigrants and their descendants played in its many facets which should become the focus of inquiry. Further research on the ethnic experience in Pittsburgh, then, should broaden and deepen the strong current of interest in the economic lives of immigrants and the political economy of their settlements and widen the scope of inquiry to include new areas of investigation. Along with fuller examinations of their jobs, the conditions of their labor, and their relations with their employers and other workers, studies need to assess more precisely the ways in which immigrants and their descendants fit into such processes as class formation, fragmentation, and polarization. Which ethnic groups, for example, benefited most from industrial expansion and which have fared the worst as the mills have shut down? How did the proliferation of white-collar and service jobs of varying skills and statuses affect the employment of particular ethnic groups? Did discrimination on the part of employers bar some ethnic groups from securing or succeeding in white-collar positions? In answering these questions, scholars must be particularly sensitive to gender and class differences within ethnic groups.

Beyond the realm of production, attention should also be paid to changing patterns of consumption. Margaret Byington's pioneering

study of household budgets, in her Survey volume, *Homestead*, remains the most thorough description of the consumption patterns of
different nationality groups in the region. Few modern studies touch
on this important subject. We know that some immigrant women,
for example, worked in the Heinz plant on Pittsburgh's North Side,
bottling pickles and canning fruit. When did these women begin to
buy the goods they manufactured rather than making them at
home? How did ethnic spending patterns for clothing, appliances,
household furnishings, and other goods vary? Over the course of the
twentieth century, did class and ethnic identity become as much or
more a matter of what one bought than of how one made a living?
Bodnar, Simon, and Weber, in *Lives of Their Own*, provide some
insight on investments in education and homeownership made by
Italians, Poles, and blacks, but what of other groups? In many ethnic
families, women were largely in charge of daily household expenditures. Hence, understanding changing patterns of consumption will
illuminate much about women's lives and about the dynamics of
ethnic families.

In hilly, riverine Pittsburgh, few immigrant families lived in
crowded tenements like those common in New York and other eastern cities. What were the consequences for immigrant families of
having slightly more privacy or a bit more open space in their neighborhoods than did immigrants in cities with flatter terrain? Did
these conditions lessen domestic violence, weaken neighborhood
cohesion, or merely allow some immigrants to supplement their
meager wages by keeping gardens or raising small livestock? On the
whole, little is known about the material culture of ethnic homes or
the intimate side of family life, from aspects of child-rearing to sexual attitudes and practices. These topics require comparison across
groups and intergenerationally.

Much has yet to be learned about the public world of Pittsburgh's
ethnic groups, as well. For some groups, not even the bare outlines of
associational life are known, and, in general, there has been little
study of the impact of these organizations on their membership.
Analysis of the ethnic press and of immigrant education have been
equally neglected. Pittsburgh's ethnic churches, synagogues, and
other religious bodies have been given some attention, but the role
of religion in the ethnic experience requires explication. How did
ethnic groups transform the sacred symbols of the homeland and
adapt them to fit the conditions of their new situation? Did commitment to religion supersede or reinforce class or national solidarity?
Examining the popular culture of Pittsburgh's ethnic groups should

be similarly worthwhile, for the realm of leisure as much as labor defined the lives of immigrants and their children.

More fundamentally, scholars of the ethnic experience in the Pittsburgh region need to assess what they mean by adjustment, acculturation, assimilation, and "Americanization." In so doing, we should remember that these processes may have had different meanings for immigrants and their descendants. The biography of one Pittsburgh Croatian immigrant, Steve Nelson, makes the point. Nelson left his ethnic community to join the Communist party, embracing an ideology that saw "America" less as land of opportunity than as capitalist empire.[75] His "Americanization" led him not toward, but away from, the values celebrated by the American mainstream. Similarly, characters in *Out of This Furnace* depict other means of "Americanization." The fictional Dobie, for instance, believes that he has become American—"Patrick Henry, Jr.," as he refers to himself—by joining with other workers to back the CIO.[76] These examples remind us that we must understand the lives of immigrants and their descendants from their own vantage points and in their own terms.

All of this implies that we must go beyond the image of Joe Magarac, beyond a preoccupation with the immigrant worker in the mill. The great task that lies ahead is to explore the transformation of ethnicity in the region. We know that in the last century immigration to Pittsburgh has declined; what has happened to ethnicity? How has ethnic culture changed, and which forces in mainstream society have most altered or undermined it? What new forms of expression has ethnicity adopted? Does it survive only as an easily consumable commodity, whether in the shape of ethnic festivals, radio broadcasts, or restaurants? Or is it more in the province of "high culture," such as in elite organizations or artifacts—the Nationality Rooms at the University of Pittsburgh, for example. Or is ethnicity more deeply and profoundly held, able to mobilize people and to shape their behavior and attitudes in enduring ways? There are hints of its continued salience that require investigation. During the early 1980s, for instance, community activists protesting the shutdown of Pittsburgh's steel mills invoked the spirit of Polish ethnicity. On one occasion, anti-shutdown groups included on their program songs by the Polish Falcon Choir, a dance by the Polish Women's Alliance, and a salute to the workers of Solidarity in Poland.[77] If such activities had resonance with the region's workers, then ethnicity, in some shape or other, may be with us still. As the capitalist transformation of the Pittsburgh region continues, so does

the metamorphosis of ethnicity; tracing that metamorphosis is the
challenge for future research.

NOTES

I would like to thank Richard McLellan for his assistance in obtaining
the material reviewed in this essay. For commenting on an earlier draft, I am
grateful to James R. Barrett, Leslie Page Moch, and the participants in the
Pittsburgh Social History Conference held at the University of Pittsburgh in
March 1987. I am especially indebted to John J. Bukowczyk for his astute
suggestions regarding both the style and substance of the essay.

1. See Clifford J. Reutter, "The Puzzle of a Pittsburgh Steeler: Joe
Magarac's Ethnic Identity," *Western Pennsylvania Historical Magazine* 63
(Jan. 1980): 31–36.
2. Margaret F. Byington, *Homestead: The Households of a Mill Town*,
vol. 4 of The Pittsburgh Survey (1910; rpt. Pittsburgh: University of Pittsburgh Press, 1974).
3. John M. Glenn, director of the Russell Sage Foundation, which sponsored the Survey, in the introduction to Elizabeth Beardsley Butler, *Women
and the Trades: Pittsburgh 1907–1908* (1909; rpt. Pittsburgh: University of
Pittsburgh Press, 1984), p. 1. See also Trisha Early, "The Pittsburgh Survey"
(seminar paper, Department of History, University of Pittsburgh, 1972), p. 4.
Early's paper traces the politics behind the writing of the Survey volumes.
4. This essay does not detail the experience of blacks in the Pittsburgh
region; this topic, related in important ways to the history of white and
other nonwhite minority groups in the region, is so extensive and particular
that it merits an essay of its own in this volume. See Laurence Glasco,
"Double Burden: The Black Experience in Pittsburgh."
5. On the Pittsburgh elite, see, for example, Thomas Kelso, "Pittsburgh's Mayors and City Councils, 1794–1844: Who Governed?" (seminar
paper, Department of History, University of Pittsburgh, 1963), and Joseph
Francis Rishel, "The Founding Families of Allegheny County: An Examination of Nineteenth-Century Elite Continuity" (Ph.D. diss., University of
Pittsburgh, 1975). For background on early German settlers see Nora Faires,
"Ethnicity in Evolution: The German Community in Pittsburgh and Allegheny City, Pennsylvania, 1845–1885" (Ph.D. diss., University of Pittsburgh, 1981), and Layne Peiffer, "The German Upper Class in Pittsburgh,
1850–1920" (seminar paper, Department of History, University of Pittsburgh, 1964). On the Irish see Victor Anthony Walsh, "Across the 'Big
Wather': Irish Community Life in Pittsburgh and Allegheny City, 1850–
1885" (Ph.D. diss., University of Pittsburgh, 1983). Notable exceptions to
this general lack of interest in the region's first immigrants include two

studies of communities which were later annexed by Pittsburgh: see John W. Larner, Jr., "A Community in Transition: Pittsburgh's South Side, 1880–1920" (seminar paper, Department of History, University of Pittsburgh, 1961), and Joan Miller, "The Early Historical Development of Hazelwood" (seminar paper, Department of History, University of Pittsburgh, n.d.).

6. Walsh, "Across the 'Big Wather'," p. 375.

7. Oscar Handlin, *Boston's Immigrants: A Study in Acculturation*, rev. ed. (Cambridge: Harvard University Press, 1957); cf. JoEllen Vinyard, *The Irish on the Urban Frontier: Nineteenth-Century Detroit, 1850–1880* (New York: Arno Press, 1976).

8. Walsh, "Across the 'Big Wather'," pp. 105–09, 18–21. See also Walsh, "Across the 'Big Wather': The Irish-Catholic Community of Mid–Nineteenth-Century Pittsburgh," *Western Pennsylvania Historical Magazine* 66 (Jan. 1983); 1–23.

9. Walsh, "Across the 'Big Wather'," pp. 210–12, 311–15. See also Walsh, " 'A Fanatic Heart': The Cause of Irish-American Nationalism in Pittsburgh During the Gilded Age," *Journal of Social History* 15 (Winter 1981): 187–204. On Irish nationalism cf. Eric Foner, "Class, Ethnicity, and Radicalism in the Gilded Age: The Land League and Irish-America," in Foner, *Politics and Ideology in the Age of the Civil War* (New York: Oxford University Press, 1980); David Brundage, "Irish Land and American Workers: Class and Ethnicity in Denver, Colorado," in *"Struggle a Hard Battle": Essays on Working-Class Immigrants*, ed. Dirk Hoerder (DeKalb: Northern Illinois University Press, 1986), pp. 46–97; and David Emmons, "An Aristocracy of Labor: The Irish Miners of Butte, 1880–1914," *Labor History* 28 (Summer 1987): 275–306. For another view of the Irish in Woods Run, see Patricia K. Good, "Irish Adjustment to American Society: Integration or Separation? A Portrait of an Irish-Catholic Parish: 1863–1886," *Records of the American Catholic Historical Society of Philadelphia* 86 (Mar.–Dec. 1975): 7–23. Good maintains that Irish ironworkers found in their church a mechanism for "acculturation and adaptation to the American environment" (p. 21).

10. John William Bennett, "Iron Workers in Woods Run and Johnstown: The Union Era, 1865–1895" (Ph.D. diss., University of Pittsburgh, 1977).

11. Bennett, "Iron Workers," chap. 3.

12. There is some discussion of these immigrants in Miller, "Early Historical Development of Hazelwood," and Joseph Johnston, "National Origins and Ethnic Groups of the People of Allegheny, Pennsylvania, in 1880" (seminar paper, Department of History, University of Pittsburgh, n.d.).

13. Faires, "Ethnicity in Evolution," pp. 126, 142–45.

14. Ibid., pp. 149, 177–79.

15. Ibid., pp. 446–50.

16. Peiffer, "German Upper Class," and Michelle Pailthorp, "The German-Jewish Elite of Pittsburgh: Its Beginnings and Background" (seminar paper, Department of History, University of Pittsburgh, 1967).

17. Marguerite Renner establishes that fewer German than Irish women

entered the growing ranks of teachers. See Renner, "Who Will Teach? Changing Job Opportunity and Roles for Women in the Evolution of the Pittsburgh Public Schools, 1830–1900" (Ph.D. diss., University of Pittsburgh, 1981), pp. 251–74.

18. Faires, "Ethnicity in Evolution," pp. 217–28, 259–64. In general, my findings regarding immigrant occupations confirm Walsh's in "Across the 'Big Wather'." See also Michael P. Weber, "Residential and Occupational Patterns of Ethnic Minorities in Nineteenth-Century Pittsburgh," *Pennsylvania History* 44 (Oct. 1977): 316–34. Comparing the mobility of German, Irish, and native-born workers in four Pittsburgh industrial wards from 1880 to 1910, Weber found that Germans trailed both native-born and, especially, Irish workers in movement up the occupational ladder. Germans were also the least likely to persist in the city.

19. Nora Faires, "Occupational Patterns of German-Americans in Nineteenth-Century Cities," in *German Workers in Industrial Chicago, 1850–1910: A Comparative Perspective,* ed. Hartmut Keil and John B. Jentz, (DeKalb: Northern Illinois University Press, 1983), pp. 37–51.

20. Faires, "Ethnicity in Evolution," pp. 422–45. See also G. Dale Greenawald, "Germans in Pittsburgh, 1850, 1880, 1930: Residency, Occupations, and Assimilation" (seminar paper, Department of History, Carnegie-Mellon University, n.d.)

21. Paul Kleppner, "Lincoln and the Immigrant Vote: A Case of Religious Polarization," *Mid-America* 48 (July 1966): 176–95.

22. See Hal Kimmins, "Joseph Barker, Mayor of Pittsburgh, 1850–51" (seminar paper, Department of History, University of Pittsburgh, 1963), and Robert Kaplan, "The Know Nothings in Pittsburgh" (seminar paper, Department of History, University of Pittsburgh, 1977). Neither Kimmins nor Kaplan offers much insight into the reaction of immigrants to Barker's popularity.

23. Kleppner, "Lincoln and the Immigrant Vote," pp. 192–95.

24. Faires, "Ethnicity in Evolution," pp. 437–39.

25. Pittsburgh's German Jewish immigrants have attracted a good deal of scholarly attention, yet aside from the documented rise of separate Jewish and Christian elites, it is not clear to what degree German Jews constituted a distinct community within the larger German immigrant group. On elites, see Pailthorp, "German-Jewish Elite," and Peiffer, "German Upper Class." The literature on German Jews includes Bruce Skud, "Ethnicity and Residence Within the Jewish Immigrant Community" (seminar paper, Department of History, University of Pittsburgh, 1975); Laurie Mizrahi, "The History of the Jewish Community of Pittsburgh, 1847–1890" (seminar paper, Department of History, Carnegie Mellon University, 1981); and Mitchell A. Nathan, "The Jewish Community of Pittsburgh: A Beginning" (seminar paper, Department of History, Carnegie Mellon University, 1982).

26. Skud, "Ethnicity and Residence," p. 4. For another view of the ethnic composition of Tree of Life congregation, see Ida Cohen Selavan, "The Founding of the Columbian Council," *American Jewish Archives* 30 (April

1978): 26–27. Selavan claims that this congregation was, from the outset, made up largely of Jews from Lithuania, Posen, and Holland.

27. Skud, "Ethnicity and Residence," pp. 11–16; Pailthorp, "German-Jewish Elite," p. 67; and Mizrahi, "History of the Jewish Community," pp. 6–9. German Christian congregations also had distinct and revealing patterns of relocation as they followed their parishioners from the downtown to various sections of the city. See Faires, "Ethnicity in Evolution," p. 428, and Greenawald, "Germans in Pittsburgh," pp. 20–26.

28. Selavan, "Founding of the Columbian Council," pp. 28–29.

29. On the Polish Women's Alliance in Pittsburgh, see Louise Misko, "A Study of Political Activities and Attitudes of Pittsburgh Poles Relative to Achieving the Independence of Poland Through Preservation of Religious, Fraternal, and Cultural Institutions" (seminar paper, Department of History, University of Pittsburgh, 1975).

30. Ida Cohen Selavan, "The Jewish Labor Movement in Pittsburgh" (seminar paper, Department of History, University of Pittsburgh, 1971), pp. 1–2.

31. The population figure for 1880 includes the inhabitants of Allegheny City, annexed by Pittsburgh in 1907. All figures are my calculations based on data in U.S. Bureau of the Census, *Thirteenth Census of the United States, 1910: Population* (Washington, D.C.: Government Printing Office, 1911), p. 212; and U.S. Bureau of the Census, *Fifteenth Census of the United States, 1930: Population* (Washington, D.C.: Government Printing Office, 1932), p. 329.

32. See also Hilda Marie Becker, "A Statistical Analysis of the Census Reports on the Distribution of the Foreign-Born White Population of Pittsburgh, Pennsylvania, 1890–1930" (MA thesis, Department of Sociology, University of Pittsburgh, 1932); Frank S. Snyder, "Spatial Distribution of Eight Foreign-Born White Groups in Pittsburgh, Pennsylvania, from 1910 to 1950" (MA thesis, Department of Geography, University of Pittsburgh, 1973); and, on McKeesport, a Monongahela River steel town, see Mary Huey, "Occupational-Nationality Structure of McKeesport, 1880" (seminar paper, Department of History, University of Pittsburgh, n.d.).

33. See Francis G. Couvares, *The Remaking of Pittsburgh: Class and Culture in an Industrializing City, 1877–1919* (Albany: State University of New York Press, 1984), pp. 88–92.

34. Bodnar, Simon, and Weber, *Lives of Their Own: Blacks, Italians, and Poles in Pittsburgh, 1900–1960* (Urbana: University of Illinois Press, 1982). See also, John Bodnar, Michael Weber, and Roger Simon, "Migration, Kinship, and Urban Adjustment: Blacks and Poles in Pittsburgh, 1900–1930," *Journal of American History* 66 (Dec. 1979): 548–65; and "Seven Neighborhoods: Stability and Change in Pittsburgh's Ethnic Community, 1930–1960," *Western Pennsylvania Historical Magazine* 64 (April 1981): 121–50.

35. Bodnar et al., *Lives of Their Own*, pp. 54–62, 67–69, 80–82, 113, 179.

36. Ibid., pp. 80–82.

37. Misko, "Political Activities," pp. 36–43.

38. Ibid. Misko provides a comprehensive examination of Polish associational life in Pittsburgh. See also Joseph Borkowski, *The Role of Pittsburgh's Polish Falcons in the Organization of the Polish Army in France* (Pittsburgh: Polish Falcons of America, 1972), pp. 18–20. On relations between Polish parishes and the Pittsburgh diocese see Daniel S. Buczek, "Polish American Priests and the American Catholic Hierarchy: A View from the Twenties," *Polish American Studies* 33 (Spring 1976): 38–39.

39. Immigrants organized ethnic fraternals throughout the United States, but Pittsburgh's hazardous mill work particularly encouraged their development. As late as 1977 fourteen ethnic fraternal associations had their national headquarters in Pittsburgh. See Margaret E. Galey, "Ethnicity, Fraternalism, Social and Mental Health," *Ethnicity* 4 (Mar. 1977): 19–53. On fraternal organizations and other societies among Hungarians see Joseph Kenneth Balogh, "An Analysis of Cultural Organizations of Hungarian-Americans in Pittsburgh and Allegheny County" (Ph.D. diss., University of Pittsburgh, 1945); and Paul Body and Mary Boros-Kazai, *Hungarian Immigrants in Greater Pittsburgh, 1880–1980*, 10 pamphlets (Pittsburgh: Hungarian Ethnic Heritage Study Group, 1981); on Finnish societies in Monessen, a steel center located thirty miles south of Pittsburgh, see Roger N. Foltz, "The Story of the Finnish Community in Monessen, Pennsylvania" (seminar paper, Department of History, University of Pittsburgh, 1964); and on institutional development among a variety of immigrant groups, including Italians, Lithuanians, Poles, Serbs, and Ukrainians, on Pittsburgh's South Side, see Larner, "Community in Transition," pp. 13–31.

40. Alexander, *The Immigrant Church and Community: Pittsburgh's Slovak Catholics and Lutherans, 1880–1915* (Pittsburgh: University of Pittsburgh Press, 1987), pp. 15–27.

41. Ibid., pp. 10–11, 20–21, 32–43, 72–73. Alexander contributes to the small body of work on intraethnic differences among Pittsburgh's immigrants. See especially Faires, "Ethnicity in Evolution" and Walsh, "Across the 'Big Wather'." Both authors are more interested in the issue of class cleavage than is Alexander. Alexander discusses the difficulty of determining the occupations and residences of working-class Slovaks in "City Directories as 'Ideal Censuses': Slovak Immigrants and Pittsburgh's Early Twentieth-Century Directories as a Test Case," *Western Pennsylvania Historical Magazine* 65 (July 1982): 203–20.

42. Tom Callister, "The Reaction of the Presbytery of Pittsburgh to the New Immigrants" (seminar paper, Department of History, University of Pittsburgh, n.d.), pp. 12–17, 23–30.

43. Ray Burkett, "Vandergrift: Model Worker's Community" (seminar paper, Department of History, University of Pittsburgh, 1972).

44. See Corinne Azen Krause, "Urbanization Without Breakdown: Italian, Jewish, and Slavic Immigrant Women in Pittsburgh, 1900 to 1945," *Journal of Urban History* 4 (May 1978): 291–306; and Krause, "Italian, Jewish, and Slavic Grandmothers in Pittsburgh: Their Economic Roles," *Frontiers* 2 (Summer 1977): 15–23.

45. Krause, "Urbanization Without Breakdown," pp. 296–98; and Krause, "Italian, Jewish, and Slavic Grandmothers," pp. 16–22. Krause contends that "Slavic" women were more likely to work for wages outside their homes, failing to distinguish among the nationalities of the women whom she interviewed. Furthermore, she refers to "Slavic culture" and "character," categories which deny the variation in the traditions and experiences of immigrants from different Southern and Eastern European nations. In collapsing diverse nationalities into the rubric of "Slavic," Krause follows in the tradition of the Survey. Other recent works are flawed by this same practice. See Paul Krause, "Labor Republicanism and 'Za Chlebom': Anglo-American and Slavic Solidarity in Homestead," in Hoerder, ed., *"Struggle a Hard Battle,"* pp. 143–69; Robert Peles, "Crisis in Johnstown: The 'Little Steel' Strike of 1937" (seminar paper, Department of History, University of Pittsburgh, 1974); and Gerald Angerman, "McKeesport: A Preliminary Study" (seminar paper, Department of History, University of Pittsburgh, 1970).

46. Byington, *Homestead.*

47. Bell, *Out of This Furnace: A Novel of Immigrant Labor in America* (1941; rpt., University of Pittsburgh Press, 1976), pp. 150–51, 173. See also S. J. Kleinberg, "Technology's Stepdaughters: The Impact of Industrialization Upon Working-Class Women, Pittsburgh, 1865–1890" (Ph.D. diss., University of Pittsburgh, 1973); Kleinberg, "Technology and Women's Work: The Lives of Working-Class Women in Pittsburgh, 1870–1900," *Labor History* 17 (Winter 1976): 58–66; and Bodnar et al., *Lives of Their Own*, p. 102.

48. Herbert G. Gutman, "The Buena Vista Affair, 1874–1875," *The Pennsylvania Magazine of History and Biography* 88 (July 1964): 250–93.

49. Krause, "Labor Republicanism," pp. 143–69. For an examination of the Homestead strike which stresses the animosity between native-born and foreign-born workers see Steven R. Cohen, "Steelworkers Rethink the Homestead Strike," *Pennsylvania History* 48 (April 1981): 155–77, esp. p. 174.

50. Carl I. Meyerhuber, "Black Valley: Pennsylvania's Alle-Kiski and the Great Strike of 1919," *Western Pennsylvania Historical Magazine* 62 (July 1979): 251–65. See also Frank Huff Serene, "Immigrant Steelworkers in the Monongahela Valley: Their Communities and the Development of a Labor Class Consciousness" (Ph.D. diss., University of Pittsburgh, 1979), pp. 194–96.

51. See Peles, "Crisis in Johnstown," and Angerman, "McKeesport."

52. Bell, *Out of This Furnace*, pp. 341–413.

53. Morawska, "The Internal Status Hierarchy in the East European Immigrant Communities of Johnstown, Pennsylvania, 1890–1930s," *Journal of Social History* 16 (Fall 1982): 75–107; and Morawska, *For Bread with Butter: The Life Worlds of East Central Europeans in Johnstown, Pennsylvania, 1890–1940* (Cambridge: Cambridge University Press, 1985).

54. Chien-shiung Wu, "The Chinese in Pittsburgh: A Changing Minority Community in the United States" (Ph.D. diss., University of Pittsburgh, 1982), pp. 8–9, 16–24, 30, 77.

55. Georgia Katsafanas and Alice Flocos, "The Greek Immigrant and the Greek Orthodox Church in Pittsburgh" (research paper, Department of History, University of Pittsburgh, 1974), pp. 4, 17, 54.

56. Alfred Ray Pannbacker, "The Levantine Arabs in Pittsburgh, Pennsylvania" (Ph.D. diss., University of Michigan, 1982), pp. 5, 36, 79, 92–93. On Arab immigrants as peddlers see Nancy Faires Conklin and Nora Faires, " 'Colored' and Catholic: The Lebanese in Birmingham, Alabama," in *Crossing the Waters: Arabic-Speaking Immigrants to the United States Before 1940,* ed. Eric Hooglund (Washington, D.C.: Smithsonian Institution Press, 1987), p. 73.

57. Skud, "Ethnicity and Residence," pp. 21–23.

58. See Patrick Lynch, "Pittsburgh, the I. W. W., and the Stogie Workers," in *At the Point of Production: The Local History of the I. W. W.,* ed. Joseph R. Conlin (Westport: Greenwood Press, 1981), p. 83; and Krause, "Italian, Jewish, and Slavic Grandmothers," pp. 18–20.

59. One of the volumes of the Survey documents the conditions in Pittsburgh's stogy industry. See Butler, *Women and the Trades,* pp. 75–97.

60. Selavan, "Jewish Labor Movement," pp. 5–12; Lynch, "Pittsburgh, the I. W. W., and the Stogie Workers," pp. 89–91.

61. Selavan, "Jewish Labor Movement," pp. 14–19.

62. Skud, "Ethnicity and Residence," pp. 24–27.

63. These population figures are based on Bodnar et al., *Lives of Their Own,* p. 187. The calculations are mine. For a concise discussion of the demography of industrial decline in the neighborhood of Hazelwood, see Joel A. Tarr and Denise DiPasquale, "The Mill Town in the Industrial City: Pittsburgh's Hazelwood," *Urbanism: Past and Present* 7 (Winter/Spring 1982): 7–14. From 1870 to 1970 the foreign-born share of Hazelwood's population decreased from 25.6 to 8.2 percent while the black share increased from 0.5 to 12.0 percent.

64. All figures derive from U.S. Bureau of the Census, population censuses for 1970 and 1980; the calculations are based on these figures. These census data obviously do not include the presumably small numbers of illegal immigrants who have settled in Pittsburgh in the last several decades.

65. Pannbacker, "Levantine Arabs," pp. 54, 79–80. Dormont was also a center of second settlement for Germans, according to Greenawald, "Germans in Pittsburgh," p. 24.

66. Wu, "Chinese in Pittsburgh," pp. 133–34, 236–37.

67. Demetrius Iatridis, "The Post-War Greek Newcomer in Pittsburgh: A Study for Community Organization" (MA thesis, School of Social Work, University of Pittsburgh, 1952), pp. 60–66, 83.

68. Bodnar et al., *Lives of Their Own,* pp. 220, 229, 232. For a description of Bloomfield as an ethnic enclave in 1980, see William Simons, Samuel Patti, and George Hermann, "Bloomfield: An Italian Working-Class Neighborhood," *Italian American* 7 (1981): 102–15.

69. Misko, "Political Activities," pp. 106, 236–39, 249–57.

70. Katsafanas and Flocos, "Greek Immigrant," pp. 32, 40, 51.

71. Marcia Chamoritz, "The Persistence of Ethnic Identity in Two Nationality Groups in a Steel Mill Community" (seminar paper, Department of History, University of Pittsburgh, 1976).

72. Myrna Silverman, "Class, Kinship, and Ethnicity: Patterns of Jewish Upward Mobility in Pittsburgh, Pennsylvania," *Urban Anthropology* 7 (Sept. 1978): 24–43.

73. Kathleen Monahan, "The Irish Hour: An Expression of the Musical Taste and the Cultural Values of the Pittsburgh Irish Community," *Ethnicity* 4 (Sept. 1977): 208–09. The studies by Monahan, Misko, Silverman, and Katsafanas and Flocos stress the durability of ethnic culture and reflect the surge in cultural pluralism during the 1970s. Another study undertaken during this period surveyed the efforts of eight ethnic groups in Pittsburgh to maintain their languages: Croatians, Greeks, Hungarians, Italians, Lithuanians, Poles, Serbs, and Slovaks. See Fazel Nur, "Language Maintenance Efforts of Several Ethnic Groups in Allegheny County, Pennsylvania" (Ph.D. diss., University of Pittsburgh, 1978). Regrettably, Nur collapses all his findings into broad categories, making it impossible to distinguish among these ethnic groups. See also Howard F. Stein, "An Ethno-Historic Study of Slovak-American Identity" (Ph.D. diss., University of Pittsburgh, 1972).

74. Marcia Chamovitz, "The San Rocco Celebration of Aliquippa: An Italian Saint in an American Setting" (seminar paper, Department of History, University of Pittsburgh, 1977).

75. Steve Nelson, James R. Barrett, and Rob Ruck, *Steve Nelson: American Radical* (Pittsburgh: University of Pittsburgh Press, 1981).

76. Bell, *Out of This Furnace*, p. 412.

77. See Staughton Lynd, "The Genesis of the Idea of a Community Right to Industrial Property in Youngstown and Pittsburgh, 1977–1987," *Journal of American History* 74 (Dec. 1987): 946–47.

Women and Class in Pittsburgh, 1850–1920

MAURINE WEINER GREENWALD

T HE WOMEN of Pittsburgh—native-born and immigrant, white and black, rich and poor—are "invisible" in the city's standard histories and folklore.[1] As a powerful steel center the Pittsburgh region has long been known as a sooty, hard-working, blue-collar, beer-drinking city. In stories, songs, and paintings, Pittsburgh figures as a "man's town" characterized by labor's muscle and brawn amidst industrial smokestacks, brilliant furnace fires, corner taverns, and rough-and-tumble sports. Most histories of the city feature male corporate leaders, male workers, and male politicians. As a result, women's contributions to the local economy and society have been largely overshadowed or overlooked. There is a presumed male quality inherent in Pittsburgh's industrial past.[2]

Steel producers excluded women from manufacturing sites except to fill severe wartime labor shortages or to comply with court-mandated affirmative action policies. Women became more visible in Pittsburgh's economy only in the 1970s and 1980s, overlapping the period of steel's sharp decline and the emergence of a more diversified regional economy. In place of steel, the driving force in Pittsburgh's economy came to depend heavily on the city's corporate headquarters and professional service sector, new high technology companies, and higher education and research institutions. Women began to move beyond "the shadow of the mills" to assume diverse new economic roles in the private and public sectors.[3]

Even in "male Pittsburgh" women have of course always had a history. To perceive the structure and content of women's lives it is necessary to shift the focus from the steel mills to home workshops and small factories, schoolrooms, telephone exchanges, and offices; from the taverns and men's social clubs to the homes of the corporate elite, middle management, and workers; from city hall to ethnic associations and community centers. Like women's history in other

33

locales, the study of women in Pittsburgh provides an important avenue for examining the intricate processes of large-scale social change.[4] Women's wage work features the complexities of sex, race, and ethnic segregation in the labor market. The study of house-wifery reveals the disparities in standards of living and consumption patterns. Women's community activism raises questions about infor-mal and formal political power. Such diverse facets of women's expe-riences collectively provide a vehicle for exploring the structure of society and the dynamics of power. The study of women in Pitts-burgh accentuates the diverse social contrasts between women and men, the working class and the middle class, and whites and blacks that have long been a hallmark of Pittsburgh.

The task of reconstructing the history of women in Pittsburgh is formidable. Knowledge of women's lives in the region deals mostly with the late nineteenth and early twentieth centuries. The available material especially illuminates a seventy-year period, a narrower time frame than for other topics in this volume. The decades before 1850 and after 1920 are virtually unexplored. Most of the existing scholarship is moreover about white working-class women. When supplemented by the pioneering Pittsburgh Survey of 1909–1914, a recently published memoir about an upper middle-class family, and scattered references to black women, what emerges is a vivid picture of social and economic inequalities in an industrial city.

This essay also explores ways in which women's lives in the second half of the nineteenth and early part of the twentieth century compared and contrasted with women in other steelmaking centers. Typical of other American cities, class and race distinctions in Pitts-burgh played major roles in determining women's employment, edu-cational, and marital experiences as well as their contributions to civic life. At the same time comparisons reveal that women's wage-earning patterns and civic life in Pittsburgh differed from those of their counterparts in other industrializing midwestern American cit-ies such as Detroit, Buffalo, and Cleveland and from what women in the English steel center of Birmingham experienced. Women in Pitts-burgh had fewer options for wage earning and fewer opportunities for civic leadership. In this sense the stereotype of Pittsburgh as a man's town had some relative meaning, but Pittsburgh's women nonetheless have a history that merits exploration in its own right.

Paid Work and Economic Development

The nature of female employment and the composition of the female labor force in Pittsburgh changed dramatically from the nine-teenth to the twentieth century. In the mid-nineteenth century

most female workers came from working-class families and were
concentrated in domestic service and manufacturing. This pattern
persisted until the closing decades of the century when white-collar
jobs in the fields of teaching, sales, telephone operating, and clerical
work attracted some middle-class women into the labor force and
widened job options for working-class women. From the 1840s to
the 1970s the female labor force in Pittsburgh and adjacent communities was characterized by extreme class, ethnic, and racial distinctions. What follows is a partial portrait of women's employment
that suggests more about these complex changes than can easily be
documented in the existing literature.

Until the Civil War women were a more important part of the
manufacturing sector in Pittsburgh than they would ever again be.
Although half the manufacturing economy in 1850 was concentrated in the iron and steel industries, female labor in the production
of cotton cloth and clothing made women a very important part of
the manufacturing sector as a whole. In the 1840s and 1850s cotton
mills in Allegheny City (now Pittsburgh's North Side) and, to a
much lesser extent, in neighboring Pittsburgh were one of the main-
stays of female employment. According to accounts of the 1848
factory riots in Allegheny City and the 1850 manufacturing census,
the cotton mills employed about fifteen hundred people, most of
whom were probably single Irish women whose families counted on
their support.[5]

The closing of the cotton mills after the Civil War and the exclusively male hiring policies in the growing iron and steel industries
relegated women to a minor role in industry. By 1910 women wage
earners constituted only 10 percent of all workers in Pittsburgh
manufacturing.[6] Rolling mills and open-hearth furnaces came to
dominate the physical landscape of the city, stretching for miles
along the Monongahela River and employing fully one-third of all
Pittsburgh workers in manufacturing in 1910. The city offered other
means of earning a living, but these jobs were fewer and less visible.[7]
While male industrial workers labored in the mills, the majority of
women in industry were needleworkers who made garments in one-
room workshops or their own homes. Scattered throughout the city
in small factories, women also produced confectioneries, crackers
and cakes, cheap cigars, paper boxes, clothbound books, and glassware. The two exceptions were the hundreds of semiskilled women
employed in large, modern commercial food firms on the North Side
and those in electrical equipment companies in East Pittsburgh, an
adjoining municipality.

As women's participation in the manufacturing sector shrank,
their contributions to white-collar employment rose. Between 1870

and 1920 women found employment in Pittsburgh's many offices, department stores, and telephone exchanges. By about 1900 the staff of the extensive, centralized public school system had become largely female. The combination of manufacturing, distribution, and communication services and the feminization of public school teaching made Pittsburgh a microcosm of employment trends throughout the country.

Pittsburgh's female labor force participation rates (LFPR) from 1890 to 1920 were similar to or only slightly lower than those of industrializing cities in the Midwest and approximated the corresponding national rates as well. The percentage of women who worked for wages outside the home in Pittsburgh was close to the rates for Detroit, Buffalo, and Cleveland. Over these decades the proportion of women working in Pittsburgh varied from 18.6 percent in 1890 to 24.8 percent in 1920. Pittsburgh's LFPRs were distinctly lower than those in textile centers, such as Lowell, which specifically recruited single and married women, and in larger cities after 1900, such as Chicago, with more diverse economies. Nonetheless, Pittsburgh's rates were virtually identical to the four national rates recorded in the respective federal censuses, as detailed in table 1.

TABLE 1
WOMEN'S LABOR FORCE PARTICIPATION RATES, 1890–1920 (percent)

	1890	1900	1910	1920
Pittsburgh	18.6	20.3	24.9	24.8
Detroit	21.3	23.7	26.7	23.7
Buffalo	19.6	20.9	24.6	24.6
Cleveland	19.5	21.3	25.4	24.5
Lowell	37.2	41.9	46.0	42.9
Chicago	16.6	23.2	31.1	32.3
National	18.9	20.6	25.4	23.7

SOURCES: Data for Pittsburgh, Detroit, Buffalo, and Cleveland, 1890–1910, are from Ileen DeVault, "Sons and Daughters of Labor: Class and Clerical Work in Pittsburgh, 1870s–1910s" (Ph.D. diss., Yale University, 1985), p. 84, n. 6. The original sources for all the cities listed above can be obtained from U.S. Census Office, *Report on Population of the United States at the Eleventh Census: 1890*, part 1 (Washington, D.C.: Government Printing Office, 1895), table 19, and part 2 (1897), table 118; and from the following publications of the U.S. Bureau of the Census in Washington, D.C.: *Special Reports: Occupations at the Twelfth Census, 1900* (1904), table 43; *Abstract of the Twelfth Census of the United States, 1900*, 3rd. ed. (1904), table 88; *Thirteenth Census of the United States, 1910: Population, Occupation Statistics*, vol. 4 (1914), table 8; *Fourteenth Census of the United States, 1920: Population, Occupations*, vol. 4 (1923), table 25; *Fourteenth Census of the United States, 1920: Population: General Report and Analytical Tables*, vol. 2 (1922), table 16.

National figures are from U.S. Bureau of the Census, *Historical Statistics of the United States, Colonial Times to 1970*, part 1, bicentennial ed. (Washington, D.C.: Government Printing Office, 1978), p. 133.

Despite the similarity between women's LFPRs in Pittsburgh and other midwestern industrializing cities, women's employment in Pittsburgh differed in one respect. The simultaneous closing of the textile mills and the extreme domination of the iron and steel industries in Pittsburgh dramatically reduced *married* women's employment options. The LFPR for Pittsburgh's married women was considerably lower than the comparable national figure. In 1900, 5.6 percent of married women in the United States worked for wages, whereas in Pittsburgh the figure was only 2.8. In 1910 the respective national and local figures were 10.7 and 3.5, and in 1920, 9.0 and 6.0.[8] Regional comparisons reinforce the view that the Pittsburgh labor market was particularly inhospitable to married women. In 1920, for example, the rate in Cleveland was 9.6 percent and in Detroit 8.2 percent. In fact, throughout the twentieth century the percentage of married women in Pittsburgh employed outside the home was significantly below the national averages and slightly below the rates of nearby cities, as detailed in table 2. Although the gap between Pittsburgh and similar midwestern cities was quite small by 1980, Pittsburgh's rate still lagged far behind the national average.

A comparison between married women's labor force participation

TABLE 2
MARRIED WOMEN'S LABOR FORCE PARTICIPATION RATES, 1920–1980 (*percent*)

	1920	1940	1960	1980
Pittsburgh	6.0	11.0	22.8	41.8
Detroit	8.2	16.2	26.5	46.6
Buffalo	5.0	12.8	25.2	43.3
Cleveland	9.6	16.7	29.4	43.9
National	9.0	16.7	31.7	51.0

SOURCES: These percentages were calculated from the following, all publications of the U.S. Bureau of the Census, Washington, D.C.: *Fourteenth Census of the United States, 1920: Population, General Report and Analytical Tables*, vol. 2 (1921), table 16; *Fourteenth Census of the United States, 1920: Population, Occupations*, vol. 4 (1923), table 25; *Sixteenth Census of the United States, 1940: Population*, vol. 4, parts 3 and 4 (1943), table 8, and vol. 3, parts 3, 4, and 5, table 8. Data for 1960 are from *U.S. Censuses of Population and Housing: 1960*, the census tracts for Pittsburgh (1962), tables P-1 and P-3; for Detroit (1961), tables P-2 and P-3; for Buffalo (1961), tables P-2 and P-3; and for Cleveland (1961), tables P-2 and P-3. Data for 1980 are from *1980 Census of Population and Housing*, the census tracts for Pittsburgh (1983), table P-10; for Detroit (1983), table P-10; for Buffalo (1983), table P-10; and for Cleveland (1983), table P-10.

The national figures for married women's labor force participation can be found in U.S. Bureau of the Census, *Historical Statistics of the United States, Colonial Times to 1970*, part 1, bicentennial ed. (Washington, D.C.: Government Printing Office, 1978), p. 133; U.S. Department of Labor, *Time of Change: 1983 Handbook on Women Workers*, Women's Bureau Bulletin 298 (Washington, D.C.: Government Printing Office, 1983), table 1-8.

in Pittsburgh and in Birmingham, England, another steel center, in the first decade of the twentieth century underscores the distinctive composition of Pittsburgh's female labor force. Married women in Pittsburgh faced at least two external barriers to employment: limited manufacturing jobs and discriminatory hiring policies. Women were excluded from the steel industry in both cities, but Birmingham offered married women jobs as industrial homeworkers and factory operatives in the jewelry, leather, brass, and metal trades. Consequently, there were three times more women in the manufacturing sector in Birmingham than in Pittsburgh, while the number of women in domestic and personal service, trade and commerce, transportation, and professional work was greater in the American than the English city. Peter Shergold correctly concludes that "Pittsburgh women served" while "Birmingham women manufactured."[9] The Pittsburgh labor market was also more restrictive because large employers, such as the public schools and electrical manufacturing firms, officially prohibited the employment of married women for many decades. If married women in Pittsburgh needed to contribute to the family income, they had to lie about their marital status or resort to taking in boarders and laundry—low-status work that paid poorly and was often undercounted by the census.

The notion that Pittsburgh women served while women elsewhere manufactured is accentuated when Pittsburgh is compared to other midwestern American industrial cities such as Detroit, Buffalo, and Cleveland. Since Pittsburgh was dominated by the manufacture of steel, only small numbers of women could find factory work of any kind. In 1920, only 14 percent of employed women in Pittsburgh worked in industry while 22 percent did so in Detroit and Buffalo and 25 percent in neighboring Cleveland. Although large numbers of women worked as seamstresses and dressmakers in all four cities, Detroit, Buffalo, and Cleveland offered women a greater number and variety of factory jobs than did Pittsburgh. Thousands of Detroit women worked in automobile factories, other iron and steel industries, and cigar and tobacco firms. In Cleveland, and to a lesser extent in Buffalo, textile and clothing manufacturers hired women as semi-skilled operatives. But in Pittsburgh, 30 percent of women wage earners in 1920 accepted employment in domestic and personal service. Only 22 percent of women workers in the other three cities were servants, laundresses, boardinghouse keepers, and the like. Detroit, Buffalo, and Cleveland even employed a somewhat larger percentage of women in the clerical occupations than Pittsburgh. Twenty-six percent of women workers in Pittsburgh were clerical workers in 1920 in contrast to 30.4 percent in Detroit, 28.2 percent in Buffalo, and 27.6 percent in Cleveland. From this comparative standpoint it is clear that Pittsburgh women faced a more limited labor market, as shown in table 3. In addition to the consistently lower incidence of married women's employment, two other features of the female labor market in Pittsburgh persisted decade after decade: the proportion of working women employed in domestic service remained higher and the proportion in manufacturing decidedly lower than in these other cities. Not until 1960 were the four cities about evenly matched in the percentage of women employed in the fields of domestic and personal service and clerical work; Pittsburgh continued to employ a lower percentage in manufacturing.[10]

A picture of women's employment in Pittsburgh in the early twentieth century indicates that gender, ethnic, racial, and class distinctions grew over time as the size and composition of the labor force changed. Segregation of the sexes and races in employment was as endemic in Pittsburgh as elsewhere in the country. Females and males as well as blacks and whites worked in different industries and in separate and contrasting occupations. Despite dramatic changes in female labor force participation and women's work settings over many decades, the practice of occupational segregation by gender and race persisted in numerous forms and promoted a large wage gap between men and women and between whites and blacks.

Numerically the most important field of employment for women in Pittsburgh, as elsewhere, was private and public housekeeping. In 1910, 21,147 (41 percent) of the 51,678 women in the Pittsburgh labor force earned their livelihoods as servants, waitresses, charwomen, cleaners, porters, housekeepers, or stewardesses. Only 5 percent (8,970) of Pittsburgh's male workers (181,959) were similarly employed. Native-born women of native-born and foreign-born parents, foreign-born women, and black women all worked as servants, but to considerably different degrees. Foreign-born women consti-

TABLE 3
WOMEN WORKERS IN SELECTED OCCUPATIONS, 1920 (*percent*)

	Pittsburgh	Detroit	Buffalo	Cleveland
Manufacturing and Mechanical Industries	14.1	22.2	22.5	25.5
Domestic and Personal Service	30.3	22.8	22.7	21.9
Clerical Occupations	26.3	30.4	28.2	27.6

SOURCE: U.S. Bureau of the Census, *Fourteenth Census of the United States, 1920: Population, Occupations*, vol. 4 (Washington, D.C.: Government Printing Office, 1923), table 2.

"White"

Separate,

tuted almost 30 percent of the total number of women in domestic and personal service; women of immigrant parents accounted for 24 percent; native-born women 19 percent; and black women 18 percent. Within this service classification black women were more concentrated in housekeeping services than any other group. Eightyeight percent of the black female wage earners in Pittsburgh worked in this field as compared to 64 percent of foreign-born working women and 27 percent of native-born women with either nativeborn or foreign-born parents.[11]

Opportunities for black women were particularly limited, a pattern that typified black women's experience throughout the United States. In the early twentieth century, when white women were entering white-collar work and black men were gaining access to the steel industry, black women could only find work as household servants, laundresses, or janitors in downtown stores and offices. Only a few black women worked in factories, in jobs that white women would not accept. Except for the holiday season, when department stores were short of help, black women were limited to the positions of stock girls and wrappers. Although the public schools educated both white and black students in the same classrooms, black teachers were, ironically, totally excluded from the Pittsburgh schools until the late 1930s. Telephone companies followed the same pattern, refusing to hire blacks as switchboard operators even though it was distinctly "women's work." The steel mills did not hire black women, even as common laborers, until the 1940s when the Fair Employment Practices Committee pressured them to comply with government regulations.[12]

Elizabeth Beardsley Butler, author of *Women and the Trades*, one of the volumes of the famous Pittsburgh Survey, found in 1909 that women's industrial and service jobs were rigidly segregated by gender from one end of Pittsburgh to another and in workplaces as different as canneries and department stores. In telephone companies women operated switchboards, whereas men installed and repaired the machines. In lamp manufacturing men shaped brass and iron, but women riveted and punched holes in metal. In cracker factories men prepared the dough that women baked and packaged. In glass manufacture men produced the glass objects that women decorated and packed. Rarely did men and women perform the same work or earn the same pay. Women's wages were usually one-half of men's.[13]

Ethnic distinctions also formed powerful barriers between groups of women. Employer and employee prejudices mutually reinforced hiring and employment patterns. Polish women dominated the female labor force in the cracker industry. In cigar factories Hungarian, Croatian, and Polish women were barred from every process except the stripping of tobacco leaves. In the metal trades two-thirds of the women making cores and winding coils were Slavic. Jewish girls and immigrant Jewish women concentrated in the stogy (cheap cigar) and garment factories and the "cheap and hustling" shops, while American-born women could be found mostly in the "better-class" stores, telegraph and telephone exchanges, book binderies, candy factories, and millinery houses.[14] According to Butler, even when women of differing ethnic identities worked in the same setting, differences of language, custom, and attitude still kept them apart.

Occupational sex, ethnic, and class segregation eventually characterized the white-collar fields of teaching and clerical work. In the case of schoolteaching, centralization increased the different forms of segregation. Marguerite Renner's case study of Pittsburgh school teachers from 1850 to 1900 corroborates the findings of other scholars, that in urban school systems women came to outnumber men by about ten to one and taught in the lower grades, while men worked in the higher grades and as managers. By the turn of the twentieth century school boards officially excluded those who did not meet the certification standards, blacks, and married women. The Pittsburgh school board specifically sought single, white, native-born, high-school educated women to teach the city's young children.[15] Women who did not fit this profile had no recourse.

The unequal distribution of women and men and the reliance on a more middle-class labor force in the Pittsburgh public schools took many decades to develop. Until 1911 the schools operated on a ward basis, allowing neighborhoods to exercise extensive control over all school policies. From the 1840s through the 1860s the ward school boards unilaterally decided the length of the school year, designed the curriculum, selected the texts and teachers, levied property taxes, and maintained school buildings and equipment. Consequently, the profile of teachers varied from ward to ward and reflected each ward's class composition—upper middle and middle-class, artisan, or unskilled. The Fourth Ward leaders, for example, selected teachers according to the rules of the patronage system, hiring the daughters of prominent citizens such as the school board president, the alderman, constable, and city councilman. The Ninth Ward board, composed of artisans, hired working-class women from unskilled and skilled families who had earned teaching certificates and fit the board's craft-based notion of a properly trained worker.[16] During the 1870s, almost 39 percent of the city's teachers came from professional, business, or white-collar families, while 55 to 60 per-

cent of the city's female teachers came from working-class backgrounds, reflecting the population of the wards.[17] Although the majority of teachers were single, a significant minority of married and
widowed women were also hired. In a few instances husband-wife
teams worked in the ward schools.

For many decades the people who hired teachers did not set high
educational standards. It was relatively easy for white, native-born
women to qualify for teaching positions in Pittsburgh and other
American cities in the nineteenth century. Until the 1870s school
board members usually required only that instructors be literate and
reasonably well versed in reading, writing, and arithmetic and that
they be respectable native-born members of the community. Pittsburghers justified their hiring of women in the same terms as common school advocates in other cities: women were ideal teachers of
little children because they were more nurturant, patient, and moral
than men and would work for half as much. In 1900 almost 70
percent of the 926 women teachers in the Pittsburgh public schools
worked as grammar-school teachers.[18] Men were chosen to be the
principals and superintendents because they were more likely to
make a career of managing educational institutions. There were few
exceptions to this rule.

As Pittsburgh extended its governmental control to adjacent communities in the 1870s, greater consideration was given to standardiz-
ing the educational requirements of students and prospective teachers. By 1909 the Pittsburgh school code had abolished ward control
of the schools. At the same time the state began to exercise more
jurisdiction over the selection of teachers through the county superintendents of education. Teachers were obliged to complete either
the normal school or the academic program of the Pittsburgh High
School. After the introduction of state certification the social origins
of female teachers began to change. In addition, the marital status of
female public-school teachers shifted after 1911 when the state of
Pennsylvania officially barred married women from teaching in the
public schools.

Urban schoolteaching was one of the best jobs open to women. It
offered better wages, steady employment for ten months a year, and
respectability. But teachers' wages were especially tailored to the
supposed needs of young single women, not to career-minded, self-
supporting wage earners. Teachers who needed or wanted higher
wages had to fight for them. After the turn of the century they
organized to improve their professional standing.

Women teachers, seasoned veterans of the classroom, were in the
forefront of the 1904 petition campaign in Pittsburgh for higher

does this include minorities?

wages, merit increases, and retirement programs. The central school board established a Teachers' Salary Commission in 1905 which raised the yearly wage scale for teachers and introduced a classification system based on experience and training. The graduated salary scale especially rewarded graduates of normal schools and academic high-school programs. Women teachers thus successfully promoted the professionalization of teaching.[19]

Professionalization ultimately changed the population of women who would become teachers.[20] Since only the most privileged working-class families could afford to send their daughters to high school, only girls from the upper strata of the working class would become teachers. Native-born women between the ages of fifteen and nineteen, for example, were more likely to attend the Pittsburgh Central High School in 1860 than any other group of female or male teens. Twenty-eight percent of all girls and boys living at home attended high school in 1860, but 36 percent of third-generation girls (native-born children of native-born parents) were enrolled. Although most female high-school students probably did not work for wages after they left school, nearly 75 percent of those who took jobs became elementary-school teachers, whether or not they completed the four-year course. The occupations of the students' fathers also underscores the special nature of this population of high-school students. At least 48 percent of the fathers of the girls who attended high school in 1855, 1860, 1871, and 1890 were either professionals, businessmen, teachers, or office workers. Skilled workers were the single most prominent group to send their daughters to high school, accounting for at least one-third of the female students.[21] It is safe to assume that the mandatory certification of schoolteachers made it more difficult for females from the lower ranks of the working class to become teachers. Their families simply could not afford to send them to high school.

Although it is not known precisely why, few male high-school students in Pittsburgh chose public-school teaching for their careers. Like their counterparts in other cities, these men may well have considered the costs of normal-school training too high for the low return on their educational investment.[22] Male high-school graduates in Pittsburgh obtained jobs in more prestigious professions, business, and clerical work. Until office employment opened to females, schoolteaching was the only higher-status job women could expect. In the end, the centralization and bureaucratization of the Pittsburgh schools standardized the selection of teachers, weakened ward control, reinforced gender segregation, and increased class homogeneity among teachers.

Race considerations in the Pittsburgh public schools add another
dimension to the meaning of segregation. Most cities segregated
students along racial lines and hired teachers accordingly: black
teachers taught in all-black schools, white teachers taught in all-
white schools. Pittsburgh pursued a different policy. The Pittsburgh
school board abolished racial segregation in 1881, bringing black and
white students together for a time while forcing black teachers from
the system for over fifty years. Although the Frick Training School
for teachers admitted black students as early as 1921, it was not
until 1937, when black professionals mobilized to integrate the
public-school teaching staff, that blacks could use their teaching
credentials in Pittsburgh. Until then they had to look for work outside their chosen profession or outside Pittsburgh. Even after blacks
vigorously protested their exclusion from public-school teaching,
integration occurred slowly. From 1937 until 1950 black teachers
were assigned only to schools whose students were predominantly
or all black. A black (woman) did not become a principal of a predominantly white elementary school until 1970, six years after the
first black (woman) was elected a member of the Pittsburgh Board of
Education.[23]

The sexual and racial division of labor in the Pittsburgh public
schools foreshadowed the gender and race line that would be drawn
toward the end of the nineteenth century in other rapidly growing
institutions such as hospitals, libraries, department stores, and offices. A study by Ileen DeVault, which examines the social origins of
Pittsburgh clerical workers from the 1870s to the 1910s, further
illuminates the process of sex and class segregation in the market-
place.[24] Despite its reputation as the city of heavy industry, Pittsburgh was a leader in business procedures as well as manufacturing
technology and organization. As industrial production increased in
size and complexity from the 1880s, the market for support services
expanded. Banking, telephone communications systems, insurance
firms, advertising companies, and mail-order houses all grew rapidly. Business consolidation and expansion increased the demand for
larger secretarial staffs. At the same time a technological revolution
in office equipment and procedures occurred. The widening use of
the typewriter and of addressograph, calculating, card punching and
sorting, and duplicating machines accelerated the speed and transformed the methods of performing office work. With the advent of
commercial education, large numbers of men and women became
available and eligible for office employment. But women and men
did not receive the same job opportunities in the clerical field.
Theories of labor supply and labor demand have been used to

explain the sexual division of labor in office employment. Labor supply theory emphasizes the value of human capital and the occupational choices people make based on their own values, aspirations, sex-role expectations, and educational credentials. According to the supply-side argument, both women and men are socialized to plan for and enter occupations that society deems suitable for them. Human capital explanations of sex segregation see the choice and preparation for a sex-typical occupation as rational planning. Since women's work histories have tended to be fragmented due to child-bearing and child-rearing responsibilities, they have usually selected occupations that offer low depreciation of skills and training while they are absent from the labor force and that permit easy reentry. Such occupations become predominantly female.[25]

Labor demand explanations argue that the exclusion of women from traditionally male jobs, especially higher paying, higher status employment, is largely a result of employer's preferences, not women's choices. According to this argument employers are unwilling to hire women for jobs that involve extensive on-the-job training and considerable responsibility because women leave their jobs to marry and raise children. So employers create segments in the labor market to maximize their profits. Women are appropriate for low-paying jobs that require few skills because they would not remain on the job for the more demanding positions. Theories of labor market segmentation or internal labor markets argue that there is a general correspondence between the sexual division of labor and the distinction between stable, high-paying jobs and low-paying jobs with high turnover. The former have historically been reserved for men and the latter for women.[26]

DeVault's analysis of clerical work can be used to support both the dual labor market and human capital explanations for the concentration of women on the bottom tier of the clerical labor force: young women eagerly sought clerical work because it suited their personal needs as women and as members of a particular socioeconomic class, but corporate managers severely restricted women to low-paying, mechanized, routinized work. Employers in Pittsburgh eventually adopted a "pyramidal" organizational structure in offices to keep up with the flood of paper transactions as inexpensively as possible. The office hierarchy was predicated on a rigid sexual division of labor. Mechanization and specialization resulted in a tri-level stratification of the office labor force: middle managers trained to deal with matters involving judgment, experience, and responsibility; lower managers with special skills, such as advanced bookkeeping; and machine operators and others who performed routine tasks.[27] Women dominated the lower tier and men the upper two. In most American offices men served as general clerks, accountants, shipping clerks, weighers and messengers, while women accounted for a majority of basic bookkeepers, cashiers, stenographers, machine operators, and filing clerks. The more labor intensive the office work the more likely it appears that women would be hired.

Businessmen vigorously promoted the exclusive employment of women as machine operators. Typewriter companies, for example, used women to demonstrate the speed and efficiency of typing correspondence. Over twenty-five thousand women were employed in 1904 through Remington Typewriter Company's "placement bureaus" in major cities across the United States. Scientific management experts also encouraged the training and employment of women in the low-paying jobs with high labor turnover. Private business schools and public high schools with commercial education programs further institutionalized the sexual division of labor. Women were especially encouraged to master the labor-intensive skills of stenography and typing for which companies particularly sought their labor. The Commercial Department of the Pittsburgh High School offered bookkeeping, stenography, and typewriting for all students but reserved managerial training for males.[28]

School attendance rates suggest that many families understood that a high-school education was a better investment for girls than boys. The national figures show that for the years from 1890 to 1928 girls outnumbered boys in high school by four to three among whites and three to two among blacks.[29] Despite the fact that girls spent only a few years in the labor force before marriage, a high-school education was essential to their occupational mobility. Without it they were restricted to domestic service, factory labor, sales, and telephone communications. Formal education offered the only opportunity for girls to acquire higher job skills, but for boys, it was only one of many avenues to a decent job.[30] Boys could train on the job for skilled manual labor and obtain other employment without completing high school. The commercial education program in the public high school consequently became an especially attractive course of study for young women. Although females were only 43 percent of the students in the Commercial Deparment of the Pittsburgh High School from 1890 to 1903, their presence in the program gradually increased over those years and far exceeded their representation in the clerical workforce in Pittsburgh.[31]

The high-school students enrolled in commercial education came from diverse backgrounds. Unskilled workers, skilled manual workers, clerical and sales workers, proprietors, and widows deliberately

chose to further their children's education. At least half the commercial education students came from working-class families. DeVault
argues that the meaning of the clerical education program was shaped
by the families' different material conditions and social status. For
unskilled workers it was "an avenue of escape from narrow and often
impoverished lives." For skilled workers it served to reinforce their
superior position within the working class. For widows clerical work
may have been seen as "a form of insurance" for daughters who might
some day face widowhood and the need to become self-supporting.[32]

The social origins and marital patterns of clerical workers offer
further insight into the importance of class identity in shaping parents' decisions to enroll their children in the public high-school program. More than any other group, skilled male workers enrolled their
sons and daughters in the special training program. These glass blow-
ers, machinists, carpenters, railroad conductors, and others turned to
the new opportunity for clerical education for their children, DeVault
contends, to protect the privileged status these families held within
the working class which was being eroded by corporate challenges to
the fathers' occupations. Employment in the clerical sector was one
way for the children of skilled workers to distinguish themselves
from other groups within the working class at a time when their
fathers' occupations were undergoing technological and social redefi-
nitions. From this point of view, native-born skilled workers from
Northern and Western European families were anxious to distance
themselves from the unskilled and semiskilled Southern and Eastern
European immigrants who "invaded" their workplaces in the late
nineteenth and early twentieth centuries. Clerical employment became an avenue by which some labor aristocrats could differentiate
their social standing from that of the new immigrants.[33]

By comparing the job options available to the sisters of women
and brothers of men enrolled in the Commercial Department of the
Pittsburgh High School, DeVault graphically demonstrates how attractive office work was to the children of skilled manual workers.
The sisters of aspiring office workers tended to be salesclerks, needleworkers, or teachers. Sales work and the sewing trades paid less than
office employment and were known for their poor working conditions. Although teachers earned more than saleswomen or seamstresses, the teachers' unemployment rates grew because job openings did not keep pace with the number of eligible applicants. As a
result, women began to look favorably upon office work since it put
them at the top of the female labor market. Similarly, the realistic
options available to aspiring male clerical workers were low-level,
white-collar jobs, principally in sales, skilled manual labor, and un-

skilled and semiskilled factory employment. In terms of wages, working conditions, and social status, clerical work was also an attractive alternative for the sons of skilled workers.[34] This portrait suggests that women did not aspire to higher-ranking office positions because they expected to marry and leave the labor force. At the same time men did not fear direct job competition from women since the demand for clerical workers outpaced supply and the gender-segregated market protected the better jobs for men.

It is difficult to interpret precisely how skilled blue-collar workers thought about white-collar work. DeVault rejects the argument that skilled workers wanted their children to leave the working class for the middle class. She contends instead that labor aristocrats were intensely proud of their privileged position in the working class. They wanted their sons to have well-paid, secure employment and their daughters to enhance their family's declining social status. By sampling the marriage records of clerical workers DeVault found that 40 percent of the labor aristocrats' daughters married clerical workers, 32 percent married manual workers, and less than 25 percent wedded professionals or retail, wholesale, or manufacturing proprietors. Since almost a third of these women married manual workers, DeVault concludes that the daughters of skilled workers shared their fathers' pride of class. Commercial education was not "solely" or "primarily" a way of attaining social mobility through marriage.[35]

There is another way to interpret these findings: almost 65 percent of skilled workers' daughters who enrolled in the commercial education program married white-collar workers of one or another sort. It could be argued, as Cindy Aron does for middle-class clerical workers, that office work fundamentally altered working-class women's behavior, attitudes, and expectations.[36] Working-class women who were employed in offices rubbed shoulders with their social equals and superiors every day. As DeVault points out, the women invested in suitable clothing to look "business-like," encouraged by "rumor, fiction, and suggestive jokes" to hope that the right attire would bring them to the attention of the boss's son.[37] Those who worked in downtown Pittsburgh probably frequented the area's fashionable restaurants, department stores, specialty shops, and movie palaces. White-collar work and marriage to white-collar workers may have signified movement into the middle class and the adoption of middle-class values. Since school and marriage records only allow us to speculate about social values, it is possible that white-collar work meant one thing to the parents of office employees and something else to the clerical workers themselves. It would be interesting to know what the educational and marital patterns were of the clerical workers' children.

No matter how one interprets DeVault's findings, it appears that clerical employment resulted in a greater distinction within the working class between those at the top and those at the middle and bottom. Teachers and clerical workers were at the top of the female labor market in the late nineteenth and early twentieth centuries. Educated working-class women had more in common with middle-class women than with women from the lower segments of their own class. Within the sexually segregated labor market, ethnicity, class, and race separated working-class women from one another. These women rarely transcended these formidable barriers to form bonds of friendship and sisterhood.[38]

Women and Labor Militancy

Patterns of women's labor militancy in Pittsburgh between the 1840s and the 1930s reflected the changing structure of the female labor market, steel's dominance of the Pittsburgh economy, and the highly segmented nature of the female labor force. Although women factory operatives participated in the labor movement, most of the Pittsburgh women who worked did so in fields outside the labor movement—domestic service, sales, schoolteaching, and clerical occupations—leaving only a tiny percentage with any opportunity for joining a trade union. Working-class housewives, who did not usually work outside their homes nonetheless strongly defended the rights of organized labor and distinguished themselves as unusually militant during numerous labor conflicts.

Women textile workers in Allegheny City were highly visible in labor conflicts of the early nineteenth century. The cotton mill strike of 1843 and the strike and riot of 1848 belie the stereotype of women as docile and passive workers. Although women earned only one-third to one-half of men's wages, they fiercely protected their rights as workers. When their livelihood was jeopardized by wage cuts, they took to the streets, sometimes as a formidable ax-wielding army capable of disorderly conduct quite contrary to the social norms governing women's public behavior. In 1848 women rioted side by side with men, oblivious to the contemporary rhetoric about female delicacy and the existence of separate spheres for men and women. A jury found the women guilty of riotous behavior, and the trial judge meted out the same sentences and fines for them that he gave to the men.[39] Other examples in which women defied gender-role conventions include the Homestead strike of 1889, the Homestead lockout of 1892, the Westinghouse strike of 1916, and the great steel strike of 1919.

During the great lockout at the Homestead steelworks in 1892, Pittsburgh women, as reported by New York and Pittsburgh newspapers, were as feisty, if not more so, than men. This time working-class housewives, who ordinarily tended only to domestic matters, urged their husbands to protect their homes just as the women had done three years earlier during a successful strike. When Henry Clay Frick hired hundreds of Pinkerton detectives to lock steelworkers out of the mills in 1892, elderly women and mothers with babes in arms and children in tow joined the crowd actions. The women released their "pent-up rage" by shouting the "vilest profanity" and jostling Pinkertons with umbrellas, brooms, and blackjacks. One woman is said to have loaded the family rifle and brought it to her husband just, it seems, as dutifully as she might have brought him his lunch. Other "boldly aggressive" women shouldered muskets and rifles themselves or helped their husbands stockpile clubs and stones to use against the Pinkertons. A newspaper headline captured the women's intense feelings in capital letters, noting that women "URGED THEIR HUSBANDS TO KILL PINKERTONS." One housewife resolutely announced, "It's either us or the mill. . . . I want my man to keep it up until we win or are destroyed." Housewives who ordinarily spent their days cooking, cleaning, and taking care of children became fierce and aggressive when they believed that their family ideals, standard of living, or the lives of their husbands were being threatened.[40]

Housewives in Homestead were not unique in this regard. The working-class women of many ethnic communities physically defended their way of life during economic crises throughout the United States. The immediate cause might differ, but housewives' public actions bore striking similarities from place to place. What one historian said about Jewish housewives' participation in the 1917 kosher food riot in New York applies as well to Homestead: "They acted together—for themselves, for their families, for their neighborhoods, and in defense of their world."[41] In contrast to New York, where grocers capitulated to the demands of the food boycotters, in Homestead the Carnegie Steel Company overwhelmed the workers so powerfully that unionization remained effectively sidetracked for thirty years.

Between the Homestead lockout of 1892 and the Westinghouse strike in East Pittsburgh in 1916, labor conflicts were less dramatic, limited to smaller groups of workers, and distinctly nonviolent. During the 1910s native-born female telegraphers worked in concert with male co-workers for better labor conditions. Of the ninety female telegraphers in the city, 42 percent joined the telegraphers' union and went on strike with men, demanding equal pay for equal work among other changes. Even though unorganized women contributed to the strike's defeat, the labor contract resulted in the abolition of wage differentials based on gender at one of the city's two telegraph companies.[42]

The stogy industry provides another example of men and women working together to improve their working conditions. In 1912 Pittsburgh's Hill district, then populated by Jewish immigrants and dotted with sweatshops, erupted with labor activity. Two hundred Hill district cigarmakers formed the Industrial Workers of the World Local 101 and went on strike in July and August. They struck again in July 1913 for higher pay and improved sanitary and safety conditions. Since entire families made stogies together, women participated in the union campaign both as family members and as workers. After an eighteen-week work stoppage, the cigarmakers won full recognition of their union, a fifty-hour workweek, better sanitary conditions, and improved pay.[43]

In 1916 a strike by thirteen thousand Westinghouse workers in East Pittsburgh involved some three thousand women. The strikers, who represented many nationalities, religions, and skills, demanded an eight-hour workday with no reduction in daily wages. Three of the strike leaders were women from the local community and another two were women organizers assigned by the American Federation of Labor. As in the case of the Hill district stogie workers, leftwing unionists played a prominent role in the strike by organizing on an industrial basis and articulating an ideal of militant unity in word and deed. In the end, the Westinghouse strikers could not match the power of the coal and iron police, company guards, railroad detectives, infantry and cavalry soldiers brought into the Turtle Creek Valley at Westinghouse's behest.[44] The memory of the crushing defeat in 1916 reduced labor's voice in East Pittsburgh for nearly twenty years, until female and male electrical workers joined the drive for industrial unions in the 1930s.

The Pittsburgh steel strike of 1919 differed markedly from the 1892 Homestead lockout. In 1892 skilled and unskilled steelworkers formed a united front, and American-born and immigrant housewives joined street actions to support their husbands. In 1919 very few of the native-born skilled steelworkers of the Monongahela Valley supported the strikers—several thousand mostly unskilled laborers from Southern and Eastern Europe. As David Brody has shown, the English-speaking steelworkers "were the soft flank in the strik-

ers' ranks," leaving the foreign-born and their wives to stand virtually alone in 1919.[45] Immigrant women attended union meetings
and street demonstrations in support of their husbands' efforts to
win union representation. Two prominent organizers for the United
Mine Workers—Mother Jones and Fannie Sellins—also lent their
considerable skills to the organization of street demonstrations to
publicize the strikers' cause. The strikers and their wives could not,
however, match the ferocity of the Pennsylvania constabulary,
which unmercifully beat strikers and supporters in their homes and
on the streets and jailed them for exercising their right to assemble
and speak. In one of the many fracases with the police, Fannie Sellins was shot to death in a nearby steel community.[46] The strikers
and wives finally yielded to the company, abandoned by their skilled
co-workers and suffering from the assaults on them by local and
state authorities.

Women in Pittsburgh throughout this period figured in labor-
management confrontations as wage earners and spouses, and sometimes as union organizers, but little is known about them. Strike
reports in daily newspapers only occasionally mention women by
name and typically give but little information about them, as witness the case of twenty-one year old Anna Katherine Bell, a seasoned
worker at Westinghouse, whose family evicted her because she participated in the 1916 strike, or the "girl in the paper mask" who
obviously feared public recognition but was determined to head one
of the many street processions of Westinghouse strikers. Such examples suggest that labor militancy meant more than an intellectual
commitment to the ideals of organized labor. At a time when police
actions against strikers and their supporters were commonplace,
labor solidarity demanded extreme courage and physical stamina,
and only exceptional women and men devoted themselves to the
cause of organized labor. Ronald Schatz discovered in studying Pittsburgh electrical workers in the 1930s that the men and women who
became labor activists had particular characteristics. They tended to
be political nonconformists and/or they possessed atypical demographic traits. Male labor activists tended to be middle-aged, skilled
workers from families of Northern European descent with long-term
commitment to union or socialist ideals, while the women typically
were young, semiskilled wage earners from Southern and Eastern
European immigrant families. The women drew support from families committed to the labor movement or from the fact that they
headed their own families and so could air their views without fear
of familial rejection.[47] Whether male or female, labor leaders stood
apart from their co-workers.

Housewifery and Social Inequality

Home life in Pittsburgh and in nearby Homestead in the early twentieth century replicated the patterns of class, ethnic, and racial stratification found in the labor force. The home, as Elizabeth Ewen has demonstrated for the Lower East Side of New York, "showed the imprint of the economic universe that structured its reality."[48] A number of studies of Pittsburgh conclude that the standard of living varied markedly within and between social classes. The result was a pyramid, with the largest group of wage earners living in unhealthy and decrepit rental property and the smallest and most privileged group living in luxurious homes of their own. During the flowering of corporate capitalism in the late nineteenth and early twentieth centuries the fruits of economic growth in Pittsburgh were distributed unevenly among the city's wage earners. Middle-level managers lived less well than the corporate elite but much better than skilled workers. The labor aristocrats in turn lived appreciably better than unskilled workers.[49] The nature of housewifery varied accordingly.

Ethel Spencer's memoir of the turn of the century reminds us that in the upper fringes of the middle class, families lived in newly built, roomy houses on spacious lots that could accommodate many children, guests, and servants. Her parents' home was decorated with marble-topped golden oak and walnut furniture. Lace curtains in the spring and heavy velours in the winter covered the tall, large windows. The rooms were lit by gas chandeliers before World War I and by electricity afterward. Except for a scarcity of bathrooms for the nine-person-plus household, the Spencers lived very comfortably.[50]

Housewifery at the Spencers reflected the family's upper middle-class social status. Ethel Spencer's mother managed the household, performing many chores herself, but she was usually assisted by a cook, chambermaid, laundress, and one of two nurses. She orchestrated the children's activities, taught and supervised the servants, planned the seasonal chores, maintained the household budget, altered and recycled the children's clothing, oversaw the children's religious training and dance lessons, and helped out in the kitchen and nursery. Housewifery was indeed a full-time, demanding job, especially in a family of seven children. Despite their large family the Spencers could afford to install home improvements that lightened the burden of domestic tasks. The family bought the latest in lighting, heating, bathing, and cooking equipment. When the house was built in the late 1880s, the newest type of gas cooking stove was installed in the kitchen, and several rooms in addition to the kitchen and bathroom had hot and cold running water. Ethel Spencer recalls

that her family ate very well because her mother and the cook spent
so much time planning and preparing meals. Mrs. Spencer's household budget easily covered the cost of the best cuts of meat, seasonal
vegetables, freshly baked bread, coffee beans, jams and jellies, fresh
eggs, butter, and cream.[51]

The nature of housework for the mass of working-class women
bore little resemblance to the Spencers' household. There were no
servants, and housewives were often responsible for boarders' laundry and meals. The lower segments of the working class rented their
housing and could not afford the household improvements available
to middle-class families. Their houses usually lacked hot and cold
running water, indoor plumbing, or gas cooking and heating stoves.
The lower standard of living made household labor much more arduous for the majority of working-class housewives than for women
from higher socioeconomic groups.

In a study of working-class housewifery in Pittsburgh in the
1870s and 1880s, Susan J. Kleinberg found that municipal services
varied from one neighborhood to another according to residents'
ability to pay for installation or use political leverage to get the city
government to do so. The many wealthy and middle-class homeowners of the newly settled suburban East End fared well in their efforts
to improve roads and access to public transportation. They were able
to persuade the city government to assume the entire cost of paving
their roads. Working-class residents in older neighborhoods could
not afford the higher rents that landlords would have charged them
for the municipal improvements, and the elected officials refused for
a time to use city revenues to pave the roads. The same was true of
the allocation of water and sewers. The City Council Water Commission decided in 1872 to use the amount of revenue from each street
as the basis for determining the installation of water pipes. Large
water mains and indoor water pipes were installed in middle-class
homes, but only small water pipes and pumps could be found in
working-class communities. The installation of sewers also varied
from one neighborhood to another. At a time when the wealthier
communities had excellent sanitary waste disposal, the industrial
neighborhoods lacked proper sewerage systems.[52]

The unequal distribution of municipal services could make routine household duties into herculean chores. In the most industrial
section of the city, the South Side, working-class women faced formidable obstacles in caring for their families. These housewives had to
haul water from backyard pumps for washing dishes and clothes,
cleaning, drinking, cooking, and bathing. In the summertime, when
the steel mills and railroads, situated on the flat lands of the area,

needed large quantities of water for their operations, the adjacent hillsides where workers lived were drained of their water supply. Women had to drag buckets of water from the pumps on the flatlands from seven in the morning until six in the evening when the increased water pressure allowed them to pump water again in their backyards. With only the occasional help of older children, the housewives became beasts of burden. On Mondays they washed clothes outdoors in the warm weather and indoors in the cold. Water had to be hauled inside and heated on top of the coal or wood stove. The stove had to be properly lit and the fire tended. With only a scrubbing board and tub, the housewife was obliged to rub, rinse, and wring the clothes by hand. The labor was made all the harder by the very dirty and sweaty garments that had to be washed every week. Although washing machines came on the market in the 1870s, only a few middle-class housewives bought them. On Tuesdays the ironing could consume half a day. Cleaning on Wednesdays and Thursdays was equally demanding because of the dirt and particles from the unpaved streets, the soot from the heating stove, and the lack of domestic appliances or help. Working-class housewives had only soap and water, rags and dusters, and their own considerable elbow grease. By the end of the week these housewives could look forward to baking and shopping. In between or during these chores these beleaguered women supervised children and prepared the daily meals.[53]

In her classic study of household budgets in Homestead in the early twentieth century, Margaret Byington found that the household routine and expenditures of steelworkers varied with the breadwinner's income and occupation. Work in the steel mills often involved weekly alternation of day and night shifts, thus disorganizing the household routine every seven days. Some family members might work in the mills while others slept. Mealtimes were often irregular, making it difficult for the workers' wives to organize their time and to provide nutritious meals. Meal preparation was additionally complicated because housewives had to prepare a lunch pail for the men to take with them or for the women to deliver.[54]

Working-class women, like their middle-class counterparts, were consummate household managers, but most had to manage on very limited resources and could buy only the daily necessities. Mrs. Spencer's careful household management allowed her to send her children to private high schools and college, take regular vacations, and outfit everyone properly. "By planning ahead, by extra labor, by wise buying," working-class women in Homestead could purchase ample food and sometimes a household utility item. Most of the women in Byington's study were lucky to make ends meet, pay the neighborhood grocer promptly, and afford commercial entertainment now and then. Working-class housewifery involved "constant watchfulness," much "patience," "practical skill," and sacrifice just to limit indebtedness for daily necessities.[55]

These two groups of women did not live in the same worlds nor were their paths likely to cross. When working-class and middle-class women met, it was as mistress and servant or social settlement worker and client, relationships that were inherently unequal. Ethel Spencer offers an insightful portrait of her family's household staff. There was a definite hierarchy among the servants: the nurse and cook ranked higher than the chambermaid and laundress. Most of the staff were single, newly arrived German or Central European immigrants; the laundress was Afro-American and married. Everyone except the washerwoman lived with the Spencers. The laundress was not only separated from the other servants by her race, marital status, place of residence, but also by the site of her work in the house. Minnie spent twenty-five years boiling clothes and ironing in the cellar. The Spencer children "were hardly aware of her departure at night." The servants seem to have gotten along with one another, although there was one instance of temporary "racial prejudice" between the German chambermaid and the newly hired Polish cook. Eventually the two became friends. Ethel Spencer remembers that her mother had very cordial relations with the servants because she appreciated their hard work and respected them as individuals. Years after the servants had married and set up households of their own, they continued to visit Mrs. Spencer. Ethel Spencer proudly reports that the servants learned to speak English, cook complicated and tasty meals, and aspire to a middle-class standard of living. None of the servants, of course, achieved a social or economic status approaching that of the Spencers.[56]

Women's Public World

Working-class women also brushed shoulders with their social superiors in the city's settlement houses. Native-born, middle-class women tried to Americanize immigrant women from Southern and Eastern Europe. According to Elizabeth Metzger, social workers in Pittsburgh's Kingsley House and Irene Kaufmann Settlement were the daughters and wives of well-to-do businessmen, attorneys, and physicians who attempted to improve the "ethical, social, and economic conditions" of the neighborhood residents.[57] To achieve these goals the settlement workers investigated immigrants' housing and health conditions, recreational patterns, diets, and cultural life. The settlement houses provided public playgrounds, bathing facilities, nursing services, libraries and reading rooms where discussion groups and debating societies could meet and where classes were held to teach English, citizenship, housekeeping, sewing, dressmaking, and millinery. As was true of other American cities in the early twentieth century, the Pittsburgh government eventually adopted many of these programs. The Board of Education assumed responsibility in 1906 for Americanization and English classes. By 1919 the city's public health association had taken over nursing services. The city gradually opened branch libraries in many communities and added playgrounds to the schools. But the settlements continued their programs in housewifery and motherhood, teaching immigrant women to check milk for contamination, sanitize their homes, cook nutritious meals, and care for their children according to American middle-class standards of proper behavior.[58] In light of the unequal distribution of municipal services, immigrant women must have found it very difficult to live up to the standards in their instruction.[59]

The history of social settlement houses in Pittsburgh also raises a key question: How important were women to social welfare reform in Pittsburgh? Roy Lubove and John N. Ingham note that women were prominent in the effort to make local and state government more responsible for improved housing, health, and labor standards.[60] The Civic Club of Allegheny County, the State Federation of Pennsylvania Women, the Consumers' League of Western Pennsylvania, the Pennsylvania and Allegheny County Child Labor Associations, and the Associated Charities of Pittsburgh depended on women to publicize their causes, lobby government officials, and coordinate legislative campaigns with other organizations. The Pittsburgh social settlements also spearheaded reform and extensively utilized the services of women. But unlike Hull House in Chicago, they did not groom women for leadership positions in social reform. Although Kingsley House, founded in 1893 by Rev. George Hodges, chose women to head the settlement from 1893 to 1902, only men directed the organization after that date. Forty-one percent of the Kingsley officers were women in 1897 (15 of 36), but less than 4 percent (2 of 52) in 1926. As Kingsley House's services grew, women staffed the teaching and nursing positions, but they lost their influential places in administration. The Kaufmann Settlement followed a similar path. The Council of Jewish Women established a Sabbath School in 1895 that became the Irene Kaufmann Settlement in 1909. In the first few years women ran the Kaufmann Settlement, but soon men outnumbered women as board members and administrative heads. In 1906 women filled three-fourths (28 of 38) of the administrative positions. By 1913, eight years after the official link with the Council of Jewish Women had ended and an election system for officers had been instituted, women held only 35 percent (12 of 34) of the administrative posts. By 1918, their percentage had dropped to 22 percent (4 of 18).[61] The development of the social settlements in Pittsburgh adopted the same sexual division of labor as other local institutions: women swelled the rank and file and disappeared from the leadership. Consequently, the Pittsburgh settlement houses did not encourage women to make, in the words of Kathryn Kish Sklar, "an indelible imprint on U.S. politics" the way that settlements in Chicago and New York did.[62] It may well be that Kingsley House and the Kaufmann Settlement were more typical of the movement nationwide than was Hull House.

Future Research

The studies of women in Pittsburgh and the scholarly debates about conceptual approaches to the study of gender suggest some particularly useful avenues for future research. White-collar employment, housewifery, and public activism all deserve further attention, but they need to be studied over a longer period of time and in a way that highlights the diverse experiences of women from different ethnic groups and races and systematically contrasts women and men.

The process of structural change in Pittsburgh, a painful transition from heavy manufacturing to a more service-oriented, high technology and a diversified manufacturing base, underscores the importance of studying the evolution of white-collar employment in Pittsburgh from the early twentieth century to the present.[63] Not only regionally, but nationally and internationally, steel production dropped markedly in the early 1980s, but Pittsburgh had begun to lose its relative ranking in national steelmaking capacity and heavy manufacturing by the third decade of the twentieth century. A longer view of the region's shift—from being a blue-collar, working-class town toward one that is more white-collar and middle class—would seem necessary.

Patterns of occupational segregation by sex, ethnicity, and race have also shifted profoundly in the twentieth century in the clerical, sales, communications, and teaching fields. In the early twentieth century, white-collar employment was the particular province of white women of Northern and Western European heritage. Later in the century, women from Southern and Eastern European families as well as Afro-American women obtained such jobs. An understanding of the timing and causes of these changes in both women's and men's employment patterns would help test the applicability of labor supply and labor demand explanations for various forms of workplace segregation.

Since entry into certain white-collar occupations depended on high-school training, it is essential to determine when secondary education became important to the lower ranks of the Pittsburgh working class. The patterns of high-school attendance probably varied significantly from one city and ethnic group to another. A study of women in New York City, for example, suggests that Italian girls did not attend high school until the 1940s because their parents depended on their labor as industrial homeworkers. When industrial homework and child labor became illegal in the 1930s, Italian families had to look for alternate ways for their daughters to contribute to the family economy. It was at this point that Italian girls were sent to high school to study office work. Until then the girls' education had been limited to elementary school and lagged far behind that of other ethnic groups.[64] The labor market and labor laws forced Italian families to evaluate their attitudes toward formal education for their daughters.

Since Pittsburgh has been home to so many different peoples, the city is an excellent laboratory for studying ethnic and racial variations in women's and men's education and employment patterns. School attendance at all levels illuminates not only the material circumstances but also the attitudes of diverse social classes and ethnic groups toward women's place inside and outside the home. A recent study of blacks, Italians, and Poles in Pittsburgh from 1900 to 1960 concludes that occupational mobility was very modest for the men in all three groups: at best they moved into skilled positions in industry, and only a few obtained white-collar jobs. The same study indicates that Polish and Italian women refrained from entering the Pittsburgh labor force until 1950, at which time they acquired jobs in offices and stores. The reasons for this delay remain unclear.[65] The roles played by economic development, state labor and education laws, and cultural tradition in shaping these patterns have yet to be analyzed.

The transformation of housewifery since the First World War is an excellent vehicle for examining several major historical developments in the twentieth century: the evolution of municipal services, the changing nature of women's domestic labor, and class distinctions in consumption patterns and household technology. Only the broadest outlines of change have been sketched for Pittsburgh (or elsewhere) to describe the dispersion of municipal services and household technology between the First and Second World Wars. Ruth Schwartz Cowan argues in her two-century history of house-work that in the 1920s and 1930s the lower ranks of the American working class could rent or buy housing with electricity, hot and cold running water, and indoor bathrooms in neighborhoods with adequate sewers, regular garbage collection, and street cleaning. Elec-tric washing machines, vacuum cleaners, and refrigeration were as yet still limited to "more comfortable" families. After the Second World War the diffusion of household amenities grew appreciably. Class distinctions in housewifery, which had been so profound ear-lier in the century, were redefined in the 1960s and 1970s. Although Cowan does not discuss the implications of her findings for the meaning of class identity in the United States, her material suggests that features of suburban housing following World War II—such as unitized kitchens, extra bathrooms, family rooms, and electric and gas appliances—may have profoundly affected the self-perceptions of blue-collar and lower-level white-collar workers and families. To what extent did the development of new housing and the adoption of new standards for housewifery redefine differences between the stan-dard of living and consumption patterns of the working class and middle class?[66]

The dramatic increase in married women's labor force participa-tion since the Second World War is partly responsible for the rising standard of living among wage earners, and it has received much attention. Scholars have studied the effect of married women's em-ployment on women's attitudes toward equal rights, voting behavior, and family roles, but its effect on class distinctions in the United States is not well understood.[67] At the beginning of the twentieth century, for women to work for wages outside the home was socially acceptable only for those white married women who had to supple-ment their family income or for women who were their family's sole support. Decade after decade controversy raged over the right of mar-ried women with employed husbands to work for wages. Married women's labor force participation continued to climb nonetheless and perhaps accelerated in the 1980s due to inflationary pressures on household budgets. Since married women have been concentrated in the white-collar sector since about 1930, it is likely that their employ-ment patterns have obscured the boundaries between the working class and middle class. The entrance of working-class women into commercial education programs at the beginning of the century was part of a gradual process that began with the upper echelon of the working class and worked its way down to the lower ranks, including more married women along the way. Women's increased employment in white-collar jobs and married women's increased labor force participation complicate class distinctions, reorganizing and perpetuating them in new patterns that differ from those which were so pronounced in 1900 and were in large part based on male occupations. Pittsburgh is an excellent setting for studying this issue since the blue-collar/white-collar occupational mix within families appears to have been prevalent.

The Pittsburgh *Business Times* published in 1987 the first issue of a quarterly supplement devoted to women in business in contemporary Pittsburgh. The entire issue celebrates female achievers serving as members of City Council, corporate vice-presidents, television news directors, judges, athletic directors, business owners, robotics researchers, bank executives, and arts promoters. As volunteers women continue to raise millions of dollars each year for charitable and cultural institutions including the opera, symphony, and ballet, as well as numerous social, health, welfare, and religious organizations.[68] Nevertheless, formidable barriers still hinder women's access to the uppermost echelons of the public and private sectors. Anecdotal comments abound about Pittsburgh's extremely conservative corporate culture and the limited opportunities available to professional women.

As women occupy increasing numbers of white-collar jobs in the Pittsburgh region, their situation calls for focused study not unlike the attention given more than seventy-five years ago by the Pittsburgh Survey to female factory operatives and housewives—only now the focus should be on clerical workers, professionals, and managers. Work along these lines would fall within the genre of scholarship which examines women's changing experiences as a way to understand such long-term social processes as the persistence of occupational sex segregation, the expansion of public schooling, the diffusion of municipal services and domestic technology, the rise of large-scale corporations, and the evolution of living standards. The studies that best expand our understanding of social change and highlight the special nature of women's experiences include female-male comparisons. It is only by contrasting women and men's schooling patterns, labor markets, job histories, volunteer activities, domestic roles, and marriage patterns that the importance of gender differences can be determined.

The interplay of gender with social and economic forces forms a complex mosaic whose history is more than a sum of discrete parts. If future studies of Pittsburgh history are designed with this thought in mind, they can better enhance our understanding of social-historical processes. The scattered literature available to date offers insights and examples but requires supplementation and synthesis.

NOTES

I wish to thank Eric Davin for research assistance and Martin A. Greenwald for editorial assistance.

1. Anne Firor Scott applies the notion of women's historical invisibility to the history of voluntary associations in "On Seeing and Not Seeing: A Case of Historical Invisibility," *Journal of American History* 71 (June 1984): 7–21.

2. R. Jay Gangewere, "The Myth of Pittsburgh," *Carnegie Magazine* January/February 1987, pp. 4–5. Photographs headlined "The Great Entrepreneurs," "The Men Who Made Pittsburgh," "Men Behind the Renaissance," and "Leaders in Industry and Business in the Seventies" are featured in Stefan Lorant, *Pittsburgh: The Story of an American City* (Lenox: Authors Edition, Inc., 1980), pp. 232–40, 366–67, 403, 526–27. Lorant's photographs of women are fewer in number than those of men and give prominence to the wives and sisters of eminent businessmen, college students, sorority sisters, club members, and writers. See pp. 245, 248, 257–58, 392, 478.

3. The phrase comes from the title of S. J. Kleinberg's book, *The Shadow of the Mills: Working-Class Families in Pittsburgh, 1870–1907,* (Pittsburgh: University of Pittsburgh Press, 1989).

4. Joan Wallach Scott, "Women in History: The Modern Period," *Past and Present* 101 (Nov. 1983): 141–57. Scott further explores the conceptual implications of studying women's history in "Gender: A Useful Category of Historical Analysis," *American Historical Review* 91 (Dec. 1986): 1053–75.

5. Monte Calvert, "The Allegheny City Cotton Mill Riot of 1848," *Western Pennsylvania Historical Magazine* 46 (April 1963): 99.

6. Peter Shergold, *Working-Class Life: The "American Standard" in Comparative Perspective, 1899–1913* (Pittsburgh: University of Pittsburgh Press, 1982), p. 70.

7. Ibid.

8. U.S. Bureau of the Census, *Historical Statistics of the United States, Colonial Times to 1970,* part 1, bicentennial ed., (Washington, D.C.: Government Printing Office, 1978), p. 133, contains the national rates of married women's labor force participation. The figures for Pittsburgh in 1900 and 1910 can be found in Shergold, *Working-Class Life,* table 17, p. 76. As Shergold indicates, the conjugal status of Pittsburgh women workers in 1910 has to be estimated from the 1900 figures which undercounted the number of women shopkeepers and boardinghouse keepers. He offers three

estimates: a minimum of 3.5 percent, a maximum of 5.8, and the mean probability of 4.6. In any case, married women's labor force participation in Pittsburgh was lower than the national rate in 1910.

9. Shergold, *Working-Class Life*, pp. 64–69.

10. The statistics on women's employment for the four cities for 1960 can be found in U.S. Bureau of the Census, *U.S. Censuses of Population and Housing: 1960, Census Tracts, Pittsburgh, Pa.* (Washington, D.C.: Government Printing Office, 1962), table P-3; ibid., *1960, Census Tracts, Detroit, Mich.*, table P-3; ibid., *1960, Census Tracts, Buffalo, N.Y.*, table P-3; ibid., *1960, Census Tracts, Cleveland, Ohio*, table P-3.

11. These percentages were computed from U.S. Bureau of the Census, *Thirteenth Census of the United States: 1910, Population, Occupation Statistics*, vol. 4 (Washington, D.C.: Government Printing Office, 1914), table 8.

12. See Peter Gottlieb, *Making Their Own Way: Southern Blacks' Migration to Pittsburgh, 1916–1930* (Urbana: University of Illinois Press, 1987), pp. 89, 104, 106–09; Dennis C. Dickerson, *Out of the Crucible: Black Steelworkers in Western Pennsylvania, 1875–1980* (Albany: State University of New York Press, 1986), pp. 161–63; Arthur J. Edmunds, *Daybreakers: The Story of the Urban League of Pittsburgh: The First Sixty-five Years* (Pittsburgh: Urban League, 1983), pp. 59–63; and Jean Hamilton Walls, "A Study of the Negro Graduates of the University of Pittsburgh for the Decade 1926–1936" (Ph.D. diss.: University of Pittsburgh, 1938), pp. 21, 42–43, 64, 70.

13. Elizabeth Beardsley Butler, *Women and the Trades: Pittsburgh, 1907–1908* (1909; rpt. Pittsburgh: University of Pittsburgh Press, 1984), pp. 287, 232, 239. For a more extensive analysis of Butler's findings, see my introduction to this edition, "Women at Work Through the Eyes of Elizabeth Beardsley Butler and Lewis Wickes Hine."

14. Ibid., pp. 62, 77, 22–23, 26.

15. Marguerite Renner, "Who Will Teach? Changing Job Opportunity and Roles for Women in the Evolution of the Pittsburgh Public Schools, 1830–1900" (Ph.D. diss., University of Pittsburgh, 1981). Cf. David B. Tyack and Myra H. Strober, "Jobs and Gender: A History of the Structuring of Educational Employment by Sex," in *Educational Policy and Management: Sex Differentials*, ed. Patricia S. Schmuck, W. W. Charters, Jr., and Richard O. Carlson (New York: Academic Press, 1981), pp. 131–51.

16. Renner, "Who Shall Teach?" pp. 257–77.

17. Ibid., table 29, p. 282.

18. Ileen DeVault, "Sons and Daughters of Labor: Class and Clerical Work in Pittsburgh, 1870s–1910s" (Ph.D. diss., Yale University, 1985), p. 97. DeVault computed this figure from Pittsburgh, Central Board of Education, *32nd Annual Report of the Superintendent of Public Schools for the School Year Ending June 1, 1900*, p. 77, and U.S. Bureau of the Census, *Special Reports: Occupations at the Twelfth Census, 1900* (Washington, D.C.: Government Printing Office, 1904), table 43.

19. Renner, "Who Shall Teach?" p. 329.

20. Marjorie Murphy, "From Artisan to Semi-Professional: White Collar Unionism Among Chicago Public School Teachers, 1870–1930" (Ph.D. diss.: University of California, Davis, 1981), pp. 98, 106, 106A.

21. Carolyn Sutcher Schumacher, "School Attendance in Nineteenth-Century Pittsburgh: Wealth, Ethnicity, and Occupational Mobility of School Age Children, 1855–1865" (Ph.D. diss., University of Pittsburgh, 1977), pp. 61, 73, 194, 192, 189.

22. Myra H. Strober and Audri Gordon Lanford, "The Feminization of Public School Teaching: Cross-Sectional Analysis, 1850–1880," *Signs: Journal of Women in Culture and Society* 11 (Winter 1986): 212–35.

23. Ralph Proctor, "Racial Discrimination Against Black Teachers and Black Professionals in the Pittsburgh Public School System, 1834–1973" (Ph.D. diss., University of Pittsburgh, 1979), pp. 28–29, 32–35, 55–56, 103, 126, 157; Edmunds, *Daybreakers, p. 59.*

24. DeVault, "Sons and Daughters of Labor."

25. Two books assess the extent of occupational sex segregation, explanations for its prevalence, its effects on workers and families, and possible remedies for it. See Barbara F. Reskin, ed., *Sex Segregation in the Workplace: Trends, Explanations, Remedies* (Washington, D.C.: National Academy Press, 1984), and Barbara F. Reskin and Heidi I. Hartmann, eds., *Women's Work, Men's Work: Sex Segregation on the Job* (Washington, D.C.: National Academy Press, 1986).

26. For excellent critiques of the theories of occupational sex segregation, see Samuel Cohn, *The Feminization of Clerical Labor in Great Britain* (Philadelphia: Temple University Press, 1985), pp. 3–35; and Ruth Milkman, *Gender at Work: The Dynamics of Job Segregation by Sex During World War II* (Urbana: University of Illinois Press, 1987), pp. 1–11.

27. DeVault, "Sons and Daughters of Labor," pp. 75–79.

28. Ibid., pp. 63, 74, 179.

29. Susan B. Carter and Mark Prus, "The Labor Market and the American High-School Girl, 1890–1928," *Journal of Economic History* 42 (Mar. 1982): 164.

30. Ibid., p. 166.

31. DeVault, "Sons and Daughters of Labor," p. 195.

32. Ibid., pp. 224, 240, 213.

33. Ibid., pp. 234–47.

34. Ibid., pp. 96–144.

35. Ibid., pp. 329, 331.

36. Cindy Sondik Aron, *Ladies and Gentlemen of the Civil Service: Middle-class Workers in Victorian America* (New York: Oxford University Press, 1987), pp. 139–61.

37. DeVault, "Sons and Daughters of Labor," p. 113.

38. For an extended discussion of the limits of sisterhood as detailed in the historical scholarship on women, see Nancy A. Hewitt, "Beyond the Search for Sisterhood: American Women's History in the 1980s," *Social History* 10 (Oct. 1985): 299–321.

39. Calvert, "Cotton Mill Riot," pp. 97–133.

40. I am indebted to Paul Krause for the following newspaper citations on working-class housewives in Homestead. For women's behavior during the 1889 strike, Pittsburgh *Post*, July 12 and 13, 1889. For the 1892 Homestead lockout, Pittsburgh *Commercial Gazette*, July 5, 6, 7, 9, 1892; New York *Tribune*, July 7, 9, 1892; New York *World*, July 6, 1892; New York *Sun*, July 6, 1892.

41. Elizabeth Ewen, *Immigrant Women in the Land of Dollars: Life and Culture on the Lower East Side, 1890–1925* (New York: Monthly Review Press, 1985), pp. 176–83, esp. 183. See also Dana Frank, "Housewives, Socialists, and the Politics of Food: The 1917 New York Cost-of-Living Protests," *Feminist Studies* 11 (Summer 1985): 255–85.

42. Butler, *Women and the Trades*, pp. 292–94.

43. Patrick Lynch, "Pittsburgh, the I.W.W., and the Stogie Workers," in *At the Point of Production: The Local History of the I.W.W.*, ed. Joseph R. Conlin (Westport, Conn.: Greenwood Press, 1981), pp. 79–94.

44. For a detailed narrative of the strike based on a survey of Pittsburgh newspapers, see Dianne Kanitra, "The Westinghouse Strike of 1916" (seminar paper, Department of History, University of Pittsburgh, 1971). See also Linda Nyden, "Women Electrical Workers at Westinghouse Electric Corporation's East Pittsburgh Plant, 1907–1945" (seminar paper, Department of History, University of Pittsburgh, 1975), pp. 16–22; David Montgomery, *The Fall of the House of Labor* (Cambridge: Cambridge University Press, 1987), pp. 317–27.

45. David Brody, *Steelworkers in America: The Nonunion Era* (1960; rpt. New York: Harper and Row, 1969), pp. 231–62, esp. 260.

46. William Z. Foster, *The Great Steel Strike and Its Lessons* (New York: B. W. Huebsch, 1920), pp. 53, 60, 146–48. For brief accounts of the police terror against the striking steelworkers and their wives, see ibid., pp. 250–51; Interchurch World Movement, *Public Opinion and the Steel Strike* (New York: Harcourt, Brace, 1921), pp. 130–31, 186–87, 212–13 (women's attendance at strike meetings is mentioned on p. 183).

47. Ronald Schatz, "Union Pioneers: The Founders of Local Unions at General Electric and Westinghouse, 1933–1937," *Journal of American History* 66 (Dec. 1979): 586–602.

48. Ewen, *Immigrant Women*, p. 127.

49. Shergold, *Working-Class Life*, pp. 207–30.

50. Ethel Spencer, *The Spencers of Amberson Avenue: A Turn-of-the-Century Memoir* (Pittsburgh: University of Pittsburgh Press, 1983), pp. 10–29.

51. Ibid., pp. 30–44.

52. Susan J. Kleinberg, "Technology and Women's Work: The Lives of Working-Class Women in Pittsburgh, 1870–1900," *Labor History* 17 (Winter 1976): 58–66. See also Kleinberg's "Technology's Stepdaughters: The Impact of Industrialization Upon Working-Class Women, Pittsburgh, 1865–1890" (Ph.D. diss., University of Pittsburgh, 1973), pp. 85–124.

53. Kleinberg, *Shadow of the Mills*, chap. 8.

54. Margaret Byington, *Homestead: The Households of a Mill Town* (1910; rpt., Pittsburgh: University of Pittsburgh Press, 1974), p. 64.

55. Ibid., pp. 75, 79.

56. Spencer, *Spencers*, pp. 30–43.

57. Elizabeth A. Metzger, "A Study of Social Settlement Workers in Pittsburgh, 1893 to 1927" (seminar paper, Department of History, University of Pittsburgh, 1974), table 36, p. 63.

58. Ibid., pp. 19–41. Similarly, educated black women taught homemaking and consumer skills to groups of black women who had migrated from the South to Pittsburgh in the era of the First World War. The Pittsburgh branch of the Urban League hired women to organize neighborhood groups of housewives to care for their homes and children as well as their ill and needy neighbors. In an effort to stabilize their labor force Pittsburgh corporations employed female and male "colored welfare workers" to help black migrants adjust to factory work and urban life. See Edmunds, *Daybreakers*, pp. 29–44.

59. Byington offered some evidence to support this comment. "The [Homestead] housewives expressed some scorn of the theoretical aspects of the problem [of economizing on food] as taught in the cooking classes of the Schwab Manual Training School, feeling that practical experience was of more value than any theory." See *Homestead*, p. 78.

60. See Roy Lubove's and John N. Ingham's essays in this volume.

61. Kathryn Kish Sklar, "Hull House in the 1890s: A Community of Women Reformers," *Signs: Journal of Women in Culture and Society* 10 (Summer 1985): 658–77. Metzger, "Social Settlement Workers," pp. 23–25, 29–30, 34–37. See also Ida Cohen Selavan, "The Founding of the Columbian Council," *American Jewish Archives* 30 (April 1978): 24–42.

62. Sklar, "Hull House," p. 658.

63. Frank Giarratani, "Perspective on Regional Structural Change: Pittsburgh and the United States," in *Perspectives on Pittsburgh: Papers for the Project on Regional Structural Change in International Perspective*, ed. Martin A. Greenwald (Pittsburgh: University Center for International Studies, University of Pittsburgh, March 1984).

64. Miriam Cohen, "Italian-American Women in New York City, 1900–1950: Work and School," in *Class, Sex, and the Woman Worker*, ed. Milton Cantor and Bruce Laurie (Westport, Conn.: Greenwood Press, 1977), pp. 120–43.

65. John Bodnar, Roger Simon, and Michael P. Weber, *Lives of Their Own: Blacks, Italians, and Poles in Pittsburgh, 1900–1960* (Urbana: University of Illinois Press, 1982), pp. 113–51, 241, 254. See also Corinne Azen Krause, *Grandmothers, Mothers, and Daughters: An Oral History Study of Ethnicity, Mental Health, and Continuity of Three Generations of Jewish, Italian, and Slavic-American Women* (New York: American Jewish Committee, n.d.), which has some suggestive information about the differing expectations of ethnic groups toward the education of women.

66. Ruth Schwartz Cowan, *More Work for Mother: The Ironies of House-*

hold Technology from the Open Hearth to the Microwave (New York: Basic Books, 1983), pp. 181–90, 193–201.

67. Robert Lynd and Helen Merrell Lynd, *Middletown: A Study in Modern American Culture* (New York: Harcourt, Brace & World, 1929); Lois Scharf, *To Work and To Wed: Female Employment, Feminism, and the Great Depression* (Westport, Conn.: Greenwood Press, 1980); Winifred Wandersee, *Women's Work and Family Values, 1920–1940* (Cambridge, Mass.: Harvard University Press, 1981); Lynn Weiner, *From Working Girl to Working Mother: The Female Labor Force in the United States, 1820–1980* (Chapel Hill: University of North Carolina Press, 1985).

68. *Woman of the Times* appropriately has a female publisher, managing editor, and art director. The first issue was released in the fall of 1987.

▪ *Double Burden:*
The Black Experience in Pittsburgh

LAURENCE GLASCO

S CHOLARLY STUDIES of black Pittsburgh are numerous but uneven in their coverage. In the 1930s the Works Progress Administration (WPA) assembled a rich body of material on the social life, politics, and even folklore of the city's blacks. But the projected general history was never completed, and its unedited pages until recently lay forgotten in the state archives.[1] The gap left by the lack of a general history, moreover, is not filled by specialized studies because these are uneven in their coverage. The nineteenth century, for example, has been especially neglected: the scholarly literature on that period consists of one article, one dissertation, and one undergraduate thesis, all of which focus on the antislavery movement of the Civil War era.

The twentieth century, in contrast, has received considerable attention. The period between World War I and World War II has been especially well covered: over one hundred specialized studies—including fifty-six master's theses and dissertations—describe the adjustment problems of black migrants and the emergence of the Hill district as a predominantly black ghetto. The years following World War II also have interested scholars: more than fifty studies—primarily doctoral dissertations—examine the racial dimensions of poverty, segregation, and governmental efforts to alleviate those conditions. Finally, black Pittsburgh from approximately 1930 to 1980 has been visually well documented in the collection of Teenie Harris, a photographer for the Pittsburgh *Courier* whose 50,000 to 100,000 photographs rival those of New York's Vander Zee collection in portraying the texture of black urban life.[2] At least for the twentieth century, then, the scholarly sources on black Pittsburgh are quite extensive—more so than for any of the city's other ethnic groups—and are sufficient to highlight the broad contours of black history in Pittsburgh.

Taken together, the historical material indicates that, in addition to suffering the same racial discrimination as their counterparts elsewhere, blacks in Pittsburgh have borne two additional burdens. The first was economic. The stagnation and decline of Pittsburgh's steel industry began just after World War I—well before that of most Northern cities—and very early closed off opportunities for black economic progress. The second burden was geographic. Whereas the flat terrain of most Northern cities concentrated blacks into one or two large, homogeneous communities, Pittsburgh's hilly topography isolated them in six or seven neighborhoods, undermining their political and organizational strength.

The record also shows that, despite these two extra burdens, black Pittsburghers accomplished much. They created a distinguished newspaper, owned two outstanding baseball teams, maintained a lively cultural life, and nurtured musicians and writers of national prominence. However, economically and politically, the community stagnated. It was unable to develop a stable working class, its middle class remained small, and its geographic fragmentation into several neighborhoods severely diluted its political and institutional strength. As a result, by the 1980s Pittsburgh blacks lagged behind their counterparts in most other cities in terms of economic and political development.

The Black Community Before World War I

The burdens of economic and racial discrimination can be traced to the earliest years of black settlement. Prior to the Civil War, Pittsburgh's black community was typical of those found in most Northern cities—small, impoverished, and victimized by racial discrimination. Although blacks arrived with the very earliest colonial settlers—as trappers, pioneers, soldiers, and slaves—the population grew slowly. In 1850 they comprised only two thousand people, less than 5 percent of the city's population, and were centered in "Little Hayti," an area just off Wylie Avenue in the lower Hill district where housing was cheap and close to downtown.

The black community was poor because racial discrimination excluded its men from the industrial and commercial mainstream of the city's economy. Barbering was the most prestigious occupation open to blacks, and they operated most of the downtown barbershops that catered to the city's elite. (To cut the hair of other blacks would have cost them their white customers.) Most, however, could find work only as day laborers, whitewashers, janitors, porters, coachment, waiters, and stewards. The men's low earnings forced their wives to seek work outside the home, typically in low-paying and demeaning jobs as servants, domestics, and washerwomen.[3]

Despite their exclusion from the city's industry and commerce, some blacks prospered. As early as 1800 Ben Richards, a black butcher, had accumulated a fortune by provisioning nearby military posts. At mid-century John B. Vashon operated a barbershop and a fashionable bath house, while John Peck was a wigmaker and barber. Most members of the black elite, however, were men of modest holdings: by 1860 Richards's fortune had been dissipated and the manuscript census listed only twelve blacks—three barbers, three stewards, a musician, a porter, a waiter, a pattern finisher, a grocer, and a "banker"—with property worth $2,000 or more.[4]

Pittsburgh's pre–Civil War black community supported a remarkable number of institutions. These included an AME (African Methodist Episcopal) church, an AME Zion church, four benevolent societies, a private school, a cemetery, a militia company, a newspaper, and a temperance society. The community also contained an impressive set of leaders, such as Vashon, Peck, and Lewis Woodson, all barbers and all active in civic affairs. The best-known leader was Martin R. Delany who, after publishing *The Mystery*, a newspaper in Pittsburgh, co-edited *The North Star*, Frederick Douglass' paper, and authored an important nationalist tract, *The Condition, Elevation, Emigration, and Destiny of the Colored People of the United States.* The race pride of Delany is indicated in a comment by Frederick Douglass that he thanked God for simply making him a man, "but Delany always thanks him for making him a black man."[5]

The community and its leaders placed great emphasis on education. They maintained their own school—the existence of which had originally attracted Delany to Pittsburgh—and avidly pursued higher education. Delany went on to become one of the first blacks to study medicine at Harvard University; Vashon's son became the first black to graduate from Oberlin College; Peck's son became one of the nation's first blacks to obtain a medical degree; and Lewis Woodson, in addition to his duties as barber and minister, taught in the community's own school.

The community stressed both cultural attainments and gentility. The accomplishments of two of its children reflect those emphases. Henry O. Tanner became an award-winning painter based in Paris, and Hallie Q. Brown became a leading elocutionist who performed throughout the United States and Europe.

Culture and gentility also became social stratifiers, often in combination with pride in place of origin. A. B. Hall, a nineteenth-century resident, recalled that the pre–Civil War community was dominated by an aristocracy of genteel families and cliques: "The Virginians, District of Columbia and Maryland folk, consorted together; the North Carolina people, thought themselves made in a special mould; those from Kentucky, just knew they were what the doctor ordered; while the free Negroes, who drifted in from Ohio and New York, didn't take a back seat for anybody."[6]

Despite their stress on culture, gentility, and education, black Pittsburghers faced daily indignities and threats to their personal and civil rights. They were excluded from, or confined to separate sections of, the city's theaters, restaurants, and hotels. Occasionally they were attacked by white mobs, although they were spared the major riots that convulsed black communities in New York, Philadelphia, and Cincinnati.

They also faced continual efforts to deny them an education. Protests by Harvard students forced Delany to leave that institution before getting his M.D. degree. Fear of student protests at the University of Pittsburgh (then the Western University of Pennsylvania) prompted professors to ask John C. Gilmer, the school's first black student, to sit in the hallway outside the lecture room.[7] And in 1834 Pittsburgh excluded black children entirely from the city's new public school system, to which blacks responded by establishing their own school and vigorously protesting their exclusion from the public system. Three years later the city provided blacks with a segregated school on Miller Street, but its wretched condition disappointed and angered black residents.

In addition to being deprived of an equal education, blacks were denied even basic rights of citizenship. In 1837 Pennsylvania disfranchised its black residents, causing enraged black communities throughout the state to hold numerous protest meetings and rallies. Even more trauma was in store when, in 1850, the federal government passed the Fugitive Slave Act. This act, which was designed to help slave catchers apprehend runaway slaves, was the most terrifying law passed in the pre–Civil War era. It made lawbreakers of Pittsburghers—black and white—who had been working to help runaway slaves, and facilitated the recapture of Pittsburgh blacks who themselves were runaway slaves. The act terrified so many that during the 1850s the city's black population dropped from 1,974 to 1,149. Those who did not flee were galvanized into action, establishing an elaborate network of spies, harassing slave catchers, and even kidnapping slaves passing through the city with their owners. The pessimism of that decade caused Martin Delany to give up hope in America and urge blacks to emigrate. Delany himself traveled to Nigeria and negotiated with Yoruba chiefs for a settlement near Abeokuta, whence his later reputation as the "Father of Black Nationalism."[8]

In resisting slavery and racial discrimination, Pittsburgh's blacks were not without white supporters. In the 1850s the city had several antislavery societies which worked to frustrate the fugitive slave law. Two local physicians who were abolitionists as well as supporters of black progress, Joseph Gazzam and Julius LeMoyne, trained Martin Delany and sponsored his entry into Harvard Medical School; another wealthy abolitionist, Charles Avery, acknowledged the social and biological equality of blacks—something rarely done even by the most ardent abolitionists. In 1849 he established the all-black Allegheny Institute (later Avery College) for their education.[9]

Following the Civil War, legislation designed to ensure the rights of Southern freedmen helped Northern blacks—including those in Pittsburgh—to secure their own civil rights. Thus, the Fifteenth Amendment to the Constitution, ratified in 1869 to secure the black vote in the South, forced Pennsylvania finally to grant the vote to all its male citizens.[10]

The fight for a desegregated school system also was won during the postbellum years. Blacks continued to protest their inferior school system until, in 1875, enrollment in the Miller Street School was so low that the Board of Education abolished it and desegregated the entire system. From that point on, Pittsburgh was one of the few large cities with a desegregated system. The victory, however, was bittersweet: cities with segregated systems provided jobs for black teachers but, because it was unacceptable for whites to study under a black, black teachers were not employed in Pittsburgh.[11]

Despite gains in education and voting rights, most other forms of racial restriction persisted throughout the post–Civil War era. In 1872 the Democratic Pittsburgh *Post* lampooned Republican hypocrisy in posing as friends of Negroes when blacks in Republican Allegheny County "could not be admitted to the orchestra, dress or family circle of the opera house, could not purchase a sleeping berth on any of the railroads that leave the city, could not take dinner at the Monongahela House, Hare's Hotel, or any A No. 1 restaurant," and could not even enter the Lincoln Club "except as a waiter."[12]

The *Post* failed to mention something even more serious: blacks could not find work—except as temporary strikebreakers—in the region's mines, mills, factories, and offices. By the turn of the century, Pittsburgh was the nation's sixth largest city, annually attracting thousands of European immigrants to its busy industries, while still confining blacks to jobs that white workers, including immi-

grants, did not want. As a consequence, most blacks still worked as
teamsters, refuse collectors, janitors, and laundresses, while the
"lucky" worked as waiters, barbers, railroad porters, butlers, maids,
coachmen, and gardeners.[13]

Although we know relatively little about the black community
between 1875 and World War I, there are several indications that
this was an important period in its development. First, it was a
period of rapid growth. Between 1870 and 1900 the rate of population growth for black Pittsburgh was greater than during any other
period, increasing from 1,162 to 20,355, and making the Pittsburgh
black community the sixth largest in the nation.[14]

During this period blacks made notable economic strides. Although the great majority worked as maids, laundresses, coachmen,
and janitors, a surprising number operated successful businesses. In
1909 an investigator counted eighty-five black-run businesses, including owners of pool rooms and print shops, plasterers, cement
finishers, pharmacists, paper hangers, and haulers—even a savings
and loan. Some businesses were quite large, such as the Diamond
Coke and Coal Company of Homestead, with one thousand employees, and two contractors employed over one hundred men each. In
addition, a caterer (Spriggs and Writt) and a wigmaker (Proctor)
served a predominantly white clientele, as did several grocers, restaurant owners, and barbers.[15]

By the end of the nineteenth century, Pittsburgh blacks had developed a set of social distinctions more elaborate than those at mid-
century. According to a contemporary observer, social lines hardened
after the Civil War as longer-term residents and property owners held
themselves aloof from Southern migrants, cultivated genteel manners, and "accepted into their 'social set' only those who could point
to parents and grandparents and say they were 'old families' with
'character'."[16]

By 1900 black Pittsburghers supported an impressive range of
clubs dedicated to social, cultural, and community uplift. The most
prestigious—Loendi, Aurora Reading Club, Goldenrod Social Club,
and White Rose Club—emphasized social and cultural activities.
Blacks were excluded from white fraternal lodges, but established
all-black chapters of their own, many of which, such as the Elks,
Masons, Odd Fellows, Knights of Pythias, and True Reformers, emphasized community improvement. They erected a Home for Aged
and Infirm Colored Women, a Working Girls' Home, and a Colored
Orphans' Home. Churches which catered to the district's elite—
Ebenezer Baptist, Grace Memorial Presbyterian, Bethel AME, Warren M.E., and St. Benedict the Moor Roman Catholic—also were

active in community activities, while over twenty-five women's clubs devoted themselves to both charitable activities and social affairs.[17]

A number of groups—including the Aurora Reading Club, the Frances Harper League, the Wylie Avenue Literary Society, the Homewood Social and Literary Club, the Emma J. Moore Literary and Art Circle, and the Booker T. Washington Literary Society—promoted the cultural life of the community through reading and discussing literature. In addition, the prestigious Loendi Club, established in 1897, invited outstanding speakers to address their members and the black community. Community interest in classical music was reflected in the creation of three black concert and symphony orchestras, one by William A. Kelly, a coal miner and graduate of Oberlin, a second by Dr. C. A. Taylor, who had played in the Toronto Civic Orchestra, and a third by David Peeler, a local contractor and builder.[18]

In sum, against all odds, and in ways not yet understood, the black community prior to World War I managed to grow substantially and to develop an impressive set of social, economic, and cultural institutions. This promising foundation, combined with the breakthrough of blacks into the city's mills and factories during World War I, should have led to a flowering of black society. As we will see, the breakthrough did energize and transform the community's social and cultural life, but economically and politically its full promise was not realized.

Between the Wars: Migration Amidst Economic Stagnation

The First World War stands as a watershed in black history, both nationally and locally. Economic expansion and the war-related cutoff of European immigration forced northern industries to open up factory jobs for the first time to black Americans. This touched off a migratory wave from the South that, between 1910 and 1930, increased the northern black population by more than five hundred thousand. Newly opened jobs at places like Jones & Laughlin Steel enlarged Pittsburgh's black population from twenty-five to fifty-five thousand, while hiring by Carnegie Steel plants in Aliquippa, Homestead, Rankin, Braddock, Duquesne, McKeesport, and Clairton raised the black population in those neighboring towns from five to twenty-three thousand.[19]

The migrants gave the community a new energy and creativity that quickly attracted attention. Wylie Avenue, Centre Avenue, and

side streets in the Hill district "jumped" as blacks and whites flocked
to its bars and night spots. The Collins Inn, the Humming Bird, the
Leader House, upstairs over the Crawford Grill, as well as Derby
Dan's, Harlem Bar, Musician's Club, Sawdust Trail, Ritz, the Fullerton Inn, Paradise Inn, and the Bailey Hotel attracted some of the
nation's finest jazz musicians. Marie's and Lola's, small and stuffy
clubs, provided spots for late night jam sessions, and helped make
Pittsburgh a center for nurturing internationally known musicians.
Jazz notables born, reared, or nurtured in the Pittsburgh area between
the wars include Lena Horne, Billy Strayhorn, Kenny Clarke, Art
Blakey, Earl "Fatha" Hines, Roy Eldridge, and Leroy Brown, in addition to such notable female musicians as Mary Lou Williams, Louise
Mann, and Maxine Sullivan. As the district's fame spread nationwide,
Claude McKay, leading poet of the Harlem Renaissance, labeled the
intersection of Wylie and Fullerton Avenues—in the heart of the
Hill—"Crossroads of the World."[20]

In addition to energizing the social and musical scene of Pittsburgh, the migrants helped create a wide range of neighborhood
sports—baseball, basketball, and football—which have been beauti-
fully described in *Sandlot Seasons* by Rob Ruck. The most famous of
such teams, the Pittsburgh Crawfords, emerged from integrated neighborhood clubs in the Hill district. Organized originally for "pick up"
games by Bill and Teenie Harris, they later were bankrolled by Gus
Greenlee, black numbers czar and owner of the Crawford Grill. The
1936 Crawfords boasted five eventual Hall of Famers—Satchel Paige,
Josh Gibson, "Cool Papa" Bell, Oscar Charleston, and Judy Johnson—
and, according to Ruck were "possibly the best baseball team [white
or black] ever assembled for regular season play." The Homestead
Grays, financed by "Cum" Posey, son of a prominent black business-
man, was also filled with premiere players. Organized around the turn
of the century as an amateur sporting outlet for black steelworkers, by
the 1930s the Grays were tough competitors of the Crawfords and,
after Greenlee resurrected the Negro National League, helped make
Pittsburgh the national center of black baseball.[21]

THE BURDEN OF ECONOMIC STAGNATION

Signs of vitality in music and sports, however, masked serious
economic problems. Black migrants entered the industrial work
force at the bottom and had almost no success in moving up. Abraham Epstein's 1918 study, "The Negro Migrant in Pittsburgh,"
found that 95 percent of black industrial workers were in unskilled
positions, and investigations five years later by two black graduate
students, Abram Harris and Ira Reid, found blacks still mired at the

bottom of the job ladder, and experiencing difficulty holding onto even those lowly jobs. In 1923, for example, some 17,224 blacks were employed in seven major industries in the Pittsburgh area, but in 1924 a recession reduced their number to just 7,636. Moreover, the building trades offered no escape; blacks were excluded entirely from more than half the trades, and the few to which they belonged confined them to positions as hod carriers and common laborers and insisted that they be paid less than union scale. A situation that was bleak during the 1920s turned disastrous during the Great Depression, when 33 to 40 percent of black adults were unemployed. A study of 2,700 black families found 41 percent destitute and another one-third living in poverty.[22]

Black economic problems were caused by more than white prejudice. Blacks had entered the city just as its heavy industry had stopped growing. Two economists writing in the 1930s noted that the expansion of Pittsburgh's iron and steel industry "began to slow up" as early as 1900, and "by 1910 the era of rapid population and industrial growth was completed." Moreover, as John Bodnar, Roger Simon and Michael P. Weber show, the steel industry was eliminating common laborers. Unskilled workers declined from 32 to 22 percent of the work force between 1900 and 1930. As a result, blacks and other unskilled workers found "increasing competition for work at the lowest levels of the occupational scale."[23]

Many white workers responded to the retrenchment in steel by leaving the city. During the 1920s the city's white population grew by only 12 percent, and it declined absolutely from 1930 to the 1980s. Blacks, with fewer options elsewhere, saw their population grow by 46 percent during the 1920s and increase in each subsequent decade. Although their numbers continued to grow, the stagnant job situation prevented blacks from establishing occupational beachheads, which had been important to white ethnics in finding employment and job mobility. Thus, even without racial discrimination— which certainly was pervasive—black migrants would have been more likely to be laid off in an economic downturn and to suffer greater occupational instability than their immigrant predecessors.[24]

As a result, work conditions for blacks remained deplorable, with few opportunities to escape into better jobs in the mines and mills. Former workers, when interviewed by Dennis Dickerson, told how they had been brought in by the trainload, housed in hastily constructed dormitories, confined to the hottest, dirtiest, and most physically demanding jobs, paid inadequate wages, harassed by prejudiced foremen and police, and excluded from most public accommodations. They became sick from the acid fumes, fainted from the

intense heat, caught penumonia in the drafty housing, and suffered
accidents at a much higher rate than white workers. They were
seldom promoted: some even trained white workers, only to suffer
the humiliation of seeing the latter promoted over them. Alienated
by the prejudice and harassment of white steel workers, they refused
to join the strike of 1919, which embittered whites but persuaded
mill owners to see blacks as a valuable and permanent part of the
labor force.[25]

Racism also pervaded the region's coal fields. The United Mine
Workers of America (UMWA), led by John L. Lewis, systematically
restricted the entry of black miners and made sure that the few who
did work were assigned only jobs that whites did not want. In 1925
this practice enabled the Mellons to destroy the union at its Pittsburgh Coal Company by creating a labor force that was half black
and assigning blacks to all jobs in the mines, including those formerly reserved for whites. The benefits proved elusive, however, and
blacks were quickly disillusioned. With the union broken, pay was
cut drastically, the number of hours increased, mine safety deteriorated, and the mine police became more vicious than ever.[26]

That blacks were not hostile to union activity, per se, is indicated
by their enthusiastic response to nonracist unions. They shunned
the UMWA, but were attracted to the Communist-led National Miners' Union (NMU), which backed up its commitment to interracial
unionism by electing a black, William Boyce, as vice-president. During the Depression of the 1930s, however, things deteriorated for
black miners, as the NMU was replaced by the UMWA, now somewhat less racist, but still an organization in which blacks had to
fight for fair treatment.

Racism and economic stagnation in coal and steel were not the
only factors complicating the economic adjustment of migrants.
Peter Gottlieb, in a recently published dissertation, *Making Their
Own Way*, argues that their unfortunate work experience was also
caused by their attitudes toward industrial labor. Using records of
the Byers specialty steel company, supplemented by interviews with
ex-steelworkers (not from the Byers company), Gottlieb found a co-
nundrum: most migrants had already experienced some industrial
work in the South, and their work performance was rated "good" by
two-thirds of the Byers foremen—but they had an astronomical turnover rate. To maintain 223 black workers in 1923, the company had
to make 1,408 separate hires! Gottlieb believes that migrants still
regarded industrial labor in the same way they had regarded it in the
South: as temporary work to be done between planting and harvest-
ing season. Consequently, many approached the Pittsburgh mills

looking for temporary work; once they had made some money they were prepared to return home. Others quit their jobs in the mill in order to return home for a visit, but planned to return later and find another job. Finally, most were prepared to quit should the work prove unsatisfactory—which often it did.[28]

CONSEQUENCES OF ECONOMIC STAGNATION

The effects of economic stagnation and racial discrimination were visible in the deplorable living conditions of the migrants. As they poured into Pittsburgh during and after World War I, the migrants settled mainly in the lower Hill district, the most densely inhabited section of the city and characterized by a housing stock already old and dilapidated. Doubling up—with men on night shifts sleeping in beds vacated by day workers—caused an unsatisfactory situation to deteriorate even further. Health conditions and crime rates reached scandalous levels. Tuberculosis, influenza, whooping cough, scarlet fever, and venereal disease plagued the community, the death rate soared from 23.4 to 31.3 per 1,000, and crime rose 200 percent.[29]

More serious, perhaps, than the physical consequences were the social and psychological. The migrants were financially unable to buy homes; by 1930 only 1 percent of the residents of the lower Hill were homeowners, compared with 24 percent of the residents of heavily Slavic Polish Hill and 47 percent of those in Bloomfield, which was heavily Italian. Lack of homeownership increased residential instability and undermined possibilities for the emergence of a stable working-class community.[30]

THE BURDEN OF GEOGRAPHIC DISPERSAL

An additional burden borne by Pittsburgh blacks was that of geographic dispersal. Pittsburgh's hilly topography created neighborhoods of small ethnic pockets. The Hill district contained the largest single group of blacks in the city, but was never the only important place of black settlement. Indeed, during World War I, only 41 percent of the city's blacks lived there, while another 17 percent lived four miles away in East Liberty, and others lived even farther away— across the Allegheny River in the North Side, south of the Monongahela River in Beltzhoover, and east of the city in Homewood.[31]

The city's terrain encouraged this process of spatial differentiation and meant that, unlike other Northern cities, growth of the black population resulted in several neighborhoods rather than in one compact ghetto. This splintering did not reduce the amount of racial segregation in Pittsburgh. Joe Darden, in his doctoral dissertation, calculated that in 1930, 72 percent of blacks in Pittsburgh

would have to move to another census tract to fully desegregate
residential patterns, and that 74 percent would have to do so in 1940.
This would put Pittsburgh among the least segregated of northern
cities in that year. However, Karl and Alma Taeuber calculated Pittsburgh's percentage as 82, which would make it average for the
North. Whichever figure one accepts, the fact is that Pittsburgh
remained a very segregated city.[32]

These numerous "mini-ghettoes," however, did have several
negative consequences. First, they delayed the time when blacks
would dominate their own "turf." By 1930 the Hill housed less than
half the city's black residents—43 percent—and, as a result, the
district was barely majority black by 1930. Between 1910 and 1930,
the district changed from 25 percent black to 53 percent black—a
majority, to be sure, but not the solid majority that would ensure
political and economic dominance.[33]

Second, neighborhood dispersal caused the separation of the
black middle class from the masses. Financially stable residents—
both black and white—tended to move away from the lower Hill and
its deteriorating conditions and settle in other neighborhoods, especially East Liberty, Homewood-Brushton, Beltzhoover, and the upper Hill (Schenley Heights or "Sugar Top"). The result was that the
lower Hill became increasingly black and poor; by the 1930s one
researcher termed it the residence of "the most disadvantaged of the
disadvantaged" in the city.[34]

Homewood is the only middle-class black neighborhood for
which we have more than fragmentary information. In that eastern
community, former residents of the Hill augmented a nucleus of
proud descendants of servants to wealthy families dating back to the
Civil War era. By 1930 these residents—described as "respectable,
working people desirous of making their homes and neighborhoods
as attractive as possible"—had a clear sense of their special residence and identity.[35]

CLASS DIVISIONS IN THE BLACK COMMUNITY

More than residence separated working-class migrants from the
middle class. Most migrants came from the deep South—especially
Georgia and Alabama—and were overwhelmingly industrial laborers. The city's more established, long-term residents, by contrast,
were born in the North or the upper South—especially Virginia and
Maryland—and worked as barbers, waiters, janitors, and (for the
lucky) postal clerks, railroad clerks, chauffeurs, and butlers. The
groups did not even share skin color: older residents tended to be
lighter skinned than the migrants.

The few blacks who attended area universities came from the middle class. Jean Hamilton Walls, the first black woman to enroll at the University of Pittsburgh, was one of those students. Her 1938 doctoral dissertation, which examined black students at Pitt, found that most had a middle-class background. Their fathers tended to be employed either in personal service, the professions, or public service; their mothers were usually housewives; and the parents of one-fourth had themselves attended college.[36]

Class differences were reflected in such institutions as the church. By 1930 the Hill district contained some forty-five churches, many of which were "storefront" institutions created by the migrants as part of their adjustment to urban life. These new churches isolated migrants from older residents and divided the community along lines of class and even color. Such divisions, moreover, could exist even within a given church: dark-skinned worshipers in one of the community's most prestigious churches reportedly sat or were seated toward the back.[37]

A church's degree of community involvement and style of service also reflected class and status differences. The more established churches tended to be involved in broader community activities. In Homestead, for example, Clark Memorial Baptist, located on the Hilltop, was organized in the 1890s by families from Virginia. The church had new buildings, educational and social programs, auxiliaries, clubs, and a foreign mission. Its ministers often had degrees from colleges or theological seminaries and gave sermons on political as well as religious topics. Physically, and probably socially as well, Clark looked down on Second Baptist Church, which began around 1905 as a storefront church located among poorer migrants in the lower "Ward." Not until the 1940s did the latter have its own building. It had few clubs and auxiliaries, no community recreation center, a membership drawn primarily from the deep South, and pastors who stuck closely to orthodox religious worship.[38]

In Pittsburgh, as in other northern cities, one response of middle-class blacks to the influx of newcomers was to withdraw into exclusive clubs. A clear example of this is provided by the Frogs, the community's most prestigious club of the time. Organized in 1910, membership in the Frogs was limited to about twenty-five, making it a smaller and more exclusive subset of the prestigious Loendi Club. With few exceptions, the Frogs were OPs, or Old Pittsburghers, residents proud of having been in the city before the Great Migration. The occupations of these club members—postal worker, railroad clerk, chauffeur, lawyer, funeral director, laundry owner—indicate how constricted was the range of jobs then open to blacks.

Because income and occupation were not convenient stratifiers, one's social life became all the more important. The Frogs was a determinedly social organization; one officer termed it a "haven for those who wish to escape from . . . noble crusades." Invitations to its social events were highly coveted, and the numerous parties and galas of its annual "Frog Week" attracted upper-class revelers from around the country.[39]

The social life of this middle class filled the "Local News" section of the Pittsburgh *Courier*, the creation of Robert L. Vann, a handsome, talented, and industrious migrant from the South. Born in North Carolina, Vann earned his undergraduate degree at Virginia Union University in Richmond. He then attended law school at the University of Pittsburgh, attracted by the availability of an Avery Scholarship for black students. The scholarship—one of several legacies of the nineteenth-century philanthropist and abolitionist Charles Avery—was only for $100, so Vann basically worked his way through law school and in 1909 became its first black graduate. He quickly gained the acceptance of Old Pittsburghers and, by 1910, had married, established his law practice, and become editor and part owner of a weekly newspaper, the *Courier*.[40] The paper covered the activities of black upper- and middle-class clubs like the Frogs, the Loendi, the Arnett Literary Society, the Aurora Reading Society, and the Girl Friends; it chronicled the social life of elite families, such as the Stantons, Jones, Monroes, Googins, Morrisons, Douglasses, Buchannons, Pooles, Leftridges, Marshalls, Byrds, Allens, and Andersons; and it informed readers about who had gone on vacation, married, and had parties.[41]

The *Courier's* masthead—"Work, Integrity, Tact, Temperance, Prudence, Courage, Faith"—reflected the values of Pittsburgh's black middle class and their faith in the American Dream. The paper endorsed the philosophy of Booker T. Washington: "Concentrate your earnings, and make capital. Hire yourselves, produce for yourselves, and sell something for yourselves." As an ultrapatriotic journal, it supported World War I, opposed socialism, had no sympathy for unions, and applauded restrictive quotas on immigrants.[42]

Although many have criticized the black bourgeoisie for its endless social affairs, Pittsburgh's black elite also pursued cultural and intellectual activities. The Centre Avenue YMCA regularly invited lecturers such as George Washington Carver, W.E.B. DuBois, Alain Locke, Walter White, and Mary McLeod Bethune. Blacks turned out to hear artists like Jesse Fausett, Countee Cullen, Langston Hughes, James W. Johnson, Marian Anderson, Roland Hayes, and Paul Robeson. They attended performances by numerous theater groups, one of which, the Olympians, in 1939 won first-place honors in the

Pittsburgh Drama League's annual competition. (The league had long banned black theater groups.) The community also supported classical music. Between 1915 and 1927, the Peeler Symphony Orchestra gave regular performances; in 1937 Dr. A. R. Taylor's symphony orchestra presented concerts at the Centre Avenue YMCA; and Aubrey Pankey, a baritone in the choir of Holy Cross Church, toured the U.S. and Europe giving recitals of lieder, opera arias, and other songs.[43]

THE BLACK MIDDLE CLASS AND THE BURDEN OF RACISM

Despite their loyalty and 100 percent Americanism, middle-class blacks bore the same burden of racism that confronted their working-class counterparts. The only position open to them in industry was that of social worker, assigned to recruit and help in the adjustment of black migrants. Even independent professionals faced difficulties: black lawyers had credibility problems with white juries, and not until 1948 was the first black physician permitted to practice at a local hospital.

Opportunities for stable work, in fact, may have declined over this period. In 1900 blacks operated most of the city's prestigious downtown barbershops; by 1930 they operated almost none. In 1900 they had driven most of the city's taxis, hacks, buses, and trucks; by 1930 they drove mainly garbage trucks. Middle-class women fared even worse. They maintained their position as domestic servants to wealthy whites, but could not clerk in "white" stores even in the Hill district. Sheer desperation drove some who could "pass" for white to do so, often with feelings of guilt that lasted a lifetime. Writing of her experiences during the 1920s and 1930s, one light-skinned woman, Martha Moore, recalled how, frustrated in not being able to find employment, she took her mother's advice and "for twenty-five working years, as a medical secretary, salesgirl, office manager, governess, and Draft Board clerk in World War II, I was any race they wanted, or designated, me to be."[44]

Especially frustrating was the continued refusal by the Board of Education to hire black teachers. Ralph Proctor shows that prior to 1937 blacks who graduated from the University of Pittsburgh's School of Education had to go elsewhere if they wanted to teach. The Board of Education often would help them find jobs in the South and in northern cities like Philadelphia and Indianapolis. Rather than leave the city, Frank Bolden—who was light enough to pass for white—went to work for the *Courier* after being told by the board's director of personnel, "It's too bad you're not white. I'd hire you immediately."[45]

Nor were opportunities much better for other college graduates.

Walls's study of black graduates of the University of Pittsburgh between 1926 and 1936 found that half of the men and 60 percent of the women lived outside the state, largely because of lack of job opportunities. Walls concluded: "Pittsburgh has suffered the loss of many of its Negro college graduates because it offered them so few opportunities to use their training in earning a livelihood."[46]

Middle-class blacks also suffered a wide range of social discrimination. In 1936 a black student at the University of Pittsburgh complained that "admission to major sports and clubs was not impossible but unbearable and often embarrassing," while another reported that his application to a student law club had been rejected because of the objection of several southern club members.[47] The Ku Klux Klan in the Pittsburgh region boasted seventeen thousand members in the 1920s, and the police tolerated crime as long as it was black-on-black. In "white" restaurants blacks found salt in their coffee, pepper in their milk, and overcharges on their bills; in department stores they received impolite service; in downtown theaters they were either refused admission or were segregated in the balcony. Forbes Field, where the Pirates played baseball, confined blacks to certain sections of the stands, and visiting blacks—even those as prominent as W.E.B. DuBois and A. Philip Randolph—could not stay in the city's hotels.[48]

Middle-class blacks had difficulty finding integrated housing. In 1911, Robert Vann (who was often mistaken for an Indian), and his wife (who looked white), bought a pleasant house in Homewood, where they got along well with the neighbors and with the other black family on the street. In 1917, after Vann bought the house next door and rented it to another black family, his neighbors began what Vann called the "Battle of Monticello Street": handbills, meetings, and unending talk about "undesirables" continued. Within ten years Vann's block was almost solidly black.[49]

Some businesses refused to sell to blacks. Vann wanted to buy a new Cadillac in 1926 and, although he could pay cash, he had to buy it in Altoona because the Pittsburgh dealer would not sell to him for fear that Cadillacs might become labeled "a nigger car." Nor were local business associations willing to accept black members, even those who were prominent businessmen and staunch advocates of capitalism. In 1925 Vann applied for admission to the Chamber of Commerce and, despite his obvious credentials, was turned down.[50]

EFFORTS AT COMMUNITY-BUILDING

Because middle-class blacks faced many of the same problems of racism as did poorer and more recent arrivals, they championed race causes in any way they could—through the pages of the Pittsburgh

Courier, through political activity, and through organizations like the Urban League and the National Association for the Advancement of Colored People (NAACP). The institution most clearly committed to creating a sense of community and promoting black interests was the Pittsburgh *Courier*, which by the 1920s was the nation's largest and most influential black paper. Vann made it a voice of protest, crusading on the issues of housing, education, job opportunities, political awareness, crime, and Jim Crow.[51]

In 1928 a close student of the black press rated the *Courier* as America's best black weekly. Its four editions—local, northern, eastern and southern—were distributed in all forty-eight states plus Europe, Africa, Canada, the Philippines, and the West Indies. By 1938 its circulation reached two hundred fifty thousand, making it the largest of all black weeklies in the United States.

Vann hired excellent writers, including George Schuyler, sometimes referred to as the "black H. L. Mencken," and Joel A. Rogers, the historian. Schuyler traveled around America reporting on blacks; Rogers traveled to Africa and Europe, researching and reporting on ancient black kingdoms and heroes. During World War II *Courier* reporters stationed in Ethiopia covered the efforts of Emperor Haile Selassie to fight off the invading Italian troops. Sports also got extensive coverage, with writers like W. Rollo Wilson, Chester "Ches" Washington, Wendell Smith, Cum Posey, Jr. (manager of the Homestead Grays), and William G. Nunn, Sr.

POLITICAL AND ECONOMIC WEAKNESSES

Despite the urgings of its politically conscious newspaper, the black community failed to develop political power commensurate with its potential. Even before the migration of World War I, black politics was characterized by voter apathy and by short-lived, ineffective organizations. Neither party offered blacks much—the Republicans because of blacks' guaranteed loyalty to the "Party of Lincoln"; the Democrats because of sensitivity to southern white prejudices. Blacks supported the openly corrupt and prejudiced Magee-Flinn machine because it provided them with a few patronage jobs and because its opponents, the reform forces, were even more prejudiced.[52]

Despite their increased numbers following the Great Migration, the political situation remained virtually unchanged. In 1919 Robert H. Logan made a promising start toward political empowerment when he became the city's first black alderman. Logan, however, was not reelected and, despite increasing numbers of potential voters, blacks suffered a string of electoral losses throughout the 1920s and 1930s. Sickened at the defeat of one especially capable candidate

(Homer S. Brown) in 1932, the *Courier* editor attributed the loss to voter apathy and disunity. The *Courier* excoriated Pittsburgh blacks as "the most backward of Negroes to be found in any large city" and considered them "absolutely in a class by [themselves] when it comes to the question of civic and political advancement."[53]

The Great Depression broke the Republican grip on the black electorate but had little effect on the political fortunes of Pittsburgh blacks. By 1932 Vann had formally switched to the Democratic party and in that year issued his famous statement urging blacks to turn "Lincoln's picture to the wall." For several reasons, however, blacks made few advances as a result of the switch. First, local whites also voted solidly Democratic in 1932, so that the party did not need the black vote to win. Also, the state's Democrats paid more attention to Philadelphia's two hundred eighty thousand blacks than to Pittsburgh's seventy thousand.[54]

White prejudice was not the only reason for black political weakness. Two others—apathy and geographic dispersal—were examined in a thesis by Ruth Simmons. Simmons noted that by 1945 blacks had only three elected officials—Homer Brown in the state legislature, plus one alderman and one constable in the Hill district—and did not control the Democratic party organization even in wards where they formed a clear majority. Simmons believed that apathy partly explained this dismal situation: "Only those persons who hold office or jobs under the patronage of a political party exhibit any political interest," she noted, and "many of the rank and file" fail to vote. A second explanation was geographic: "If Negro voters were not so widely scattered throughout the city, they might obtain a few more elective offices."[55]

Other community institutions also were ineffective as voices of black protest. The NAACP, after an enthusiastic founding in 1915, soon declined in membership and spirit. A later director, Rev. J. C. Austin, spoke of finding the organization "just about as Christ found Lazarus after the fourth day of his death, not only dead but buried," and national headquarters informed the Pittsburgh branch that it was "weaker than any other unit we have in a city of the same size." By the late 1920s Homer Brown and Daisy Lampkin had revived the organization, and through the late 1930s membership stabilized at about one thousand. But by the late 1930s, with Brown in the state assembly, the branch once again became dormant.[56]

The Urban League was more active, but was oriented toward social welfare rather than political protest. Created in 1918 as the outgrowth of the interracial Pittsburgh Council of Social Services Among Negroes, the officers and employees of the league were drawn from the elite of the black community. To fulfill their primary goal of promoting the welfare of black migrants and industrial workers, they established a travelers' aid booth, provided homemakers to advise on nutrition and family matters, placed welfare workers in some of the larger mills, and commissioned an important study by Abraham Epstein on the plight of black migrants in the Hill district. The league had less to offer middle-class job seekers, but did provide jobs for some of them as social workers in its own offices, helped train young women in secretarial skills, raised scholarship money for college expenses, and urged the employment of black teachers by the public schools.[57]

Inexplicably, although the black community had little political or institutional strength at the local level, it elected a politician of influence at the state level. Constance Cunningham has described the career of Pittsburgh's most famous black politician, Homer S. Brown. Born in West Virginia, in 1923 Brown became the third black to graduate from the University of Pittsburgh's law school. After establishing a law practice with Richard F. Jones, one of his classmates, Brown became president of the local NAACP, a position he held until 1949. In 1934—two years after the loss so greatly lamented by Robert L. Vann—Brown was elected to the state legislature. Why Brown succeeded at the state level, while he and others regularly failed at the local level, is not clear, although Simmons attributes his subsequent reelections to the respect and popularity he enjoyed among white as well as black voters.[58]

Brown had a distinguished legislative career. Respected for his knowledge of constitutional law, he helped David Lawrence and Richard King Mellon craft legislation to launch the "Pittsburgh Renaissance." He also was applauded by *Time* magazine for astuteness and integrity, and achieved international recognition as a member of the U.S. National Commission to UNESCO when he helped draw up the Convenant on the International Rights of Man.

Brown's most dramatic achievement during the 1930s was the convening of a state legislative hearing on the refusal of the Pittsburgh Board of Education to hire black teachers. After the schools were desegregated in the 1870s, the board adamantly refused to hire black teachers because of objections to having them teach white students. As more schools became predominantly black, pressure to hire black teachers increased—if only for the purpose of controlling the students. Leadership to change board policy, however, had to come primarily from the state level because black Pittsburgh lacked the political activism to force a change. Moreover, the community was divided on which unpleasant choice to campaign for—integrated schools with white teachers or segregated schools with black. Faced with that dilemma, the community fell silent. However, the board, in response to the legislature's finding of discrimination, promised to promote its one black professional, Lawrence Peeler, a part-time teacher hired in 1933 to teach music in the Hill. However, the board did not begin hiring black teachers in more than token numbers until the late 1940s.[59]

In sum, racism, economic decline, and geographic dispersal frustrated the benefits promised by blacks' entry into the city's industries during World War I. By World War II, black Pittsburgh was socially and culturally vibrant, but was politically and economically stagnant. Moreover, as we will see, the continuing decline of Pittsburgh's industrial base devastated the black working class, while a continuation of the neighborhood divisions and apathy prevented blacks from attaining the political influence that their numbers would permit. On the other hand, the shift to a service economy, combined with the lowering of racial barriers to employment and public accommodations, opened up new opportunities to the middle class.[60]

After World War II: Middle-Class Gains and Working-Class Frustrations

DIMINISHED OPPORTUNITIES FOR THE WORKING CLASS

Because the period following World War II witnessed a number of very important developments, it is unfortunate that it has not received much scholarly attention from historians. One of the most notable trends was a bifurcation of the black experience along class lines, caused by the deindustrialization of the region. Between the end of World War I and the early 1980s, Pittsburgh's share of national steel production fell from 25 percent to 14 percent, and its manufacturing employment declined from three hundred thirty-eight to two hundred forty thousand, a loss that proportionally was larger than in all but one of the nation's major manufacturing centers.[61]

The consequences of this decline were serious for most Pittsburghers, but were devastating for working-class blacks. By the 1980s black families lived on an income that was only 57 percent that of local whites and suffered an unemployment rate of 35 percent, 3.5 times the white rate. Pittsburgh blacks also lost ground relative to blacks elsewhere: in 1987 a national consultant ranked Pittsburgh forty-first out of forty-eight metropolitan areas in terms of the comparative economic status of local blacks and whites. These dismaying economic developments were reflected in other areas of black life. A 1987 study by the University of Pittsburgh's Institute for the Black

Family documented the disintegration of many local families. As was true in black communities throughout the nation, the black marriage rate was falling while rates of teenage motherhood and the proportion of female-headed households were rising.[62]

Compounding this economic disaster were a series of other setbacks. The first was the destruction of the lower Hill—home to thousands of residents—for the sake of "urban renewal." Surprisingly, this was accomplished with the cooperation of the community's leading politician, Homer Brown, who, like other Pittsburghers, apparently regarded the area as a slum not worth saving. While in the state assembly in the late 1940s, Brown worked with Lawrence, Mellon, and other business leaders to pass legislation permitting the condemnation of the area and its renewal as part of an expanded central business district. After helping to pass such legislation in the late 1940s, Brown, as the first black judge on the Allegheny County Court, authorized its demolition in the early 1950s. What urban historian Roy Lubove has called "bulldozer renewal" uprooted fifteen hundred black families in order to make way for a domed Civic Arena and luxury apartments. That it was accomplished without protest is simply another sign of the community's passivity. Not until the mid-fifties, when developers began to eye the middle and upper Hill—the residence of middle-class blacks—was a protest mounted and the process halted.[63]

Because the uprooting was sudden and no provision was made for their resettlement, the scattering of these families did not produce better housing for them. They simply crowded into integrated neighborhoods such as East Liberty and Homewood-Brushton, beginning the abandonment of those areas by the middle class, both black and white.[64]

A second setback for the local black poor was their inability to get government agencies to respond effectively to their needs. The federal War on Poverty of the 1960s often failed because of poor management, politics, and insufficient funds, but Pittsburgh's agency, CAP (Community Action Pittsburgh), won praise from the local press and was cited by the federal government as a model agency. A study of CAP argues that it failed because of the geographic dispersal of its clients into seven neighborhoods. Competing leaders, priorities, and strengths of those neighborhoods deprived them of the unity needed to force the bureaucracy to respond.[65]

The failure of the War on Poverty led to the riots of the late 1960s, which some saw as a "political" statement by the poor to call attention to their continuing plight. A further effort was made in 1969, the year after the riots, when Nate Smith and the Black Construction

Coalition halted work on ten building projects, put 800 marchers on
the North Side and, after two weeks of demonstrations, produced the
Pittsburgh Plan, hailed as a national model to train blacks for construction jobs, but which was only marginally effective.[66]

Nor were opportunities opening up for black children to escape
the poverty of their parents. Throughout the postwar period Pittsburgh's schools became increasingly segregated because of housing
patterns. The percentage of black pupils in the Pittsburgh school
system rose from 19 percent in 1945 to 37 percent in 1965; over
approximately the same years, the percentage of black pupils in
schools with overwhelmingly black enrollments rose from 45 percent to 67 percent in the elementary grades, and from 23 percent to
58 percent in the secondary grades. Even more serious, academic
performance in the segregated schools was substandard, with elementary students in predominantly black schools reading as much
as three grade levels below their counterparts in predominantly
white schools.[67]

Part of the failure to force federal agencies and the local school
board to effectively address black needs stemmed from continued
political weakness. Following the war, tentative steps toward political power were only partially successful. In 1954 blacks elected their
first City Council member, Paul Jones, and in succeeding years maintained at least one member on the council. In 1958 they elected
K. Leroy Irvis to the Pennsylvania legislature, and over the next
three decades of a distinguished career, Irvis served as minority
whip, majority whip, and Speaker of the House. Black political presence probably peaked in 1970, with the rise of Lou Mason to council
president. But its fragility was revealed in 1985 when, due largely to
political in-fighting among black ward leaders and low voter turnout, blacks wound up with no representation on city council. Some
blamed the at-large electoral system, without explaining how during
the previous three decades blacks had consistently elected one,
sometimes two, councilmen. Rev. Junius Carter, Episcopal priest of
Pittsburgh, echoed the lament of Robert L. Vann fifty years earlier
when he complained: "Pittsburgh is lagging behind every other city
in the nation in terms of the political power of blacks."[68]

Unable to change the system, blacks were forced simply to cope.
Melvin Williams provides an ethnographic study of that adjustment
in Belmar, a neighborhood of middle-class blacks located near Homewood that was "invaded" by lower-class persons displaced from the
lower Hill. Williams documents three major life styles by which
residents adapted to the stresses of ghetto life. One he labels "mainstream" because it is pursued by the few who have both the values

and the economic means of the majority population, and whose behavior is "difficult to distinguish . . . from their white counterparts anywhere in America." Another style, which he labels as "spurious," characterizes those with mainstream values who lack the income, networks, and "stylistic codes" needed to live a mainstream life. Resentful of the mainstreamers and contemptuous of the "no-good" among whom they are forced to live, they lead an isolated, bitter life, "in but not of the ghetto." Finally, there is the group called "no-good" by the spurious but, curiously, labeled "genuine" by Williams. Proud of their flamboyant life style and contemptuous of the values of mainstream society, they live by their wits— the women on welfare and the men by hustling and various illegal and semilegal activities.[69]

There is perhaps a fourth adaptive life style, pursued by a group which one might label the "churched." Williams, in another study, describes people whose entire lives revolve around church activities and who have as little as possible to do with this world, which they regard as "irredeemably wicked." Although his Zion Holiness Church (not a real name) is located in the lower Hill, its counterparts exist throughout the ghetto, as poor blacks attempt to create a community of morality and order amidst a society characterized by immorality and disorder.[70]

EXPANDING OPPORTUNITIES FOR THE MIDDLE CLASS

If opportunities were closing off for the black working class, they were opening up for the middle class. The latter had long suffered from occupational exclusion. During the 1920s, when black workers were finding jobs in the city's factories and mills, middle-class blacks were still excluded from offices, stores, schools, corporations, and most professions. The postwar period was to be their period of opportunity.

A major breakthrough occurred in the late 1940s when the Board of Education increased its hiring of black teachers. As late as 1948 only fifteen black teachers had been hired, but as schools became increasingly black, the board finally changed its policy. Hiring increased steadily in the 1950s and accelerated in the 1960s, so that by 1970 over four hundred black teachers were employed in the Pittsburgh school system, and blacks made up over 10 percent of the system's professional staff (although 37 percent of the student population). As Pittsburgh closed the hiring gap with other cities, blacks overcame barrier after barrier to equal employment in the schools. With pressure, protest, and demonstrated competence, they moved from being elementary teachers in the Hill (1937), to secondary teachers in the Hill (late 1940s), to elementary teachers outside the Hill (mid-1950s), to secondary teachers outside the Hill (mid-1960s). A similar progression occurred for principals, who were hired first in the Hill (1955), and then outside the Hill (1968), until by 1970 the system employed 48 black principals and assistant principals. One argument for hiring black teachers and administrators was that they could help raise the academic performance of their students. And, indeed, Doris Brevard of Vann, Janet Bell of Westwood, Vivian William of Madison, and Louis Venson of Beltzhoover—all principals of elementary schools composed primarily of students from "deprived" backgrounds—created what one researcher has termed "An Abashing Anomaly" by raising their schools' reading and mathematics scores consistently above national norms.[71]

Occupational breakthroughs in other fields were aided by the city's shift to a service economy. Indeed, the massiveness of that shift was emphasized by two economists, who noted that by the 1980s Pittsburgh's industrial mix had become more heavily oriented toward nonmanufacturing activity than in the United States as a whole. As steel production shrank, medicine, business services, education, and banking increasingly became the engines of Pittsburgh's employment growth. Success in making this transition to a service economy was partly responsible for the rating of Pittsburgh as the nation's "most livable city" in the 1985 Rand McNally Atlas.[72]

Most black Pittsburghers would have disputed the city's livability ranking, but some middle-class blacks enjoyed expanding job opportunities. Gladys McNairy was appointed to the Board of Education and in 1971 was elected board president. At the University of Pittsburgh, a sit-in by black students at the computer center produced a Black Studies Department, with Jack Daniel (then a graduate student) as chairman, and Curtiss Porter as associate chairman. This was followed by a program to increase the number of minority students and faculty, such that by 1980 there were 2,233 black students and 103 black faculty on the Oakland campus, comprising, respectively, 7.6 percent of the students and 4.9 percent of the faculty. As moderate as those figures may seem, by one calculation the University of Pittsburgh had become "the institution with the largest degree of overrepresentation" of black undergraduates among the nation's major research universities. In addition, during the 1970s, blacks at the University of Pittsburgh were promoted to several positions of responsibility: Lawrence Howard became dean of the School of Public and International Affairs, David Epperson dean of the School of Social Work, and James Kelly dean of the School of Education.[73]

Opportunities also opened up in the private sector. The city's Urban league, which in the 1920s had been involved in finding employment primarily for working-class blacks, now was negotiating to place blacks in clerical, managerial, and professional positions. In the 1940s the league's "young firebrand" staff member, K. Leroy Irvis, led the picketing of downtown stores that resulted in the hiring of black salesclerks—the first such incident in the nation. In the 1950s, under the leadership of Joe Allen, the league pressed for increased black hiring, especially in the public sector. In the 1960s, under the leadership of Arthur Edmunds, it worked with a handful of black leaders and enlightened corporate executives— notably Fletcher Byrom of Koppers—to get the city's many corporations to hire blacks for more than janitorial duty. They began by hiring secretaries, and, to provide a pool of future black managers, supported NEED, a scholarship fund for black college students.[74]

In the 1980s, expanded corporate hiring, affirmative action requirements, and an enlarged pool of black college graduates led to occupational breakthroughs of blacks into middle-management positions. A few of those hired gained professional recognition and moved into upper management, including William Bates, 1987 president of the Pittsburgh chapter of the American Institute of Architects; Darwin Davis, senior vice-president of Equitable Life Insurance Company; Jack Burley, vice-president of finance at Heinz Corporation; and Lawrence Moncrief, vice-president and general counsel of the H. K. Porter Company.

The postwar period also saw the fall of Jim Crow patterns of segregation that had humiliated blacks for over a century. After the war, blacks—and their white allies—mounted attacks on segregated public facilities, including hotels, parks, pools, restaurants, and theaters. Typical of these efforts was the protracted campaign in the 1950s, led by the Urban League's executive secretary, Joseph Allen, to desegregate city swimming pools. During the 1950s and 1960s the barriers continued to fall, until by the 1980s even the city's most prestigious clubs opened to blacks: Darwin Davis, for example, integrated the exclusive Duquesne Club; Milton Washington the Pittsburgh Athletic Association; and Eric Springer, David Epperson, and James Kelly the University Club.

The fall of such barriers, ironically, undermined some black institutions. Following the integration of major league baseball, professional black teams withered away, including the Pittsburgh Crawfords and the Homestead Grays. The Pittsburgh *Courier* followed a similar trajectory: as white papers included more news about blacks, the *Courier*'s circulation dropped until it was bought by Chicago interests in the mid-1960s. New businesses opened, however, that benefited from the improved racial climate. These included Beacon, a construction firm owned by Milton Washington with over one hundred employees; Sheradan Broadcasting, owned by Ronald Davenport, which in the 1970s was one of the few black-owned stations in America; and ALBA Travel, created by Audrey Alpern and Gladys Baynes Edmunds, one of the larger female-owned businesses in Pittsburgh.[75]

Black Pittsburgh also witnessed a minor flowering of cultural life. August Wilson, a playwright born and reared in the Hill district, won the New York Drama Critics award in 1985 for *Ma Rainey's Black Bottom*, and *Fences*, based on life in the Hill district in the 1950s, won the Pulitzer Prize in 1987. John Wideman, a basketball star and Rhodes Scholar reared in Homewood-Brushton, authored a well-received trilogy of novels set in the neighborhood of his youth— *Damballah, Hiding Place*, and *Sent for You Yesterday*, the last of which won the PEN/Faulkner Award in 1984. Another work, *Brothers and Keepers*, published in 1984, was featured on the TV show *60 Minutes*, and a subsequent novel, *Reuben*, also received national attention. In 1964, Patricia Prattis Jennings, daughter of the Pittsburgh *Courier* editor, P. L. Prattis, broke the color barrier at the Pittsburgh Symphony when she became the orchestra's pianist.[76] One year later Paul Ross became one of the symphony's violinists. In jazz, Pittsburgh continued its prewar creativity, producing such international stars as Billy Eckstine, Erroll Garner, and Mary Lou Wiliams in the 1940s; Ahmad Jamal, Stanley Turrentine, Dakota Staton, and George Benson in the 1950s and 1960s; and (though not from Pittsburgh) Nathan Davis in the 1970s and 1980s.

In addition, the University of Pittsburgh's Kuntu Repertory Theatre, directed by Rob Penny and Vernell Lillie, and Bob Johnson's Black Theater Dance Ensemble brought excellent theater and dance to the area, while Roger Humphries and Walt Harper continued the city's notable jazz tradition. Local artists such as Harold Neal and Thad Mosely continued to receive local and occasionally national recognition. In 1987 Bill Strickland's Manchester Craftsmen's Guild opened on the North Side, a multimillion-dollar building which quickly became a showplace for black and white photographers, artists, and musicians.

Conclusion

This survey has shown that, like their counterparts elsewhere, blacks in Pittsburgh have borne a heavy burden of racial discrimination. Disfranchised in 1837, they regained the vote only after the Civil War. Long barred from the city's economic mainstream, they entered local industry only after the First World War. And until World War II they were even excluded from most forms of public accommodation.

The study also suggests that, in addition to racial discrimination, blacks here have borne two further burdens. The first burden was economic. As blue-collar workers in a city whose industrial base declined earlier than most, they had fewer opportunities for economic mobility and fewer resources with which to support a middle class. The second burden was geographic. As members of a community dispersed across seven or more neighborhoods, they faced unusual difficulties developing community-wide social, economic, and political institutions.

Although the literature on black Pittsburgh is surprisingly rich, there remain important gaps in our knowledge. We especially need investigations of black neighborhoods. Although the existence of several black residential areas inhibited black unity and delayed black economic and political progress, the precise ways in which this operated remain unclear and need to be specified. Furthermore, many people derived a great degree of satisfaction living in smaller areas where they knew each other and interacted on a regular basis. Many older black residents remember friendly interaction with their white neighbors, and maintain a strong sense of identification with their old neighborhoods.

The role and position of black women in Pittsburgh also needs to be explored. Despite the numerous studies of the black community, none specifically address the experience and contribution of black women to the community. In Pittsburgh, as elsewhere in America, black women have been instrumental in the development and sustenance of numerous community organizations, the church in particular. They maintained an active club life and engaged in numerous activities of benefit to the community. Sororities and clubs such as the Business and Professional Women's Organization would be excellent topics for research. Moreover, Pittsburgh had a surprisingly large number of women-owned businesses; the entrepreneurial role of these women would be a fascinating and important area of study.

We also need an examination of the pre–World War I community, in order to understand how a people with so few job opportunities managed to create such vibrant social, cultural, and economic institutions by the turn of the century. We especially need an investigation of the leading members of that early community, the Old Pittsburghers, in order to understand the failure of them (and their

children) to maintain their entrepreneurial beginnings and to develop political leadership after World War I. In addition to understanding the black elite, we need studies of the black middle class of
Pittsburgh, particularly of their neighborhoods, such as Homewood,
East Liberty and Beltzhoover—the destination of many "migrants"
who were local and middle class rather than Southern and blue
collar. Moreover, we need studies of race relations. Rob Ruck's study
of black sport documents a remarkable amount of informal integration among neighborhood sandlot teams, and there may have been
other areas of interractial cooperation, such as schools and family
visiting. The presence of so much "high culture" within the community also needs to be investigated, particularly the active theater life
and the symphony orchestras.

For the period following the Second World War, virtually all aspects of community life need to be examined. We have no studies of
community leaders or achievers during this period. In particular, we
need a study of K. Leroy Irvis to place alongside that of Homer
Brown, explaining how both those men continued to be reelected to
an important statewide post, despite relative political weakness at
the local level. We need a study of the remarkable dislocation of
thousands of blacks, without protest, from the lower Hill district
during the early 1950s. Finally, we need studies of such community
institutions as the leading churches and the NAACP. The research
agenda is set far into the future.

The big question that needs to be explored regarding the history
of black Pittsburgh is why so much political passivity prevailed here
even following the dramatic increase of black numbers during and
after the 1920s. The hypothesis of a special double burden invites
comparative research, both to test its validity and to give it broader
significance. First, we need comparative histories of black political
development, in order to locate the political development of black
Pittsburgh somewhere along a continuum. Second, we need a comparative economic history of American cities in order to assess how
special was the economic burden of Pittsburgh's blacks. A comparison with Detroit, whose economy declined some forty years after
Pittsburgh's, would help establish whether racism or overall economic opportunities were more important in blocking the development of a stable black working class. Third, we need comparative
studies of the social development of black communities nationally
in order to assess the institutional strengths and weaknesses of
black Pittsburgh. A comparison with Cincinnati, another hilly, industrial city with dispersed black communities, would tell us
whether geographic dispersal inevitably undermines the entrepre-

neurial, institutional, and political development of a community. Finally, a comparison with cities that had segregated school systems but employed black teachers would clarify the trade-offs between segregation and jobs for the middle class.

Through such comparative studies, the experience of black Pittsburghers can take on more than local significance. Whether the Pittsburgh experience was unique—or an advanced stage of a national experience—their lives and experiences can provide a message for blacks throughout the nation.

NOTES

1. The WPA project was part of the Federal Writers Project Ethnic Survey, ca. 1940, microfilmed by the Pennsylvania State Archives in Harrisburg (reel 2). The typescript was discovered in the early 1970s by Professor Rollo Turner of the Black Studies Department of the University of Pittsburgh.

2. Harris's photographs are being identified and catalogued by Dennis Morgan and Professor Rollo Turner of the University of Pittsburgh's Department of Black Community Education, Research and Development (Black Studies). See Rollo Turner, "The Teenie Harris Photographic Collection," *Pennsylvania Ethnic Studies Newsletter* (University of Pittsburgh, University Center for Interational Studies), Winter 1987. Turner's article estimates the collection contains 50,000 items; Morgan, in a personal communication with the author, reports having located perhaps another 50,000 items in Harris's possession, plus photographs taken by other *Courier* photographers.

3. There were also black communities on the north side of the Allegheny River in Allegheny, as well as in Minersville (today Schenley Heights), a few miles east of the Hill. The pre–Civil War black community is described in Ann G. Wilmoth, "Pittsburgh and the Blacks: A Short History, 1780–1875" (Ph.D. diss. Pennsylvania State University, 1975); Ernest J. Wright, "The Negro in Pittsburgh," an unpublished typescript for the WPA; and Janet L. Bishop, "In the Shadow of the Dream: The Black Pittsburgh Community, 1850–1870" (undergraduate paper, Department of History, Princeton University, 1982). Regarding Minersville, see Bradley W. Hall, "Elites and Spatial Change in Pittsburgh: Minersville as a Case Study," *Pennsylvania History* 48 (Oct. 1981): 311–34.

4. Wilmoth, "Pittsburgh and the Blacks," pp. 9–16, table 27.

5. Wright, "The Negro in Pittsburgh," chap. 3, pp. 29ff.

6. Ibid., chap. 12, pp. 2–7, 14–20; Rayford Logan and Michael Winston, eds., *Dictionary of American Negro Biography* (New York: Norton, 1982), pp. 67–68, 577–80. The genteel life style of Pittsburgh's black community in the 1850s is recalled by Brown in *Tales My Father Told and Other Stories* (Wilberforce, Ohio: Homewood Cottage, 1925). In the 1890s two Tanner descendants, Parthenia Tanner and Mary Tanner Miller, were the first

blacks to be hired as switchborard operators by the Bell Telephone Company in Pittsburgh.

7. Wright, "The Negro in Pittsburgh," pp. 52ff.; Wilmoth, "Pittsburgh and the Blacks," p. 19; *University Times* (University of Pittsburgh), Feb. 19, 1987.

8. Wilmoth, "Pittsburgh and the Blacks," chaps. 5 and 6. It was always exceptional for free blacks in southern cities to be able to vote. Even in the North, however, free blacks were increasingly restricted in their access to the franchise. Boston was the only major city in which, during the first half of the nineteenth century, they were allowed to vote on equal terms with whites. See Leonard P. Curry, *The Free Black in Urban America, 1800–1850: The Shadow of the Dream* (Chicago: University of Chicago Press, 1981), p. 88; Wright, "The Negro in Pittsburgh," chap. 4; Richard J. M. Blackett, "Freedom, or the Martyr's Grave": Black Pittsburgh's Aid to the Fugitive Slave," *Western Pennsylvania Historical Magazine* 61 (Jan. 1978): 117–34; Wilmoth, "Pittsburgh and the Blacks," chap. 4 and pp. 52ff. The phrase "Father of Black Nationalism" was used by Theodore Draper in *The Rediscovery of Black Nationalism* (New York: Viking, 1969), p. x.

9. Wilmoth, "Pittsburgh and the Blacks," chap. 3.

10. Ibid., chap. 5.

11. Ibid., chap. 6.

12. Wright, "The Negro in Pittsburgh," chap. 5, pp. 1–67. Quote is from pp. 65–66.

13. However, the few who were employed in steel apparently were promoted and reasonably well treated. See R. R. Wright's survey of "One Hundred Negro Steel Workers," in Paul U. Kellogg, ed., *Wage-Earning Pittsburgh*, The Pittsburgh Survey, vol. 6 (New York: Survey Associates, 1914), pp. 97–110. Also see Wright, "The Negro in Pittsburgh," chap. 12.

14. This simply reflected overall growth of the city's population, which by 1900 was the seventh largest city in the nation. Not until 1920 did blacks make up more than 5 percent of the city's population.

15. These businesses were not confined to the Hill district, but were dispersed in several sections of the city—the Strip, Hill district, North Side, and Lawrenceville. Helen A. Tucker, "The Negroes of Pittsburgh," *Charities and the Commons*, Jan. 2, 1909, reprinted in Kellogg, *Wage-Earning Pittsburgh*, pp. 424–36.

16. Wright, "The Negro in Pittsburgh," chap. 12, pp. 2–7, 14–20, 28–49.

17. Andrew Buni, *Robert L. Vann of the Pittsburgh Courier: Politics and Black Journalism* (Pittsburgh: University of Pittsburgh Press, 1974), p. 26. Of necessity they also had their own cemeteries. See Joe T. Darden, "The Cemeteries of Pittsburgh: A Study in Historical Geography" (MA thesis, Department of Geography, University of Pittsburgh, 1967). Between 1896 and 1924 community activities were listed in "Afro-American Notes," which appeared each Sunday in the Pittsburgh *Press* newspaper.

18. Wright, "The Negro in Pittsburgh," chap. 12, pp. 28–49; chap. 13, esp. pp. 5–7.

19. For an overview of black migration during World War I, see Florette Henri, *Black Migration: Movement North, 1900–1920* (New York: Doubleday, 1976). Frank Bolden, former reporter for the Pittsburgh *Courier*, feels that the earlier migrants were a more sober, family- and church-oriented group, whereas those who came during the 1920s were more likely to be single and "problem-prone." Moreover, he argues, this partly reflected the type of recruiter. Rev. Nelson of East Liberty, for example, recruited the former type, whereas "Reverend" Van Harty of the North Side was much less selective. For an overview of the evolution of the Hill district, see Ralph L. Hill, "A View of the Hill: A Study of Experiences and Attitudes in the Hill District of Pittsburgh, Pa., from 1900 to 1973" (Ph.D. diss., University of Pittsburgh, 1973).

20. Wright, "The Negro in Pittsburgh," p. 75. The McKay quote is from Ishmael Reed, "In Search of August Wilson," *Connoisseur*, March 1987, p. 95. Also see William Y. Bell Jr., "Commercial Recreation Facilities Among Negroes in the Hill District of Pittsburgh" (MA thesis, Department of Social Work, University of Pittsburgh, 1938).

21. Rob Ruck, *Sandlot Seasons: Sport in Black Pittsburgh* (Urbana: University of Illinois Press, 1987), pp. 61, 114, 154–55. See also the same author's "Soaring Above the Sandlots: The Garfield Eagles," *Pennsylvania Heritage* 8 (Summer 1982): 13–18, for an examination of a neighborhood football team during the same period; and David K. Wiggins, "Wendell Smith, the Pittsburgh *Courier-Journal* and the Campaign to Include Blacks in Organized Baseball, 1933–1945," *Journal of Sport History* 10, no. 2, (Summer 1983): 5–29.

22. Published as Abraham Epstein, *The Negro Migrant in Pittsburgh* (Pittsburgh: National Urban League, 1918), p. 32. Abram Harris, "The New Negro Worker in Pittsburgh" (MA thesis, Department of Economics, University of Pittsburgh, 1924); Ira DeA. Reid, "The Negro in Major Industries and Building Trades of Pittsburgh" (MA thesis, Department of Economics, University of Pittsburgh, 1925), pp. 10, 35ff., and Ira Reid, *Social Conditions of the Negro in the Hill District of Pittsburgh* (Pittsburgh?: General Committee on the Hill Survey, 1930); James H. Baker, "Procedure in Personnel Administration with Reference to Negro Labor" (MA thesis, Department of Economics, University of Pittsburgh, 1925); Alonzo G. Moron and F. F. Stephan, "The Negro Population and Negro Families in Pittsburgh and Allegheny County," *Social Research Bulletin* 1 (Apr. 20, 1933); John T. Clark, "The Migrant in Pittsburgh," *Opportunity* 1 (Oct. 1923): 303–07, and "The Negro in Steel," *Opportunity* 4 (Mar. 1926): 87–88; W. P. Young, "The First Hundred Negro Workers," *Opportunity* 2 (Jan. 1924): 15–19.

See also, John N. Rathmell, "Status of Pittsburgh Negroes in Regard to Origin, Length of Residence, and Economic Aspects of Their Life" (MA thesis, Department of Sociology, University of Pittsburgh, 1935), p. 30; John V. Anderson, "Unemployment in Pittsburgh with Reference to the Negro" (MA thesis, Department of Economics, University of Pittsburgh, 1932); Thomas Augustine, "The Negro Steelworkers of Pittsburgh and the

Unions" (MA thesis, Department of Sociology, University of Pittsburgh, 1948); Floyd C. Covington, "Occupational Choices in Relation to Economic Opportunities of Negro Youth in Pittsburgh" (MA thesis, Department of Economics, University of Pittsburgh, 1928); Howard D. Gould, "Survey of the Occupational Opportunities for Negroes in Allegheny County" (MA thesis, Department of Economics, University of Pittsburgh, 1934); Betty Ann Weiskopf, "A Directory of Some of the Organizations to which People in the Hill District of Pittsburgh Belong" (MA thesis, Social Administration, University of Pittsburgh, 1943).

23. Glenn E. McLaughlin and Ralph J. Watkins, "The Problem of Industrial Growth in a Mature Economy," *American Economic Review* 29 (Mar. 1939), pt. 2, supplement, pp. 6–14, cited in Roy Lubove, ed., *Pittsburgh* (New York: New Viewpoints, 1976), pp. 112–21; John Bodnar, Roger Simon, and Michael P. Weber, *Lives of Their Own: Blacks, Italians, and Poles in Pittsburgh, 1900–1960* (Urbana: University of Illinois Press, 1982), pp. 117, 186.

24. Bodnar et al., *Lives of Their Own,* pp. 55–82, esp. p. 59. See also their "Migration, Kinship, and Urban Adjustment: Blacks and Poles in Pittsburgh, 1900–1930," *Journal of American History* 66 (Dec. 1979): 548–65.

25. Dennis C. Dickerson, *Out of the Crucible: Black Steelworkers in Western Pennsylvania, 1875–1980* (Albany: State University of New York Press, 1986), *passim.*

26. Linda Nyden, "Black Miners in Western Pennsylvania, 1925–1931: The National Miners Union and the United Mine Workers of America" (seminar paper, Department of History, University of Pittsburgh, 1974), pp. 3, 41, and Nyden, "Black Miners in Western Pennsylvania, 1925–1931: The National Miners Union and the United Mine Workers of America," *Science and Society* 41 (Spring 1977): 69–101. See also Gerald E. Allen, "The Negro Coal Miner in the Pittsburgh District" (MA thesis, Department of Economics, University of Pittsburgh, 1927).

27. Nyden, seminar paper, pp. 39–46.

28. Peter Gottlieb, *Making Their Own Way: Southern Blacks' Migration to Pittsburgh, 1916–1930* (Urbana: University of Illinois Press, 1987), pp. 185, 192, 118, 205ff. The dissertation was completed in 1977. See also Gottlieb's "Migration and Jobs: The New Black Workers in Pittsburgh, 1916–1930," *Western Pennsylvania Historial Magazine* 61 (Jan. 1978): 1–16, and Russell Bogin, "The Role of Education in Black Employment in the A. M. Byers & Co." (seminar paper, Department of History, Carnegie Mellon University, 1981).

29. Reid, *Social Conditions,* pp. 39–41, 58–63. See also Maryann Brice, "Vocational Adjustment of 101 Negro High School Graduates in Allegheny County" (MA thesis, Department of Social Work, University of Pittsburgh, 1938); Elsie R. Clarke, "A Study of Juvenile Delinquency in a Restricted Area of Pittsburgh" (MA thesis, Department of Sociology, University of Pittsburgh, 1932); Wiley A. Hall, "Negro Housing and Rents in the Hill District of Pittsburgh, Pa." (MA thesis, Department of Economics, Univer-

sity of Pittsburgh, 1929); Ruth M. Lowman, "Negro Delinquency in Pittsburgh" (MA thesis, Social Work, Carnegie Institute of Technology, 1923); Anna C. Morrison, "Study of 141 Dependent Negro Children Placed in Boarding Homes by the Juvenile Court of Allegheny County" (MA thesis, Department of Sociology, University of Pittsburgh, 1935); Donald J. Richey, "The Legal Status of Education for Colored People as Determined by Court Decisions" (MA thesis, Department of Sociology, University of Pittsburgh, 1932).

In addition, see Dean Scruggs Yarbrough, "Educational Status of Negro Public School Children as Reflecting Economic and Social Problems" (MA thesis, Department of Economics, University of Pittsburgh, 1926); Marion M. Banfield, "A Settlement House and Its Negro Neighbors" (MA thesis, Social Administration, University of Pittsburgh, 1941); Horace W. Bickle, "A Study of the Intelligence of a Group of Negro Trade School Boys" (MA thesis, Department of Psychology, University of Pittsburgh, 1930); Celia R. Moss, "Social and Economic Factors Affecting the Health and Welfare of a Group of Migrant Families: A Study of 29 Negro Families Migrating to Aliquippa, Pa., from Americus, Georgia, in 1936, for the Purpose of Employment at the Jones & Laughlin Steel Corporation" (MA thesis, Social Administration, University of Pittsburgh, 1943); Shirley A. Butler, "Attitudes of a Selected Group of Negro Women on the Selection of Medical Facilities in Pittsburgh" (MA thesis, Department of Social Work, University of Pittsburgh, 1948); Juliata Martinez, "The Role of the Group Worker in the Area of Interracial Education" (MA thesis, Social Administration, University of Pittsburgh, 1944).

30. Bodnar et al., *Lives of Their Own*, p. 256. For a description of Bloomfield, see William Simon, Samuel Patti, and George Herrmann, "Bloomfield: An Italian Working-Class Neighborhood," *Italian Americana* 7 (1981): 102–16. For other neighborhoods, see Alexander Z. Pittler, "The Hill District in Pittsburgh: A Study in Succession" (MA thesis, Department of Sociology, University of Pittsburgh, 1930); Delmar C. Seawright, "Effect of City Growth on the Homewood-Brushton District of Pittsburgh" (MA thesis, Department of Sociology, University of Pittsburgh, 1932); Everett Alderman, "Study of Twenty-Four Fraternal Organizations and Clubs of the Larimer Avenue District, East Liberty, Pittsburgh" (MA thesis, School of Education, University of Pittsburgh, 1932).

31. Alonzo G. Moron, "Distribution of the Negro Population in Pittsburgh, 1910–1930" (MA thesis, Department of Sociology, University of Pittsburgh, 1933), pp. 28–30.

32. See Joe Darden, *Afro-Americans in Pittsburgh: The Residential Segregation of a People* (Lexington, Mass.: D. C. Heath, 1973), p. 13; Karl and Alma Taeuber, *Negroes in Cities: Residential Segregation and Neighborhood Change* (Chicago: Aldine, 1965), table 4.

33. Moron, "Distribution of the Negro Population," p. 29. Curiously, the overall index of racial segregation in Pittsburgh was still quite high.

34. The quotation is from Bell, "Commercial Recreation Facilities."

Whites also moved out of the Hill district at this time. Italians and Jews moved to Bloomfield and Squirrel Hill, respectively. For the recollections of a Jewish resident, see M. R. Goldman, "The Hill District of Pittsburgh as I Knew It," *Western Pennsylvania Historical Magazine* 51 (July 1968): 279–95. Also see Kurt Pine, "The Jews in the Hill District of Pittsburgh, 1910–1940: A Study of Trends" (MA thesis, Social Administration, University of Pittsburgh, 1943).

35. Bodnar et al., *Lives of Their Own*, p. 197; Steven Sapolsky and Bartholomew Roselli, *Homewood-Brushton: A Century of Community-Making* (Pittsburgh: Western Pennsylvania Historical Society, 1987). See also Geraldine Hermalin, "Recreational Resources for the Negro in the Homewood-Brushton Area, 1945" (MA thesis, Social Administration, University of Pittsburgh, 1945); Helen Judd, "A Study of Recreational Facilities for Negroes in Manchester" (MA thesis, Social Administration, University of Pittsburgh, 1945); Carole T. Szwarc, "Manchester: A Study in Contrast, 1930–1968" (seminar paper, Department of History, Carnegie Mellon University, n.d.); Hilda Kaplan, "Recreational Facilities for the Negro in the East Liberty District, with Special Emphasis on Tracts 7G, 12 and 12E" (MA thesis, Social Administration, University of Pittsburgh, 1945); Ruby E. Ovid, "Recreational Facilities for the Negro in Manchester" (MA thesis, Department of Social Work, University of Pittsburgh, 1952); Pauline Redmond, "Race Relations on the South Side of Pittsburgh as Seen Through Brashear Settlement" (MA thesis, Department of Social Work, University of Pittsburgh, 1936); Gertrude A. Tanneyhill, "Carver House and the Community: A Study of Twenty-two Families Served by Brashear Association in December 1940 and 1941" (MA thesis, Social Administration, University of Pittsburgh, 1942).

36. Jean Hamilton Walls, "A Study of the Negro Graduates of the University of Pittsburgh for the Decade 1926–1936" (Ph.D. diss., University of Pittsburgh, 1938), p. 18 and *passim*.

37. Based on personal communications from several Pittsburghers to the author. This seating pattern might have been the result simply of dark-skinned newcomers joining an older, predominantly light-skinned congregation in which the established members continued to occupy seats toward the front.

38. Gottlieb, *Making Their Own Way*, p. 267. For the description of a storefront pentecostal church in the Hill district, see Melvin D. Williams, "A Pentecostal Congregation in Pittsburgh: A Religious Community in a Black Ghetto" (Ph.D. diss., University of Pittsburgh, 1973), published as *Community in a Black Pentecostal Church: An Anthropological Study* (Pittsburgh: University of Pittsburgh Press, 1974). See also Noshir F. Kaikobad, "The Colored Moslems of Pittsburgh" (MA thesis, Department of Social Work, University of Pittsburgh, 1948); Dennis C. Dickerson, "Black Ecumenicism: Efforts to Establish a United Methodist Episcopal Church, 1918–1932," *Church History* 52 (Dec. 1983): 479–91, and also his "The Black Church in Industrializing Western Pennsylvania, 1870–1950," *West-*

ern *Pennsylvania Historical Magazine* 64 (Oct. 1981): 329–44. Also see Haseltine T. Watkins, "The Newer Religious Bodies Among Negroes in the City of Pittsburgh" (MA thesis, Social Administration, University of Pittsburgh, 1945).

39. Marjorie Allen, "The Negro Upper Class in Pittsburgh, 1910–1964" (seminar paper, Department of History, University of Pittsburgh, 1964), p. 5 and *passim.*

40. The *Courier* had been established several years previously by Edwin Nathaniel Harleston, a guard at the H. J. Heinz food-processing plant. See Buni, *Vann*, pp. 42–43.

41. *University Times*, Feb. 19, 1987; Buni, *Vann*, pp. 37–38, 80–81. Buni's excellent biography is rich in detail on the social and political history of black Pittsburgh from about 1910 to 1940. Frank Bolden, former reporter for the *Courier*, is a storehouse of information and insight. See also Miriam B. Rosenbloom, "An Outline of the History of the Negro in the Pittsburgh Area" (MA thesis, Social Administration, University of Pittsburgh, 1945); and Charles C. Berkley, "The Analysis and Classification of Negro Items in Four Pittsburgh Newspapers, 1917–1937" (MA thesis, Department of Sociology, University of Pittsburgh, 1937).

42. Buni, *Vann*, pp. 107ff.

43. Wright, "The Negro in Pittsburgh," chap. 13, pp. 3, 62.

44. Charles J. Burks was the first black to be admitted to the University of Pittsburgh's medical school; he gained staff privileges at Montefiore Hospital. *University Times*, Feb. 19, 1987; Dr. M. R. Hadley was the first on the staff of any Allegheny County hospital, having joined McKeesport Hospital in 1945. See *Pittsburgh Press*, Roto Section, Oct. 17, 1982, p. 22. Bodnar et al., *Lives of Their Own*, p. 242; Martha B. Moore, *Unmasked: The Story of My Life on Both Sides of the Race Barrier* (New York: Exposition Press, 1964), p. 59.

45. Ralph Proctor, Jr., "Racial Discrimination Against Black Teachers and Black Professionals in the Pittsburgh Public School System, 1834–1973" (Ph.D. diss., University of Pittsburgh, 1979), table II-5, pp. 46–48. See also Barbara J. Hunter, "The Public Education of Blacks in the City of Pittsburgh, 1920–1950: Actions and Reactions of the Black Community in Its Pursuit of Educational Equality" (Ph.D. diss., University of Pittsburgh, 1987).

46. Walls, "Negro Graduates," p. 65.

47. Ibid., p. 105. Track and field was the only major sport open to blacks in the 1930s; those who did participate distinguished themselves. Everett Utterback was chosen team captain in 1931 (causing an uproar); John Woodruff and Herb Douglass won medals in the 1936 and 1948 Olympics. See Jim O'Brien, ed., *Hail to Pitt* (University of Pittsburgh, 1982), pp. 246–47. Not until the 1940s were blacks allowed on the football and basketball teams.

48. Buni, *Vann*, pp. 74, 176; Wright, "The Negro in Pittsburgh," chap. 5, pp. 70ff.

49. Buni, *Vann*, pp. 62–63.

50. In 1939, one year before his death, Vann finally was invited to join the Chamber of Commerce. Ibid., pp. 122, 315.

51. Ibid., chap. 7.

52. Ibid., pp. 91ff.

53. Ibid., p. 357, n. 75. By the late 1920s the *Courier* gave less attention to Pittsburgh, apparently reflecting Vann's disillusionment with local politics and his increasing interest in securing a position with Roosevelt's New Deal (ibid., pp. 142ff.). Walter Tucker in 1930 was elected the first black state legislator from Western Pennsylvania (ibid., p. 356, n. 75). Also see James Brewer, "Robert L. Vann and the *Pittsburgh Courier*" (MA thesis, Department of History, University of Pittsburgh, 1941); James S. Galloway, "The *Pittsburgh Courier* and FEPC Agitation" (MA thesis, Department of History, University of Pittsburgh, 1954).

54. Buni, *Vann*, pp. 194, 272–73.

55. Ruth Simmons, "The Negro in Recent Pittsburgh Politics" (MA thesis, Department of Political Science, University of Pittsburgh, 1945), pp. 38, 43, 140, and *passim*. Also, James S. Galloway, "The Negro in Politics in Pittsburgh, 1928–1940" (seminar paper, History Department, University of Pittsburgh, n.d.).

56. Henry A. Schooley, "A Case Study of the Pittsburgh Branch of the NAACP" (MA thesis, Department of Sociology, University of Pittsburgh, 1952), pp. 8, 58–59, and *passim*; Buni, *Vann*, pp. 58–59, 200; Frank Zabrosky, "Some Aspects of Negro Civic Leadership in Pittsburgh, 1955–1965" (seminar paper, Department of History, University of Pittsburgh, n.d.), p. 9.

57. Arthur J. Edmunds, *Daybreakers: The Story of the Urban League of Pittsburgh: The First Sixty-Five Years* (Pittsburgh: Urban League, 1983), pp. 29ff. See also Ruth L. Stevenson, "The Pittsburgh Urban League" (MA thesis, Department of Social Work, University of Pittsburgh, 1936); Antoinette H. Westmoreland, "A Study of Requests for Specialized Services Directed to the Urban League of Pittsburgh" (MA thesis, Department of Social Work, University of Pittsburgh, 1938); Katherine R. Wilson, "A Personnel Guidance of Negro Youth in the Urban League of Pittsburgh" (MA thesis, Department of Psychology, University of Pittsburgh, 1940).

58. "Homer S. Brown: First Black Political Leader in Pittsburgh," *Journal of Negro History* 66 (Winter 1981–82): 304–17; Simmons, "Negro in Recent Pittsburgh Politics," p. 39.

59. Proctor, "Racial Discrimination," pp. 50–62, 75–79, and *passim*. In fact the board's hiring also reflected a greater sensitivity to problems of interracial relations. See, for example, Anthony J. Major, "An Investigation of Supervisory Practices for the Improvement of Instruction in Negro Public Schools" (Ed.D. diss., School of Education, University of Pittsburgh, 1940); Kathryn R. Witz, "A Study of the Dispositions Given Negro and White Juvenile Delinquents at the Juvenile Court of Allegheny County" (MA thesis, Department of Sociology, 1949); Margaret D. Thomson, "The Role of

the American Service Institute's Consultant on Interracial and Intercultural Problems for Summer Camps in Allegheny County" (MA thesis, Department of Social Work, University of Pittsburgh, 1949); Ignacia Torres, "Methods of Developing Interracial and Intercultural Practices in the Pittsburgh YWCA: A Descriptive Study of Three Departments" (MA thesis, Department of Social Work, University of Pittsburgh, 1947); Kenneth R. Hopkins, "Some Factors Involved in the Discontinuance of Participation of Selected Junior-Age Negro Boys in the Program of the Irene Kaufmann Settlement" (MA thesis, Department of Social Work, University of Pittsburgh, 1952); Lillian B. Hotard, "Problems of Implementing Agency Policy: A Study of Children in Inter-Racial Activity Groups in Wadsworth Hall Who Discontinued Participation During 1949–1950" (MA thesis, Department of Social Work, University of Pittsburgh, 1951); Elizabeth B. Jackson, "The Pittsburgh Council on Intercultural Education: A Study in Community Organizations" (MA thesis, Department of Social Work, University of Pittsburgh, 1948); Albert G. Rosenberg, "Interracial Developments in Hill District Agencies and Schools, 1944 to 1948" (MA thesis, Department of Social Work, University of Pittsburgh, 1949).

See also, Arthur M. Stevenson, "Analysis of Five Agencies of Pittsburgh Providing Intercultural Services" (MA thesis, Department of Social Work, University of Pittsburgh, 1949); Carl Birchard and William B. Weinstein, "The Curtaineers: A Study of an Interracial Dramatic Project of the Irene Kaufmann Settlement of Pittsburgh from Its Inception" (MA thesis, Department of Social Work, University of Pittsburgh, 1948); Goldie Gibson, "The Change of Policy in SoHo's Afternoon Program and Its Effect on Interracial Relationships in Groups" (MA thesis, Department of Social Work, University of Pittsburgh, 1950); Cuthbert G. Gifford, "The Hill District People's Forum: A Study of Social Action as Carried Out by the Forum from Its Inception in 1942 to March, 1947" (MA thesis, Department of Social Work, University of Pittsburgh, 1947). Anne M. Barton, "Ethnic and Racial Groups in Rankin, Pa.: A Study of Relations Between Them as Expressed Through Various Social Forces" (MA thesis, Department of Social Work, University of Pittsburgh, 1947). That black students had special needs is obvious from the studies of Helen M. Simmen, "A Study of the Graduates of the Fifth Avenue High School" (MA thesis, School of Education, University of Pittsburgh, 1928); Irene A. Thompson, "Everyday English of a Ghetto Group" (MA thesis, School of Education, University of Pittsburgh, 1929); J. Allen Figurel, "The Vocabulary of Underprivileged Children" (Ph.D. diss., School of Education, University of Pittsburgh, 1948).

60. Race relations received increasing attention during the 1940s and were the subject of numerous master's theses. See Celia Bach, "The Pittsburgh Council on Intercultural Education: A Descriptive Study" (MA thesis, Social Administration, University of Pittsburgh, 1946); Bernard M. Shiffman, "The Development of Interracial Camping in Allegheny County" (MA thesis, Department of Social Administration, University of Pittsburgh, 1946); Marylyn E. Duncan, "Areas of Racial Tension in a Public Housing

Project: A Study of Relations Between Building Location in Terrace Village and Racial Tensions, with Special Reference to Play Areas" (MA thesis, Social Administration, University of Pittsburgh, 1946); Thelma W. Smith, "The Selection of Foster Homes for Negro Foster Children of the Juvenile Court of Allegheny County" (MA thesis, Social Administration, University of Pittsburgh, 1942); Frances A. Zorn, "Substitute Parental Care of the Negro Child" (MA thesis, Social Administration, University of Pittsburgh, 1941); Alberta W. Brown, "A Descriptive Study of the Negro Students Received in Transfer from the South at Herron Hill Junior High School from the Fall of 1943 Through the Fall of 1945" (MA thesis, Social Administration, University of Pittsburgh, 1946); Francis J. Costigan, "Employment Opportunities for Students Leaving Herron Hill Junior High School" (MA thesis, Social Administration, University of Pittsburgh, 1944).

61. For an analysis of this trend nationally, see William J. Wilson, *The Declining Significance of Race* (Chicago: University of Chicago Press, 1978). Frank Giarratani and David Houston, "Economic Change in the Pittsburgh Region," typescript, Mar. 27, 1986, p. 12; *U.S. Census of Manufactures, 1947: vol. 3, Statistics by States* (Washington, D.C.: U.S. Government Printing Office, 1950); and *U.S. Census of Manufactures, 1977: vol. 3, Geographic Area Statistics: Pt. 2 General Summary, Nebraska-Wyoming* (Washington, D.C.: U.S. Government Printing Office, 1971).

62. Giarratani and Houston, "Economic Change," p. 27; L. Haley, president, Urban League of Pittsburgh, press conference, Jan. 22, 1987. William O'Hare, a Washington, D.C., consultant, ranked metropolitan areas by comparing the economic status of local blacks to that of local whites; see the Pittsburgh *press*, Feb. 8, 1987, B5. Also see Institute for the Black Family, "Final Report of the Black Community Profile Study" (typescript, University of Pittsburgh, 1987), p. 8; Barry Wellman, "Crossing Social Boundaries: Cosmopolitanism Among Black and White Adolescents," *Social Science Quarterly* 52 (Dec. 1971): 602–24; P. S. Fry and K. J. Coe, "Achievement Performance of Internally and Externally Oriented Black and White High-School Students Under Conditions of Competition and Co-operation Expectancies," *British Journal of Educational Psychology* 50 (June 1980): 162–67.

63. Roy Lubove, *Twentieth Century Pittsburgh: Government, Business, and Environmental Change* (New York: John Wiley, 1969), pp. 130–32.

64. However, the exodus by whites, already underway before urban renewal, was swifter than that of middle-class blacks. Homewood-Brushton, for example, went from 22 percent black in 1950 to 66 percent black in 1960. Destruction of the lower Hill probably increased residential segregation, although indexes of residential segregation show remarkably little change from 1930 to 1970. Darden, *Afro-Americans in Pittsburgh*, calculated that for each of those years, 70–75 percent of blacks (or whites) would have to change wards to desegregate the city. He also presents calculations showing that the segregation was due primarily to racial rather than economic factors. See also M. Ruth McIntyre, "The Organizational Nature of an Urban Residential Neighborhood in Transition: Homewood-Bruston of Pittsburgh" (Ph.D. diss., University of Pittsburgh, 1963); Suk-Han Shin, "A Geographical Measure-

ment of Residential Blight in the City of Pittsburgh, Pa., 1950 and 1960" (Ph.D. diss., University of Pittsburgh, 1975); William S. J. Smith, "Redlining: A Neighborhood Analysis of Mortgage Lending in Pittsburgh, Pa." (MA thesis, Department of Geography, University of Pittsburgh, 1982); Jacqueline W. Wolfe, "The Changing Pattern of Residence of the Negro in Pittsburgh, Pa., with Emphasis on the Period 1930–1960" (MA thesis, Department of Geography, University of Pittsburgh, 1964); Brenda K. Miller, "Racial Change in an Urban Residential Area: A Geographic Analysis of Wilkinsburg, Pa." (Ph.D. diss., University of Pittsburgh, 1979); Stanley Lieberson, *Ethnic Patterns in American Cities* (New York: Free Press, 1963).

See also R. Jeffrey Green and George M. von Furstenberg, "The Effects of Race and Age of Housing on Mortgage Delinquency Risk," *Urban Studies* [Great Britain] 12 (Feb. 1975): 85–89; Arnold J. Auerbach, "The Patterns of Community Leadership in Urban Redevelopment: A Pittsburgh Profile" (Ph.D. diss., University of Pittsburgh, 1960); Edmund M. Burke, "A Study of Community Organization in Urban Renewal: Social Work Method in a Non-Social Work Setting" (Ph.D. diss., University of Pittsburgh, 1965); Carol Bauer and Joan Scharf, "A Study of Residents' Attitudes Toward an Interracial Private Housing Development" (MA thesis, Department of Social Work, University of Pittsburgh, 1959); Gayle H. Dobbs, "Support Systems and Services Utilized by Parents in Beltzhoover" (MA thesis, Department of Child Development, University of Pittsburgh, 1980); Rodney A. Pelton, "The Value System of a Large Voluntary Negro Civic Organization Within a Poverty Area: The Homewood-Brushton Committee Improvement Association" (Ph.D. diss., University of Pittsburgh, 1968).

65. Neil Gilbert, "Clients or Constituents? A Case Study of Pittsburgh's War on Poverty" (Ph.D. diss., University of Pittsburgh, 1968), pp. 23, 119, 125. See also his "Maximum Feasible Participation? A Pittsburgh Encounter," *Social Work* 14 (July 1969): 84–92. Regarding other efforts, see David H. Freedman, "An Analysis of the Institutional Capability of the City of Pittsburgh to Provide Effective Technical Assistance to Black Businessmen" (Ph.D. diss., University of Pittsburgh, 1971). Also, Colin De'Ath, "Patterns of Participation and Exclusion: A Poor Italian and Black Urban Community and Its Response to a Federal Poverty Program" (Ph.D. diss., University of Pittsburgh, 1970); Barbara B. Jameson, *A Study of the Social Services Program of the Housing Authority, City of Pittsburgh* (University of Pittsburgh: School of Social Work, 1975); Norman J. Johnson and Peggy R. Sanday, "Subcultural Variations in an Urban Poor Population," *American Anthropologist* 73 (Feb. 1971): 128–43; Shirley A. Ali, "A Case Study of Interracial Child Care Relations in a Residential Institution" (MA thesis, Department of Child Development, University of Pittsburgh, 1975); Larry V. Stockman, "Poverty and Hunger: A Pittsburgh Profile of Selected Neighborhoods" (Ph.D. diss., University of Pittsburgh, 1982); Earl D. Hollander, "Utilization and Views of Available Health Care Among Residents in Three Poverty Neighborhoods in Pittsburgh, Pa." (Ph.D. diss., University of Pittsburgh, 1970).

66. Stanley Plastrik, "Confrontation in Pittsburgh," *Dissent* 17 (Jan.–Feb. 1970): 25–31; Irwin Dubinsky, "Trade Union Discrimination in the

Pittsburgh Construction Industry: How and Why It Operates," *Urban Affairs Quarterly* 6 (Mar. 1971): 297–318. Also Daniel U. Levine, "Research Note: Are the Black Poor Satisfied with Conditions in Their Neighborhood?" *Journal of the American Institute of Planners* 38 (May 1972): 168–71; Norman D. Randolph, "Perceptions of Former Trainees of the Effectiveness of the Comprehensive Employment Training Act (CETA) Training Programs in Allegheny County" (Ph.D. diss., University of Pittsburgh, 1983).

67. Pittsburgh Board of Public Education, *Annual Report for 1965: The Quest for Racial Equality in the Pittsburgh Public Schools*, pp. 7, 11, 14, 19, 21. David H. Elliott, "Social Origins and Values of Teachers and Their Attitudes to Students from Poverty Backgrounds" (Ph.D. diss., University of Pittsburgh, 1968); Louis H. MacKey, "The Pennsylvania Human Relations Committe and Desegregation in the Public Schools of Pennsylvania, 1961–1978" (Ph.D. diss. University of Pittsburgh, 1978).

68. Rollo Turner, "Citizen, Freeman, or Elector: A History of Blacks in the Political and Electoral Process in Pittsburgh, 1837–1975," typescript, March 1987; Pittsburgh *Press*, Oct. 27, 1986. Carter's quote appeared locally; my source is the Dayton (Ohio) *Daily News*, Nov. 30, 1986, p. 3B.

69. Melvin D. Williams, "Belmar: Diverse Life Styles in a Pittsburgh Black Neighborhood," *Ethnic Groups* 3 (Dec. 1980): 23–54; for a fuller description, see the same author's *On the Street Where I Lived* (New York: Holt, Rinehart & Winston, 1981).

70. Williams, *Community in a Black Pentecostal Church*, pp. 4, 21, *passim*.

71. Board of Education, *Annual Report for 1965*, p. 7 and *passim*; Proctor, "Racial Discrimination," chap. 3, tables, 3, 4, 5, 9, and 1; pp. 149–64; table IV-1. Proctor documents efforts by these teachers to promote black pride and improve race relations. See also Philip E. McPherson, "Coordination of Efforts for the Improvement of Race Relations in the Pittsburgh Public Schools" (Ph.D. diss., Harvard University, 1966); Theodore Vasser, "A Historical and Phenomenological Study of Black Administrators in the Pittsburgh Public School System Inclusive of the Years 1950–1977" (Ph.D. diss., University of Pittsburgh, 1978); Richard D. Gutkind, "Desegregation of Pittsburgh Public Schools, 1968–1980: A Study of the Superintendent and Educational Policy Dynamics" (Ph.D. diss., University of Pittsburgh, 1983); Chapman W. Bouldin, "An Analysis of How Black Americans Are Depicted in Eleventh Grade United States History Textbooks Used in the Secondary Public Schools from 1930 to 1979" (Ph.D. diss., University of Pittsburgh, 1980); Edward A. Suchman et al., *The Relationship Between Poverty and Educational Deprivation: Final Report* (University of Pittsburgh: Learning Research and Development Center, 1968); Edward S. Greenberg, "Black Children, Self-Esteem and the Liberation Movement," *Politics and Society* 2 (Spring 1972): 293–307; Jack L. Palmer, "A Case Study in School-Community Conflict Over Desegregation" (Ph.D. diss., University of Pittsburgh, 1974); Barbara Sizemore, *An Abashing Anomaly: The High-Achieving Predominantly Black Elementary School* (University

of Pittsburgh: Department of Black Community Education, Research, and Development [Black Studies], 1983).

72. Giarratani and Houston, "Economic Change," p. 11.

73. Homer Brown was the first black appointed to the Board of Education, in 1943. At the University of Pittsburgh, during the 1980s the growth of black student enrollment leveled off, and the number of black faculty actually declined. See the annual *Fact Book*, published by the Office of Institutional and Policy Studies. For an analysis of the impact of UCEP, the special admissions program for minority students, see Diane D. Eddins, "A Causal Model of the Attrition of Specially Admitted Black Students in Higher Education" (Ph.D. diss., University of Pittsburgh, 1981); Wilma R. Smith, "The Relationship Between Self-Concept of Academic Ability, Locus-of-Control of the Environment, and Academic Achievement of Black Students Specially Admitted to the University of Pittsburgh" (Ph.D. diss., University of Pittsburgh, 1972). Barbara K. Shore, "New Careers: Myth or Reality? A Study of the Stirrings of Indigenous New Professionals" (Ph.D. diss., University of Pittsburgh, 1971). The universities also became more attuned to nearby black neighborhoods. See Ralph H. Holmes, "The University of Pittsburgh's Trees Hall–Community Leisure-Learn Program: 1968–1982" (Ph.D. diss., University of Pittsburgh, 1983); Sheila L. Johnson, "The Ethnomusicology Program at the University of Pittsburgh: A Case Study" (Ph.D. diss., University of Pittsburgh, 1983); Dorothy Burnett, "Black Studies Departments and Afro-American Library Collections at Two Predominantly White Universities: A Comparative Analysis" (Ph.D. diss., University of Pittsburgh, 1984). The relative performance of the nation's research universities was documented by Alexander Astin, *Minorities in American Higher Education* (San Francisco: Jossey-Bass, 1982), p. 136. The ranking was based on a comparison of the percentage black enrollment in a specific university to the proportion black enrollment in all universities in the state. Penn State came out near the bottom nationally.

74. Edmunds, *Daybreakers*, pp. 112–37. See also "Blacks in White: Affirmative Action in Health Services [in Pittsburgh]," *Urban League Review* 4 (Winter 1979): 65–71. James O. F. Hackshaw, "The Committee for Fair Employment in Pittsburgh Department Stores: A Study of the Methods and Techniques Used by the Committee in Their Campaign to Secure a Non-Discriminatory Hiring Policy in the Department Stores of Pittsburgh" (MA thesis, Department of Social Work, University of Pittsburgh, 1949).

75. We very much need studies of black businesses in Pittsburgh. For one of the few case studies, see Kem R. Kemathe, "Selected Characteristics of Black Businesses in the U.S., 1969–1972, with Two Case Studies of Black-Owned Businesses in Pittsburgh as Special References" (Ph.D. diss., University of Pittsburgh, 1978).

76. Patricia Prattis Jennings also was probably the first black to play for the orchestra in any capacity when she performed in 1956, with William Steinberg conducting. Personal communication, Mar. 29, 1988.

Working-Class Formation, Development, and Consciousness in Pittsburgh, 1790–1960

RICHARD OESTREICHER

 M OST OF ITS smokestacks now rust idly; local fans who drink Iron City Beer as they watch the Steelers on television are an ironic reminder of the city's past. Through most of its history Pittsburgh has symbolized modern industry and blue-collar America, perhaps more than any other city in the United States. To understand the social reality behind that symbolism is to understand some of the most fundamental processes of social change in American history. As Fort Pitt became the Iron City, the United States was changing from an agrarian republic to the world's leading industrial power. In the agrarian republic, most people worked on family farms, slave plantations, or in artisanal workshops. In the Iron City, most people were either wage earners or bosses. As the Iron City became the Steel City, the bosses became corporate executives. As the executives decided to invest their capital elsewhere, their employees, the "mill hunks," became unemployed. The creation, evolution, and decline of the Iron City is in large part the story of the formation, development, recomposition, and conflicts of the city's working class.

By class *formation* I mean the economic and structural changes in a society that lead to the emergence of large numbers of manual wage earners who are unprotected and unencumbered by the social relationships of such earlier labor systems as serfdom, slavery, or traditional artisanship. The result is an economic class which may or may not have other common cultural and demographic characteristics. Class formation will be a continuing feature of an expanding capitalist economy, but the particular way in which initial class formation takes place in a society tends to shape the process in the future.

Class *development* means the gradual development of common values, symbols, modes of thought and language, behavior, day-to-

111

day living patterns, and formal institutions among the members of an economic class as a result of association or common experiences; that is, the development of a working-class culture or a social class. Such class development is partially (but never simply) a product of the simultaneous development of a business class in antagonism to workers. If this sense of antagonism becomes highly developed, working-class culture will display distinctly oppositional qualities which implicitly or explicitly challenge prevailing social institutions and power relations.

Class *recomposition* refers to the dynamic nature of class; a working class is never simply made, but is constantly in the process of being remade as individuals are leaving working-class occupations while others are entering them, as these occupations are qualitatively changing, and as the institutional structure of society is changing.

Class *conflict* may include both covert and individual forms of resistance, such as limitation of output or purposefully inefficient work, and overt, collective forms such as strikes, boycotts, demonstrations, and political activities. It is also often manifested in cultural and social struggles which do not directly appear to concern class interests, such as conflicts over public morality, educational policies, allocation of urban services, and uses of parks and other public places.

As I survey the scholarly literature and selected primary documents about Pittsburgh's working class, I have three objectives. First to use Pittsburgh as a test case for the interpretive framework I am developing for a larger social history of American workers; second, to place the historiography of local labor history within that framework, summarizing what we know and what we still need to learn more about; and third, to suggest what has been distinctive or different about this region's labor history as compared to other places.

Initial Class Formation in the Pittsburgh Region, 1790–1850

By 1850 Pittsburgh, already known as the Birmingham of America, had established its preeminence as America's Iron City. Rolling mills and iron foundries lined the riverbanks of Pittsburgh and its surrounding industrial towns, and a diverse array of secondary metals firms turned pig iron into pipes, nails, tools, machines, and virtually every other iron product. In the next decade the continuing rise of these metals industries would be linked to the region's coal supply. Pittsburgh sat astride one of the largest soft coal seams in the world. Other industries complemented iron and coal, including glass, cotton textiles, and light consumer goods intended both for the local population and other towns downstream.[1]

This manufacturing base made Pittsburgh, and nearly all the surrounding towns of the metropolitan area, an overwhelmingly blue-collar community. Michael Holt estimated that 33 percent of Pittsburgh's-adult white male work force were unskilled workers in 1850, 30 percent were skilled workers, and another 8.7 percent were small manufacturers, some of whom were self-employed artisans.[2] Inclusion of blacks and women and workers in the satellite mill towns would increase the estimated proportion of workers in the metropolitan work force to over three-fourths, substantially above Holt's estimate for Pittsburgh's white males.

Until someone undertakes a detailed study of the antebellum labor history of Pittsburgh, we will know little about the lives and work of these people, how they came to be wage laborers in Pittsburgh, or what would become of them in the next decade or two.[3] Any explanation of initial class formation in Pittsburgh must analyze two sets of changes: first, the development of a local manufacturing economy organized on the basis of wage labor, and second, the social and demographic processes which produced a local labor force and determined how its labor would be allocated. The broad explanations for the city's economic development have already been described by local historians: its location at a natural transportation nexus at the junction of its three rivers; the easy access to coal, iron, and other natural resources; the natural barrier of the Allegheny Mountains, which limited distribution of the products of more developed eastern and foreign manufacturers to the rapidly expanding agricultural population of the Ohio Valley; the stimuli of trade interruptions during the War of 1812 and tariff increases in the 1820s; the improvements in road, canal, and rail transportation from the 1820s through the 1840s. It is much less clear how the organization of work changed as local industries developed or how changes in labor processes determined class formation.

Certainly the beginnings of a class structure date from the very beginning of white settlement in the region. As Robert Harper demonstrated in his careful analysis of surviving tax records of the late 1700s, Western Pennsylvania was never an egalitarian society, even in its frontier phase. In 1800, a smaller proportion of the region's population owned taxable wealth than the average for the entire state. In Fayette County (which Harper studied most thoroughly) over half (53.2 percent) the taxpayers and nearly two-thirds of the adult white males were landless in 1796, a substantially higher rate of landlessness than the national average (50.6 percent in 1798).

Other forms of wealth were even more unequally distributed than land.[4] The occupational structure of Fayette County taxpayers in 1796 breaks down as follows:

Professionals, merchants, and land speculators	10.5%
Yeomen	30.8
Artisans	16.1
Landless farmers (renters and tenants)	21.3
Laborers	17.1
The poor	4.2

While the tax lists include no more than half the local work force, they provide a starting point for understanding the region's occupational structure.[5] In Fayette County in 1796, landowning farmers constituted less than a third of the taxpayers, artisans about another sixth. Close to half belonged to what contemporaries called "the dependent classes": landless tenant farmers and renters, agricultural and other laborers, and "the poor" (mainly widows and the aged). The tax lists include no servants or apprentices, very few women (except a few landowners—mainly widows), and none of the working population under the age of twenty-one. If we added estimates for these groups, the percentages of professionals, merchants, and farmers would fall sharply, and the proportion of laborers and other wage earners would be much higher than the taxpayer total shown above.

Harper's reconstruction demonstrates that even in a frontier agricultural region, like Western Pennsylvania in the 1790s, wealth was unequally distributed, society was hierarchical, and perhaps one-quarter of the work force earned their livings in small-scale manufacturing, craft production, or other nonagricultural manual pursuits. Much of this nonagricultural manual labor force, especially the unskilled, was young and mobile. Some of the unskilled laborers were children of local farmers, but there also seems to have been a substantial pool of transients. In one township, two-thirds of the laborers present in 1796 had not been there three years before. Judging from local landownership records, few acquired land, the most likely path to upward mobility. It seems probable that this rural proletariat (a national phenomenon) eventually served as a major labor pool for expanding industries, but so far we do not know enough about the backgrounds of the workers in early local industries to say for sure.[6]

Unfortunately, we cannot document the distribution of wealth in Pittsburgh in the same detail as surrounding counties, but scattered references to early local manufacturing and lists of occupations do allow us to sketch the rough outlines of the rise of Pittsburgh's manual labor force.[7] As early as 1792 one observer listed the occupations of thirty-six Pittsburgh "mechanics" in a community of 130 families. In that year George Anshutz, an Alsatian immigrant, set up Pittsburgh's first iron furnace just outside the city limits. The first glass factory was established in 1797 and the first rolling mill in 1803. By 1803 the total value of the city's industrial production was estimated at $358,908.[8]

The scale of these enterprises was very small, and all or nearly all must have been organized around the traditional craft hierarchy of masters, journeymen, and apprentices. An 1802 almanac lists 163 craft proprietors, while the total production figures for the following year suggest a labor force of about three hundred, and thus an average firm size of two workers per firm. Possibly some of the individuals in the 1802 list were journeymen, and not masters, but even if the average firm size was slightly larger than two, the majority of local craft producers could not yet be described as proletarians. Certainly close to half were self-employed, and they worked in an economy and labor system in which the hope of future proprietorship seemed high. There were also dock hands, day laborers, servants, sailors, and carters whose employment status resembled later wage laborers much more closely, but we know nothing about their numbers.[9]

The number of workers, range of industries, and scale of operations all expanded rapidly over the next decade. Tench Coxe listed 494 industrial enterprises functioning in Allegheny County in his national survey of manufactures conducted in 1810, while a local survey in 1817 found that 259 industrial establishments in the city of Pittsburgh employed 1,637 hands to make aggregate products worth $2.3 million.[10]

By the teens, we can already discern a marked tendency toward the proletarianization of craft producers. As shown in table 1, mean firm size had increased to more than six workers per firm. Fourteen relatively large firms in iron, glass, and other industries employed between twenty and forty workers each. Together the employees of these fourteen firms comprised 26 percent of the industrial work force. While the size of such firms was still modest by later standards, they were already large enough and highly capitalized enough that few journeymen could hope to become proprietors of similar enterprises.[11]

Even in less capitalized industries, with smaller firms still organized around traditional craft methods, the paternalistic and familial relationships between masters and journeymen common to earlier craft production were giving way to more impersonal class relationships between employers and employees. By 1817, for example, fourteen local firms employed a total of 109 shoemakers. Local journey-men had been fighting a running battle with their employers over terms of employment for more than a decade. Shoeworkers first struck for higher wages in 1804, and the record of the 1815 conspiracy trial suggests that they had achieved a de facto union shop with a negotiated wage scale by 1809. It was this organizational power which led the shoe manufacturers to meet together and undertake prosecution of their workers for criminal conspiracy in 1814. An initial trial appears to have ended in a negotiated settlement without a verdict, but the following year shoe manufacturers met again, assessing themselves financial contributions to undertake a more vigorous prosecution, and importing Jared Ingersoll, the state attorney general, to assist in presentation of their case.[12]

The trial record reveals the market relationship between employ-

TABLE 1

GROWTH OF PITTSBURGH'S INDUSTRIES AND INDUSTRIAL WORK FORCE, 1803–1850[a]

	Value of Production ($millions)	Number of Workers	Number of Firms	Mean Firm Size
1803	0.4	300[b]	163[c]	1.8[c]
1810	1.0	750[b]	—	—
1815	2.6	1,960	—	—
1817	2.3	1,637	257	6.4
1826	2.6	2,997	—	7.9[d]
1839	14.0	—	—	—
1850	16.7[e]	8,899 (city)	756	11.8
		15,017 (county)	1,209	12.4

SOURCES: Catherine Elizabeth Reiser, *Pittsburgh's Commercial Development, 1800–1850* (Harrisburgh: Historical and Museum Commission, 1951), pp. 7, 14, 19–28, 203; Michael Holt, *Forging a Majority: The Transformation of the Republican Party in Pittsburgh, 1848–1860* (New Haven, Conn.: Yale University Press, 1969), pp. 17–25; Frank C. Harper, *Pittsburgh of Today: Its Resources and People* (New York: American Historical Society, 1931), pp. 206–08; Barbara Turner, "Organized Labor in Pittsburgh, 1826–1837" (seminar paper, Department of History, University of Pittsburgh, 1973), p. 11; James C. Holmberg, "The Industrializing Community: Pittsburgh, 1850–1880" (Ph.D. diss., University of Pittsburgh, 1981), pp. 70, 73, 75.

a. All these figures should be used with caution. Totals from year to year and source to source vary widely. Some totals may include only industries within the city limits, while others may include some surrounding towns. Sources rarely identify the basis of inclusion precisely.

b. Estimate based on projecting the ratio of workers to value of total product in 1815, where both figures are known, back to early product totals.

c. Based on possibly inaccurate listing of number of master craftsmen in 1802.

d. Based on selected industries for which sufficient data is available in the 1826 city directory.

e. Allegheny County. Based on tabulation of the 1850 manuscript census of manufactures, Holmberg estimates $17.3 million for the county and $11.5 million for the city.

ers and employees. For example, the boarding system had once given such crafts a familial atmosphere; young journeymen and apprentices lived in their master's households, taking their meals with the master's family. By 1815 boarding had been placed on a cash basis with deductions from wages for those workers still boarding with their masters and negotiated cash equivalents paid to workers, apparently now the majority, who chose not to.

We do not have a comparable documentary record of any other local craft in these years, but the patterns of employment, firm size, and capitalization suggest that shoemaking was not the only craft in which relationships between employers and employees were changing. Over the next forty years we can see a steady progression of tendencies toward increased firm size, concentration of manufacturing employees in larger, more highly capitalized enterprises, and the consequent transformation of virtually all manual workers, including even most artisans, into wage laborers.

One contemporary observer estimated the total proportion of workers in the iron industries in 1860 (including secondary metals production such as metal casting shops and machine shops) at 49 percent of the Pittsburgh area's workers, while glass factories employed 13 percent and textile firms 16 percent. In all three industries, large, highly capitalized firms predominated, with a combined average of fifty-eight workers per firm in primary iron and steel, glass, and textiles, nearly five times the average for all manufacturing firms in the county. By mid-century, perhaps three-quarters of the Pittsburgh region's industrial workers found themselves in the most industrialized and technologically advanced sectors of the industrial economy, and close to half of all industrial workers (43.8 percent) worked in firms of more than one hundred employees.[13]

The early predominance of highly capitalized, large industrial firms in this region was distinctive. Only in the much less economically diversified textile mill towns of New England could one find similar concentrations of large, heavy industrial firms in the antebellum period. Equally important, the key antebellum industries in Western Pennsylvania—iron, secondary metals, glass, cotton textiles, and coal—were all industries which had never been organized around the craft hierarchy of the preindustrial artisanal workshop. While skilled craft workers played crucial roles in iron smelting, metal fabricating, glass production, and textile manufacturing, in each case their skills were products of the industrial revolution, not holdovers from a preindustrial craft heritage. In this regard Pittsburgh stood in marked contrast to the largest industrial cities of the East, such as New York and Philadelphia, where traditional craft production still played a much larger role in the industrial economy and class formation had revolved far more around reorganization of the traditional crafts. This difference may explain local peculiarities in the evolution of working-class culture, the relatively greater salience of class in Pittsburgh's political and cultural life over the next generation compared to many other cities, and the early emergence in Pittsburgh of some of the country's most powerful craft unions.[14]

Reconstructing the demographic basis of this initial working-class formation in Pittsburgh is far more difficult than suggesting its economic and social roots. Of the three possible sources of labor—natural population increase, migration of rural laborers to the city, and immigration—the first must have been the least significant. From the 1790s through the late 1830s population grew so rapidly that only a small proportion could be accounted for by natural increase. After the depression of the late 1830s population growth slowed, but even then only a small fraction of Pittsburgh's workers could have been Pittsburgh-born. In a sample of 200 Pittsburgh workers drawn out of the 1850 manuscript census, only 25 percent had been born in Pennsylvania, and presumably many of those had migrated from other parts of the state.[15]

After 1830, and especially after 1845, national immigration totals shot up, and immigrant workers dominated in most eastern and midwestern cities. That was true in Pittsburgh, where immigrants made up 69.5 percent of the workers in my 1850 census sample: 33.5 percent of the total sample were Irish, 25 percent German, 10 percent British (English, Welsh and Scots), 1 percent all other immigrants. It seems likely that native rural migrants had been more important to working-class formation in Pittsburgh before 1840, but we cannot say for sure.[16]

Class Development to 1850

To what extent did these workers develop common habits, values, and day-to-day living patterns? To what degree did this emerging economic class also became a social class? We await Pittsburgh's antebellum labor historian to answer those questions, but references in existing works to early activities of the labor movement and to class appeals in local politics provide some hints.

Throughout the first half of the nineteenth century, from the first shoemakers' strike in 1804 to the large cotton and iron rolling mill strikes in 1848 and 1850, local historians note several dozen strikes and union organization in more than a dozen different crafts. One student of the Jacksonian era found evidence of craft unions in sixteen Pittsburgh crafts between 1826 and 1837. In a relatively small town with a population of only 21,000 as late as 1840, the number of strikes and labor organizations seems impressive. By the 1820s and 1830s craft organizations played major roles in the city's public ceremonial life such as when five mechanics' societies sponsored an 1826 procession honoring the memory of John Adams and Thomas Jefferson. During the 1848 and 1850 cotton and rolling mill strikes, the two published analyses indicate that strikers enjoyed widespread public support.[17]

Class sentiment also seems to have played some role in political culture. In the early 1830s, for example, a group of local politicians organized a Pittsburgh Workingmen's Party in emulation of the highly publicized workingmen's parties in New York and Philadelphia. According to William Sullivan's analysis, the organizers were a disgruntled faction of Whig politicians with no genuine working-class credentials. They were trying to use the "Workingmen's" label to manipulate part of the local electorate for factional purposes, but their attempt demonstrates that, at least in their judgment, class imagery had significance for the electorate.[18]

Again in the 1840s, a faction of Democratic politicians and newspaper editors employed explicitly pro-working-class rhetoric in their appeals to voters—even condoning working-class violence during the 1848 and 1850 strikes. Their purposes may also have been manipulative. Whigs dominated local politics until the 1850s, and the Democrats depended on Irish Catholics as their most consistent political base. Democratic politicians may have defended the rights of labor in an effort to broaden their political base, assuming that those voters who would be offended by a pro-worker bias would be Whig voters anyway.[19]

Although the primarily ethnoreligious basis of partisanship certainly limited the political potential of class appeals, Michael Holt's statistical analysis of Pittsburgh elections around mid-century confirms that class sentiments played some role in local political cultures. In most elections between 1847 and 1850 the correlation (Pearson's r) between Democratic vote by wards and the percentage of skilled workers in each ward was between .4 and .6; and between Democratic vote and the percentage of unskilled workers .2 to .3. The Whig vote correlated negatively with both groups. Thereafter, during the political realignment of the 1850s, the pattern of correlations between the working class and the major parties was much less consistent but still significant in some elections.[20]

We should not make too much of a few only moderately strong ecological correlations, but the statistics on voting behavior are consistent with what we know about workers in other cities and in Pittsburgh after 1860. For workers confronting the reality of an emerging class relationship, class imagery had some appeal, but class represented an ideological frame of reference competing with other powerful sources of identity.

Working Class Recomposition, 1840–1880

The economic forces promoting class formation in Pittsburgh did not change qualitatively from the first industrial enterprises which superseded traditional craft production in the 1810s until the emergence of large-scale, corporate, mass production at the end of the nineteenth century. By 1880, when 79 percent of the adult male work force in Pittsburgh and Allegheny City was working class, 39.5 percent of manufacturing employees worked in primary iron and steel production, 13.4 percent worked in secondary iron production, and 13.3 percent worked in glass. The mean firm size of all industrial firms was 27.9, while blast furnaces averaged 71 hands in 1869 and 176 in 1899, and steel mills 119 and 1869 and 412 in 1899. Before the 1880s, there were no technological innovations in the key industries—iron, glass, and coal—fundamental enough to alter much of the tempo and organization of labor or its shopfloor cultures.[21]

This steady progression of an economy revolving around a few key heavy industries, and increasing firm size in the industries, promoted not only class formation but also the development of common habits, cultural practices, and attitudes among the majority of Pittsburgh's working people. Economic tendencies were reinforced by the physical realities of industries which depended so completely, even for the most skilled workers, on dangerous, heavy labor. The ecology of these industries, combined with the natural terrain of steep hills surrounding narrow valleys bisected by the three rivers, similarly promoted the development of densely packed and self-contained mill towns and working-class neighborhoods. In Pittsburgh's South Side, for example, 85 percent of household heads were working-class in 1880.[22]

A number of historians, most notably Francis Couvares, John Bennett, Shelton Stromquist, Susan Kleinberg, and Paul Krause, have reconstructed life in these working-class communities in the generation after the Civil War with skill and sensitivity. Bennett and Krause focus on particular communities: Bennett on Woods Run on Pittsburgh's West End and Krause on Homestead. Stromquist focuses on skilled ironworkers, Couvares on the impact of skilled workers on the wider cultural life of the city, and Kleinberg on working-class women and the working-class family. All describe a mutualistic ethos grounded in work cultures, reinforced by face-to-face familial, kin, and neighborhood social networks, maintained by powerful labor organizations, and defended with an egalitarian and universalist republicanism symbolized by union logos like clasped hands superimposed on the globe, or slogans like the Knights of Labor's, "An injury to one is the concern of all!"[23]

Male skilled workers, such as iron puddlers or glassblowers, played leadership roles in the formal organizations of this culture, in the informal defense of its moral precepts on the shop floor, and in its wider public expressions. This leadership, expressed in the code of manly obligation, helped to dampen potential conflicts between skilled and unskilled who were otherwise separated by a wide economic gulf. Skilled workers who accepted the mantle of moral leadership saw themselves as the defenders of the weak and protectors of the entire community. Potential conflicts between the skilled and unskilled were also mitigated by life-cycle relationships between them. Older skilled workers in the iron foundries and glass houses supervised crews of generally younger unskilled and semi-skilled workers whom they trained as future craftsmen, and they tended to recruit apprentices and helpers among their sons, other relatives, or younger neighbors (and fellow countrymen in the case of immigrants).

Much of the strength and vigor of this culture depended on the assertive masculinity of these workers for whom craft pride and identity flowed from their sense of physical power and ability to overcome fatigue and danger. As Kleinberg energetically argues, this male cultural dominance probably meant that working-class families in the industrial communities of this region were even more authoritarian and male-dominated than elsewhere in American society, while working-class women probably had even fewer opportunities for economic independence.

Neither Kleinberg nor any of the other historians of working-class life in the region consider how gender relationships may, at the same time that they foreclosed opportunities for women, also have fostered working-class mutualism. To what extent did the mutualistic ethos depend on the emotions and positive associations generated by a family economy which expected each member of the household to sacrifice personal needs in order to perform specific tasks, prescribed both by gender and by stage of life cycle, for the greater good of the entire family? To what extent did successful defense of larger community interests and even trade union goals depend on the informal social networks among women? To what extent were women as child-rearers almost exclusively entrusted with transmission of working-class values across generations?

Nor do the labor historians, for the most part, consider how the same economic forces that stimulated development of working-class culture and solidarity also stimulated countervailing tendencies toward working-class fragmentation. The steady expansion of the industrial economy and the growing scale of industrial firms, while marking workers off more clearly from employers, also created new hierarchies within firms. Skilled workers could become supervisors; internal job ladders developed as tasks once performed by individual craftsmen were subdivided and differentiated. Craft unions in both iron and glass recognized the phenomenon and attempted to respond by amalgamating separate organizations representing these subcrafts, but conflicts among subcrafts continued to be a problem both for the Amalgamated Association of Iron and Steel Workers and for the Window Glass Workers of the Knights of Labor. The vitality of an expanding industrial economy also opened up opportunities for a variety of small businesses that catered to the needs of working-class communities: grocery stores, funeral parlors, taverns, and so forth. Few workers could ever hope to own a steel mill or a glass factory, but we need to consider more carefully how other opportunities for social mobility shaped relationships among workers and created internal stratification within the working class.[24]

We do not have any studies of social mobility in Pittsburgh before the twentieth century, but data uncovered by social historians on schooling and marriage patterns are at least indirectly suggestive. First, studies by Carolyn Schumacher and Ileen DeVault demonstrate that by the late nineteenth century significant and increasing minorities of skilled workers' families decided to keep their children in school beyond the elementary grades. This decision involved a major financial sacrifice for families who heretofore had counted on the earnings of teenage children to augment household income. The sacrifice appears to have paid off for their children, who did increase their probabilities, compared to other working-class children, of entering more highly paid jobs.[25] Second, data on choices of marriage partners demonstrate that a segment of male skilled workers and daughters of skilled workers sought upward mobility through marriage; as a group the probability that they would choose marriage partners out of the lower middle class was higher than a random choice, while the probability of an unskilled spouse was much lower than random (table 2).

Such data only hint at the tensions working people must have felt as they weighed the conflicting implications of different life strategies, tensions between the collective vision and solidarity of the mutualistic ethos, and the fragmentation implicit in strategies for individual improvement. These tensions between solidarity and fragmentation would have been inherent in the varying possibilities open to different individuals in an expanding market society. However, two major upheavals of mid-century America, the Civil War and the rise of mass immigration, enormously aggravated such tensions and stimulated awareness of what have been two of the most fundamental crosscutting pressures to class solidarity ever since—race and ethnicity.

Emancipation added four million propertyless individuals to the free population of the United States at a time when the total number of nonagricultural wage earners in the country was only slightly over three million.[26] Equally important, the Civil War helped to crystallize reinterpretations of the classical republican heritage which had been slowly fracturing along class lines for more than a generation. Artisanal republicans, since the 1820s and before, had emphasized the sanctity of labor and the necessity to protect industrial liberty as a

TABLE 2

Marriage Choices of Male Skilled Workers and Daughters of Skilled Workers

| | Class of Spouse | | | |
	Unskilled	Skilled	White Collar	Upper
1880	0.65[a]	2.0	1.1	0.19
1900	0.88	1.5	1.1	0.14

Sources: 1880: Susan Kleinberg, "Technology's Stepdaughters: The Impact of Industrialization Upon Working-Class Women, Pittsburgh, 1865–1890" (Ph.D. diss., University of Pittsburgh, 1973), p. 238; 1900: Dan Tuden, "The Impact of Class on Marriage Patterns in the 1900s in Pittsburgh," unpublished paper, 1986, on file in Department of History, University of Pittsburgh, p. 14.

Note: In each case the figure is the ratio of the number of spouses of a given class chosen by male skilled workers or daughters of skilled workers to the number of potential spouses of that class available in the marriage pool; e.g., in 1880 male skilled workers and daughters of skilled workers chose unskilled spouses at 65 percent of the rate expected if marriage partners were assigned at random. They chose skilled spouses at double the rate expected with random assignment.

a. For 1880 data are from a citywide sample, daughters of widows were excluded (since class cannot be determined), and the distribution of occupations in the work force was taken as the estimated marriage pool. To the extent that potential marriage partners had a different occupational distribution from the rest of the population, the ratios may be skewed. For 1900, the results are based on a sample of three Pittsburgh wards, and the potential marriage pool is estimated as equivalent to the actual distribution of classes in the entire sample.

prerequisite to political liberty. The Civil War and Reconstruction helped to focus and popularize this critique, and the analogy between black slaves and white factory slaves became a standard ploy in the rhetoric of labor organizers. "Labor's spokesmen insisted that social reconstruction be extended northward. 'So must our dinner tables be reconstructed,' demanded the Boston Labor Reform Association in 1865, 'our dress, manners, education, morals, dwellings, and the whole Social System.' "[27]

Race had relatively little direct impact on the recomposition of Pittsburgh's working class in the nineteenth century since the number of black workers remained low until the migrations of the World War I era. The indirect ideological impact of the Civil War, however, was enormous. For the next generation, local workers and labor organizations consistently drew on republican imagery to justify nearly all their protests and actions. Labor organizers consciously laced their speeches with language designed to invoke Civil War memories. Pittsburgh's *National Labor Tribune,* for example, declared in the mid-1870s that the local labor movement "must drill our soldiers from the corporal's squad up to the division drill" to storm the "Vicksburgs of Labor's enemies," and the men who "carried the banners of Grant and Sherman" now needed "another Lincoln" in their fight for "industrial independence."[28]

Ethnicity played an even greater role in shaping the cultures of Pittsburgh's workers. As early as 1850, when an estimated 70 percent of the city's working class were immigrants, that influence was obvious. Over the next several decades the proportion of immigrants actually declined slightly, but ethnic influence was also felt in the second generation as the children of the initial cohorts of Irish and German immigrants reached maturity in the post–Civil War decades. In 1880, immigrants and their children made up 76 percent of the adult male working class in Pittsburgh and Allegheny City.[29]

Some scholars have treated ethnic and class consciousness as polar opposites, but the relationships between them are more complex. Certainly some types of ethnic feelings, like the intense nativist crusade which swept through the city in the late 1840s and early 1850s and remained a continuing theme in urban politics throughout the century, not only divided workers but actually led to physical conflicts and pogroms. But Michael Holt's analysis of Pittsburgh nativism revealed (as scholars have for other cities) that even nativism was often expressed in class terms. It appealed disproportionately to native workers who saw mass immigration as a plot by their class antagonists to use immigrants to destroy the rights and standards of American labor.[30]

We can find evidence of the contradictory meanings of ethnic sentiments among every other nationality of Pittsburgh's workers. Ethnic particularism and memories of Old World hostilities almost always interfered, to some extent, with cooperation between workers of different ethnic backgrounds, but at the same time ethnic and class feelings could often mesh and reinforce each other, enormously strengthening individual commitment to group solidarity. This was especially true since the combined impact of chain migration and ethnic hierarchies in hiring often meant that particular nationalities concentrated in specific occupations, particular factories, and sections within factories. Workers often found themselves united with fellow ethnics against Anglo-Saxon employers. While some scholars have assumed that acculturation and assimilation was a prerequisite to working-class solidarity, many nationalities of immigrant workers have viewed defense of ethnic community culture as part of working-class loyalty and have seen assimilation to American culture as a form of embourgeoisification.[31]

On balance, how had these contradictory tendencies—mutualism and individualism, solidarity and internal conflicts—shaped the culture of Pittsburgh's workers by the end of the 1870s? Despite tensions and variations, workers could see that they had a great deal in common. They faced similar antagonisms with employers in the large factories where the majority of them worked. Although there were variations between nationalities and skill levels, and the presence of a very prosperous minority amongst them, the majority of all working-class families had to struggle to stay barely above a poverty income. Within each nationality, the economic gap between the average wealth of middle-class households and skilled workers' households was at least three times greater than the difference between the averages for skilled and unskilled households.[32]

Working people lived together in overwhelmingly working-class neighborhoods where they could see the physical similarities in their circumstances. In 1880, 69 percent of Pittsburgh's workers lived in neighborhoods which were more than 75 percent working class.[33] The non-working-class minority clearly lived better than they did. In Pittsburgh's Third Ward in 1880, Kleinberg found that 68 percent of working-class households lived in dwellings of under 800 square feet of floor space (90 percent under 1000 square feet), while 59 percent of the non-working-class households lived in dwellings of greater than 800 square feet (47 percent over 1000 square feet). While some middle-class households were already starting to acquire the household appliances that were to become one of the hallmarks of consumer culture in the twentieth century, such acquisitions were beyond the means of nearly all workers. For example, of thirty-nine households listed as owning washing machines in Pittsburgh and Allegheny in an 1870 advertisement, only eight were working class. None of the 162 households in Pittsburgh with private telephones in 1886 were working class.[34]

Working-class children in the 1870s could expect to live lives similar to those of their parents. While there were significant variations among nationalities, between skilled and unskilled workers' families, and between boys and girls, most working-class children would leave school by the age of fifteen when most of the boys and about 20 percent of the girls went to work. In the next few years, although a few of them might hope to improve their prospects by marrying up, over 80 percent would marry children of other workers like themselves, in over half the cases from their immediate neighborhood.[35]

The tensions between races, religions, ethnic groups, and differing skill levels, the conflicting pulls of alternative life strategies were all very real, but by the 1870s the commonalities among Pittsburgh's workers, the values of their work cultures, and the personal networks of their neighborhoods were sufficient to sustain intense class conflicts over the next fifteen years.

Class Conflict, 1877–1892

On the morning of Sunday, July 22, 1877, Pittsburgh photographer S. V. Albee hitched his horse up to his wagon to carry his cameras and dark tent down to the smoldering Pennsylvania Railroad yards which ran for nearly two miles through what is today Pittsburgh's Strip district. There, in a series of photographs which he aptly titled "The Railroad War in Pittsburgh," Albee recorded images of what indeed was the ruins of a war zone. Crowds of men, women, and children, some still picking through the wreckage, watched Albee as he moved systematically from the remnants of Union Station, along rows of burned-out warehouses, down tracks littered with thousands of singed car wheels and fragments of locomotives, past the foundations of what had been two giant roundhouses. The destruction reached the fringes of Lawrenceville, the Irish working-class neighborhood where many of the workers lived who had fought the Pittsburgh and Philadelphia troops trying to protect the railroad's property the day before.[36]

Fifteen years later on another July morning, a few miles over the ridges which separate the Allegheny and Monongahela river valleys, three thousand Homestead steelworkers fought a pitched battle on the riverfront with three hundred heavily armed Pinkertons who had attempted to land in Homestead to defend the Carnegie Steel Works where the workers had just been locked out. Three Pinkertons and nine workers died from wounds they sustained before the Pinkertons surrendered to the angry crowd. A few days later, a Washington photographer traveled to Homestead, just before the state militia arrived, and made a series of negatives which he sold to the largest national distributor of parlor stereographs.[37]

These two dramatic events, both of which drew international attention, bracket an era in which Pittsburgh's ironworkers, glassworkers, and rail crews repeatedly confronted their employers in conflicts which sharply divided their communities along class lines. For both sides, these recurring conflicts demonstrated the underlying differences not only in their economic interests, but also in their worldviews, in their understanding of the republican heritage that many of their grandparents had once held in common.

For employers and their supporters, who reacted with self-righteous moral outrage, workers' behavior in such conflicts constituted nothing less than terrorist assaults on the sanctity of property, the most fundamental prerequisite for a just social order. As Judge Edward Paxson (the chief justice of Pennsylvania, who had come to Pittsburgh to oversee the trial of thirty-three Homestead workers accused of treason against the Commonwealth) argued in his summation to the grand jury, the workers were guilty of "resisting the law and resorting to violence and bloodshed in the assertion of imaginary rights." The significance of such crimes, he concluded, went far beyond this particular case. "We have reached the point in the history of the state when there are but two roads left us to pursue; the one leads to order and good government, the other leads to anarchy. The great question which concerns the people of this country is the enforcement of the law and the preservation of order."[38]

But to workers, Carnegie, Frick, and their agents were the real anarchists. Carnegie's policies were "subversive of the fundamental principles of American liberty ... unconstitutional, anarchic and revolutionary." The workers at Homestead had been defending the republic against this subversive attack, their lawyer explained. "Within two days of the anniversary of the Declaration of Independence," Homestead's workers had faced "an invasion of armed men employed by a man named Frick; it was an assault upon this grand old commonwealth; ... upon the state in which the Declaration of Independence was signed. ... an attack upon the people of Homestead. ... [who] attempted to defend their rights," and several had died heroically doing so.[39]

The differences between these visions gave the successive strikes and political conflicts of the 1870s and 1880s the character of skirmishes in a guerrilla war. The superior economic resources of employers, their ability to draw on the powers of the state militia and the courts, meant that employers would win their share of victories, but until the 1890s, each time workers were defeated, they were able to respond and win back much of what they had lost. Iron puddlers first organized in the 1850s as the Sons of Vulcan. By the 1870s the puddlers combined with other craftsmen in the Amalgamated Association of Iron and Steel Workers. They first established a wage scale governing the entire region's industry in a regionwide strike in 1867, defended it successfully in the strikes of 1874 and 1879, but lost decisively in 1882. They regained their wage scale in the mid-1880s and sustained it against Carnegie's first attempt to crush the union in 1889—the dress rehearsal for the events of 1892.[40] Glassworkers fought similar battles in 1867, 1873–1874, 1877–1879, and 1882.[41] Railroad workers lost their 1877 strike, although they had won the military struggle in the rail yards. Many workers responded by deserting the Republicans and Democrats for the Greenbackers in local elections in 1877 and 1878: the Greenbackers received nearly a third of the votes in Pittsburgh and Allegheny City and carried many of the most heavily working-class wards and outlying mill towns. Workers' political influence was also revealed in the successful political campaigns of such trade unionists as William McCarthy, elected mayor of Pittsburgh in 1865 and 1875.[42]

In some respects the events of 1892 were the culmination of these displays of working-class power. Workers won the shoot-out with the Pinkertons. Their calls for solidarity were heeded as thousands contributed to their strike fund, and thousands of other iron and steel workers joined them in sympathetic strikes at other Carnegie properties in Duquesne, Lawrenceville, and Beaver Falls. The Allegheny County Democratic Committee endorsed the strikers' actions in a unanimous resolution. And in a succession of trials, Frick, Judge Paxson, and Carnegie's lawyers found, much to their disgust, that local juries consistently refused to find the Homestead strikers guilty of any wrongdoing.[43]

But this time workers faced a new antagonist. Carnegie Steel had been leading the technological transformation of the iron and steel industry for the last decade and no longer needed the skills of the Amalgamated's craftsmen. The company, which was soon to become the centerpiece of the giant merger into U.S. Steel, the world's first billion-dollar corporation, had virtually limitless resources compared to the workers.[44] The shoot-out at Homestead proved to be the final chapter in the struggle to maintain workers' power through community-based artisanal republicanism. The strike was lost, and the union never recovered. Homestead, which had been a workers' town where union officials ran the local government, became a company town for more than forty years. When Margaret Byington arrived there in 1907 to study working-class life, many workers refused to talk to her, fearing that the company would somehow find out what they had said. The defeat at Homestead symbolized the dynamics of a new political economy emerging out of massive technological change at the end of the nineteenth century.

The Second Industrial Revolution, 1890–1930

As late as 1880, despite the steady growth in firm size and capitalization over the previous sixty years, most of Pittsburgh's factories were still relatively small by twentieth-century standards. Carnegie's Edgar Thomson works, the largest factory in Allegheny County, had 1,500 employees;[45] most of the iron foundries and rolling mills had no more than a few hundred, while few factories in other industries were as large as a foundry or rolling mill. Even the larger firms were mostly organized internally as congeries of smaller shops where artisans directed work crews of a half dozen to a dozen workers at a work pace dictated by custom, by union contract, and by the limits imposed by production processes, such as the time it took for a glass pot or iron furnace to reach its prescribed heat.

Beginning slowly in the 1880s, and with an accelerating pace after 1890, a second industrial revolution transformed the local industrial economy by World War I. Bureaucratic corporations, who invested their capital by the tens of millions rather than tens of thousands and counted their employees by the thousands rather than the dozen or hundreds, superseded the smaller individual proprietorships and partnerships who had directed the nineteenth-century economy. Large-scale corporate mass production enjoyed decisive advantages not only because of the enormous financial resources it concentrated, but also because of the efficiencies gained by reorganizing work processes with flow-through and machine-paced production methods.[46]

From Wheeling and Weirton in the west to the string of mill towns lining the Monongahela River in the south and east, steel companies built the mammoth plants which now stand as rusting hulks along the region's riverbanks. Steel mills, still by far the most important industries, were complemented by new glass houses, oil refineries, secondary metals firms, and the giant Westinghouse Electric factory which transformed the Turtle Creek Valley from a wooded glen to the largest of all the mill complexes. By 1925 there were thirty-eight factories and industrial firms in Allegheny County with more than 1,000 employees each. The two largest, Westinghouse (20,000) and the Carnegie Steel Homestead Works (10,000), together had as many employees as all of Pittsburgh's 1,112 industrial establishments had had in 1880.[47]

Industrial transformation created social transformation, altering work rhythms and shop-floor experiences, economic and residential geography, neighborhood structures and institutions, demography and ethnic composition, family life and consumption patterns. Both skilled and unskilled workers faced new work regimes with machine-paced production, time-and-motion studies, and the constant push to work faster and more efficiently.

The earlier industrial economy had been concentrated along the riverbanks in the older sections of the city. There was no room there for giant mills which needed thousands of acres of cheap, undeveloped real estate. Industrialists looked outward, beyond the city, and as their new mills went up in places like Braddock, Homestead, Clairton, and Aliquippa, new towns sprang up around them. Company officials dominated local government in these towns. Local merchants and professionals looked to the companies for leadership. Companies concerned with maintaining low overhead costs kept local tax rates low, insuring that these communities provided a minimum of social services to their residents. Well into the twentieth century, muddy unpaved streets lined with small, closely packed houses lacking indoor plumbing or running water ran through some working-class residential neighborhoods in Homestead, Braddock, and Duquesne.[48]

While immigrants and their children made up three-quarters of the working class in Pittsburgh in 1880, almost all came from Germany, Ireland, and Great Britain. Southern and Eastern Europeans made up less than 1 percent of the working class. But by the 1890s, social and economic changes in Southern and Eastern Europe had started to undermine the livelihoods of marginal peasants, agricultural laborers, and rural handcraft workers. Pittsburgh's mills became magnets for virtually every nationality from that region. By 1907, 81.4 percent of the unskilled laborers in Carnegie Steel's Pittsburgh mills were Eastern Europeans.[49]

The influx of these new immigrants altered cultural relationships within the working class. A new ethnic hierarchy emerged with greater clarity than the hierarchy of a generation earlier. Upper management and professional positions were occupied disproportionately by old stock, white, Anglo-Saxons; jobs in the skilled trades and as immediate production supervisors were filled disproportionately by the old immigrants and their children; and unskilled laborers were almost exclusively new immigrants and (especially after World War I) blacks.[50]

Chain migration, kin networks, and company employment practices facilitated residential and occupational clustering among the new immigrants, but the demographic characteristics of the population initially inhibited development of stable communities. New immigrant workers were disproportionately young, transient males who intended to work in Pittsburgh for a few years and then return home. As many as half may eventually have done so. This transience is reflected, for example, in the personnel records of the A. M. Byers Corporation, a large wrought-iron pipe manufacturer studied by Michael Santos, where 52.5 percent of the unskilled workers between 1916 and 1930 (mainly Eastern Europeans and blacks) left in less than one month and 97.7 percent left in less than one year. The mobility data of John Bodnar, Roger Simon, and Michael P. Weber shows that only 17.8 percent of the Poles in a 1900 census sample still lived in the city only five years later.[51]

How did this vast sociological transformation affect the consciousness of workers and their sense of personal and collective identity? If both economic changes and ethnic loyalties had simultaneously united and divided workers in the previous period, after 1890 the more varied ethnic mix, the more far-reaching social changes, and, for some people at least, the greater economic opportunities in a fast-growing city must have produced an even more complex set of contradictory pressures.

Concentration of workers in the corporate core of the industrial economy did provide a functional basis for bridging differences in workers' backgrounds and motivations. Drawn literally from the four corners of the earth, divided between transients and industrial veterans and by economic differentials between the skilled and unskilled, workers nonetheless found themselves herded together by the thousands into giant industrial complexes where they faced common dangers, similar pressures of regimented and systematic work discipline, and a common antagonist in the authority of the corporation. Even though such corporations had vast economic resources and political influence to use against workers, the enormous overhead costs of their new technologies made them vulnerable to the consequences of disruptions in production.

How could such potential opportunities for workers be translated into something more than erratic and informal resistance to the imposition of work discipline? On what basis, with what language, what symbols, what programs could organizers appeal to workers who did not share common values like those of the artisanal communities of the previous era? Certainly each nationality of new immigrants *did* create its own churches, its own fraternal societies, its own informal social networks. Especially among many of the Eastern European nationalities, peasant village traditions emphasized communal identity and loyalty. But effective organization and protest now demanded, at the very least, forging alliances among many separate national communities and overcoming the desire of transients to avoid conflict.

The language of nineteenth-century artisanal republicanism still served as a primary ideological framework for native and old immigrant skilled workers. While their unions had been almost completely driven out of the corporate sector of the economy, they maintained and even increased their hold in construction and in many small-scale consumer industries. Bricklayers and masons, for example, were 100 percent organized in Pittsburgh in 1920.[52] But increasingly, where such republican traditions survived, they took on far more conservative meanings than they had in the Gilded Age. Most craft unions virtually abandoned universalist interpretations of republicanism (for example, rights for all, or that all men are created equal) and ambitions for universal organization of all workers, moving toward jealous defenses of their jurisdictional turf heavily overladen with nativist hostility toward new immigrants and militant racial exclusion.

There were important exceptions, especially in trades like the machinists which had extensive and increasing employment in large corporations. Nor should we underestimate the continuing positive inspiration of earlier republican rhetoric for natives and old immigrants. In 1919, for example, the local organizational secretaries of the National Committee for Organizing Iron and Steel Workers in Homestead, Braddock, Clairton, and McKeesport were all second-generation Irish workers.[53] But the dominant tendency is perhaps best symbolized by the differing relationship between skilled and unskilled, old and new immigrant, in the 1892 and 1919 steel strikes. In 1892, the skilled natives and old immigrants in the Amalgamated lodges led a struggle in which they conceived of themselves as agents and protectors of the entire community, seeking and receiving the cooperation and support of the unskilled new immigrants. In 1919, in large part, skilled natives and old immigrants reluctantly acquiesced to a strike they considered a "Hunky" affair, sometimes echoing the nativist and antiunion rhetoric of their employers.[54]

The inspirational potential of nineteenth-century artisanal repub-licanism was limited by more than just the tendency to interpret it in narrow and defensive ways. To unskilled new immigrants work-ing in giant corporations, the language of personal independence and freedom from external authority must have had an air of unreality. But if workers were to be moved on some other ideological basis, what other set of symbols had sufficient resonance in their back-ground and experiences to have any credibility?

Significant minorities in most of the new immigrant nationali-ties had prior exposure to socialist appeals in the Old Country, and they attempted to inspire their fellow country people with a combi-nation of direct-action tactics and a long-term socialist vision. The impact of the former is most evident in the series of mass strikes, led by radicals, beginning in 1909 with the McKees Rocks Pressed Steel Car Company, in which IWW (Industrial Workers of the World) activ-ists served as organizers and spokesmen. The IWW also organized a successful strike of more than one thousand, primarily Jewish, ci-garmakers in 1913, while a group composed mainly of ex-IWW's led even larger strikes which shut down the Westinghouse Works in 1914 and 1916.[55] The 1916 Westinghouse strike ended after strikers who had walked out on April 21, 1916, attempted to convince steel-workers in Braddock and Rankin to join them on May 1 in response to a call by the American Federation of Labor for a national eight-hour-day strike on that day. U.S. Steel officials closed their plants on May 1, when a mass march of Westinghouse workers arrived at the plant gates, but assembled 1,000 armed company guards at the plant who fired into the crowd the next day, killing three workers and wounding several dozen. The governor sent 1,000 National Guard troops to Westinghouse, the company broke off negotiations, and when twenty-three strike leaders were arrested for inciting and as accessories to the murder of the workers killed by U.S. Steel guards, the strike collapsed.[56]

In less dramatic ways socialists attempted to inculcate their ideas by linking the traditions of communal self-help organized by the fraternal societies of many Eastern European nationalities with their long-term vision of a cooperative society. Hungarian socialists in East Pittsburgh, for example, organized a Hungarian Working-men's Sick Benefit Federation in 1908 which claimed 10,000 na-tional members by World War I. The connection between immedi-ate communal aid and a larger socialist vision was also clearly evident in the socialist-led Jewish fraternal, the Arbeiter Ring, or Workmen's Circle, which enrolled, according to one student, 2,000 members in Pittsburgh in a Jewish population of 30,000. The cover of a 1905 convention report declared, "We Fight Against Sickness, Premature Death, and Capitalism."[57]

Socialist influence is also indicated by the comparative success of the Socialist party in local elections just before World War I. Eugene V. Debs received 15.5 percent of the vote in Allegheny County in 1912 and averaged 25 percent in sixteen Western Pennsylvania steel towns, compared to a national average of 6 percent. Between 1911 and 1917 socialists elected local officals in Turtle Creek, New Castle, North Versailles, Pitcairn, and McKeesport.[58]

Scholars disagree whether these strikes, left-wing influence in fraternal associations, and political campaigns are only the most visible indicators of much more widespread sentiments or are essentially a sideshow for communities of workers with generally more cautious and conservative values. David Montgomery, writing from a national perspective but using many examples from this area, argues the former; John Bodnar, also looking broadly at working-class consciousness, but drawing heavily on the extensive oral history research he has directed in the Pittsburgh area, argues the latter.[59]

Depending on where and when we look, both may be true. Workers' attitudes, and their sense of what was possible, shifted in response to changing circumstances, but this almost bipolar debate—protosocialist workers' control versus conservative, security-conscious pragmatic realism—strikes me as an intellectual blind alley for another reason. Neither addresses how the social and cultural changes accelerated by the second industrial revolution had started to alter the meaning of working-class life. As large-scale corporate mass production also became a system of mass consumption, it slowly undermined the foundations of the producer-based culture of the nineteenth century, changing how culture was transmitted, how politics was organized, how people spent their (increasingly significant) leisure time, how they defined their identities and ambitions. These changes began before 1890, and their full impact was not yet evident in 1920, but already we can see glimmerings of social trends which would become dominant after the Depression when neither Montgomery's worker's control of production nor Bodnar's security-conscious realism would be as crucial to the cultural and political agenda of workers as they had been before.

Unfortunately the social history of Pittsburgh workers after 1920 (just as for most of the rest of the country) consists mainly of a few studies of unionization in the 1930s and 1940s and then ceases in 1949–1950 when the CIO (Congress of Industrial Organizations) purged its left wing.[60] Lacking a body of research which addresses social and cultural change comparable to the studies which allow us to reconstruct working-class culture between 1850 and 1920, I am left with some speculations and a research agenda for future historians. Three themes stand out. First how did the settling in of the new immigrants intersect with rising incomes and mass consumption to alter cultural boundaries among workers and between workers and other classes? Second, how did this affect the ideology and political orientation of workers? Third, how (and why) did the increasing cultural and political significance of gender and race affect working-class culture?

Economic and ethnic stratification within the working class had always been important, but in the late nineteenth century the economic and cultural boundaries between the working class and the middle class were sharp and distinct. By the 1920s, both as a result of real improvements in the incomes of some workers, and the ways in which mass consumption homogenized cultural differences, the boundaries between classes seem to have become less distinct. For example, children of skilled workers, especially those of native and old immigrant backgrounds, took disproportionate advantage (compared to other workers) of opportunities opened up by expansion of the high schools (and later colleges) and by white-collar employment.[61] Did they mingle socially with middle-class children in the growing commercial departments of the high schools? Did marriage patterns change, as a result, from the largely class-based pattern of the previous generation? Did significant numbers of more prosperous working-class families move out of older working-class neighborhoods to the new suburban tracts starting to fill the South Hills and East End on the city's fringes? Was the assertive "Americanism" and hostility to the "Hunkies" displayed by many native and old immigrant workers during the 1919 steel strike an indication of these underlying social changes?

The 1919 steel strike also demonstrates how new immigrants had settled in, developed stable and cohesive social networks, and meshed among themselves. Frank Serene found evidence of this emerging stability in the fluctuations in Catholic church membership totals, marriages, and baptisms in the immigrant churches of the Monongahela Valley. Before World War I, membership followed the business cycle, rising in periods of increasing employment, falling sharply during downturns and layoffs, but after 1914 this pattern was decisively broken. During production declines in 1918–1919 and again in 1920–1921, church membership went up rather than down.[62] The contrast, only suggested by these church statistics, is more clearly demonstrated after 1930 where the mobility data of Bodnar, Simon, and Weber, and Michael Santos's analysis of A. M.

Byers personnel records, reveal extremely high residential and occupational persistence among new immigrants, a nearly total reversal
of the transience of the previous generation.[63]

As the new immigrants and their descendants settled into the
mill towns, as sons followed their fathers into the same mills, their
communities developed institutions, informal social networks, and
mutualist loyalties. The cultural basis of these loyalties may have
been quite different from the ethos which had characterized the
artisanal-republican culture of the Gilded Age, but the social impact
was similar. The stability of neighborhood and employment patterns and the continuing overlap of class and ethnic hierarchies reinforced and strengthened loyalties over time.

More speculative, on the basis of current research, is the extent
to which such networks and loyalties were defined in class as well
as ethnic terms. Although individual nationalities continued to
maintain separate formal institutions, there was considerable meld-
ing of new immigrant workers, especially in the second generation,
both demographically and in their attitudes toward each other. By
the 1930s, weak ethnic boundaries within the new-immigrant working class were evident in union organizing campaigns and in voting
behavior.[64] A harbinger of what was to come was the exceptionally
strong performance of Robert M. La Follette in the 1924 presidential campaign in Pittsburgh (36 percent compared to a national
average of 16 percent), especially in new-immigrant, working-class
neighborhoods.[65]

Thus, at the same time that skilled native and old-immigrant
workers were distancing themselves from an earlier class identification, class consciousness seems to have been increasing in Pittsburgh
among the descendants of new-immigrant workers. By the 1930s and
1940s, what we often think of as the characteristic indicators of
working-class consciousness in modern America—militant mass production unionism, staunchly Democratic political loyalties, and a
positive orientation toward state action to redress problems of economic inequality—were disproportionately centered among workers
of new-immigrant stock.

After 1919, memories of strike defeats, family economic obligations, and limited expectations all induced caution, but it is a mis-
reading of the attitudes and values of new-immigrant workers to
equate this cautiousness with ideological conservatism or a narrow,
market-oriented pragmatism. That workers of new-immigrant stock
avoided risks, and once they settled in, placed a high premium on
keeping mill jobs which paid better than almost any other jobs they
could hope to get, may indeed, as John Bodnar argues, indicate "real-

ism," but it does not tell us very much about their moral vision—what they defined as just and proper—or what they were willing to fight for when they saw reasonable chances of winning.

A Mon Valley Polish steelworker's explanation of why he went on strike in 1919 may provide some clues to those values: "Just like a horse and wagon. Put horse in wagon, work all day. Take horse out of wagon—put in stable. Take horse out of stable, put in wagon. Same way like mills. Work all day. Come home. Wife say, "John, children sick. You help with children." You say, "Oh, go to hell"—go sleep. Wife say, "John you go town." You say, "No"—go sleep. No know what the hell you do. For why this war? For why we buy Liberty bonds? For the mills? No, for freedom and America—for everybody. No more horse and wagon. For eight-hour day."[66]

"For eight-hour day," a limited and practical goal, but its justification suggests, in primitive form, what was to become by the 1930s a much more fleshed-out and explicitly articulated rationale for unionization and for a variety of public policies supported by unions, by urban liberal politicians, and by working-class Democratic voters. There is no hint in the Polish steelworker's speech, as there rarely would be in the language of 1930s unionism, of the visionary social reconstruction, or self-directed workers' control, that was implicit in the more radical versions of earlier artisanal republicanism. There is instead an unconditional and universalist assertion of human rights based on a new conception of ethnic democracy.[67] Working people were entitled to humane treatment, to freedoms of speech and assembly necessary for organization, to the economic resources and time essential to a decent family life. They were entitled to these things not because they had earned them by being good workers, or learning English, or demonstrating good citizenship, or adopting Yankee ways, but because *in America* these things were guaranteed to all people simply because they were human beings.

What is, nonetheless, surprising is the extent to which this broad moral vision was tied to a narrow economic program. Visions of workers' control in earlier versions of artisanal republicanism, we now recognize, had their roots in workers' actual experiences—the ways in which, for example, iron puddlers and glassblowers already did supervise and control so much of the labor process. To what extent does the absence of such a vision among unskilled mass production workers also reflect the realities of work experiences where they had little control over their labor beyond the ability to resist increases of pace, and almost no understanding of the overall production process?

What experiences could they draw upon as positive examples of

how their work situations might be altered? The most widespread
and fundamental source of change in work situation for workers in
large-scale corporate mass production between 1915 and 1935 was
the introduction of the cluster of policies known as corporate
welfarism and managerial reform. In Pittsburgh's steel mills, the
most important of these changes included formal and written personnel policies, regularized promotion ladders, shortened work
days, profit-sharing plans, fringe benefits and pensions, company-
sponsored social and recreational activities, grievance procedures,
and employee representation plans.[68]

Most historians have treated corporate welfarism as a transpar-
ent subterfuge aimed at reasserting social control over workers in
the face of World War I era challenges.[69] The social control motive
was unquestionable, and rarely denied by corporate officials, but we
need to consider the broader meaning and impact of these programs.
How did the introduction of corporate welfarism and managerial
reform alter the day-to-day experiences of workers? How did these
changes implicitly legitimate workers' beliefs that they were entitled to better conditions, suggest models of how to make decisions
about labor policies, and kindle new resentments when company
programs failed to deliver on their promises? Was it only a tactical
ploy when union organizers in many mills in the 1930s entered
company employee representation plans and used them to mobilize
workers? The program, policies, and even much of the bureaucratic
style of the Steel Workers Organizing Committee (SWOC) after 1936
(and many other CIO unions) closely paralleled the agenda of corporate welfarism. SWOC's seniority systems for example, literally
took over existing company promotion ladders, rarely proposing any
alterations in how seniority was to be defined or how it was to be
applied to decisions about promotions or layoffs.[70] It is only a slight
exaggeration to argue that unionization in steel amounted to trans-
ferring administration of company controlled welfarism to the
union. The limited and bureaucratic nature of mass production
unionism has provoked a sustained critique of union officials by
New Left labor historians, but how much of this outcome was ultimately a product of the previous experiences of the workers they
organized?

Finally, Pittsburgh's labor historians need to consider the impact
of gender and race on the working class in the Progressive era and
after. In part, that consideration is historiographic affirmative action. The working class has always been half female. And after the
mass migrations of the World War I era, which doubled Pittsburgh's
black population, blacks made up a much larger proportion of Pitts-

burgh's working class than ever before. But labor historians also
need to give far more attention to gender and race because changes
in gender and race relations were central to the most basic changes
in working-class culture. Increasing employment of women, especially married women, transformed working-class family and community life, facilitated upward mobility for working-class children
who were now freer to continue their education, and provided the
additional income to working-class families which helped to blur
cultural boundaries between classes. Class-based resentments of
white workers about the disappointments and difficulties in their
own lives have often been expressed as racial conflict when fear of
black incursions crystallized working-class community solidarity.
Racial attitudes, and issues like welfare policy which often serve as
publicly acceptable proxies for these attitudes, have been the most
significant crosscutting pressure against working-class political solidarity in much of the twentieth century.

The New Deal Compromise, 1930–1960

The social changes already evident by the 1920s formed a back-
drop for working-class life for the rest of the twentieth century. But
after 1933, the impact of social change was felt in a new political
economy which stimulated class loyalties among workers even as
long-term changes may have been undermining some of the traditional sources of class identity. In Pittsburgh this new political economy emerged more clearly, more completely, and more enduringly
than in most of the rest of the United States.

At a broad descriptive level the story is clear. Three things happened: unionization of mass production, political realignment, and
sustained prosperity after 1940 with national doubling of real household income by the late 1960s.[71] Between 1892 and 1930, except in
parts of the craft periphery, Pittsburgh and its surrounding mill
towns had been an open-shop region politically dominated by Republicans. By the 1940s, steel and other mass production industries
were virtually 100 percent unionized, and local politics became
nearly as strongly Democratic as in the Deep South. Unions played
major roles in local politics in many steel towns, electing union
leaders in Duquesne, Clairton, and Aliquippa starting in 1937, and
all local politicians in the region, if they hoped to be elected, had to
maintain a respectful posture toward workers.[72]

Stability replaced transience in working-class neighborhoods.
Among mass production workers, especially the children of the new
immigrants, there was little upward social mobility. Few left the

working class. In Bodnar, Simon, and Weber's sample of second-generation Poles and Italians, almost none of the Poles and less than a quarter of the Italians had entered non-working-class occupations by 1960. Yet, by the end of the era, most working-class households could afford many of the symbols of middle-class consumption. By 1960 in four Polish and Italian working-class neighborhoods they studied, between 58 and 74 percent of residents owned their homes, compared to averages of 20 to 33 percent among a sample of first-generation Poles and Italians in 1930.[73]

The impact of this new political economy based on unionization, Democratic politics, occupational and neighborhood stability, and rising incomes was both profound and yet quite restricted in scope. The *New York Times Magazine* sent a reporter to Homestead on the eve of the 1949 industrywide steel strike to gauge the mood of rank-and-file labor in postwar America. The reporter interviewed Henry J. Mikula, a thirty-six-year-old son of an immigrant steelworker who had already put in twenty years in U.S. Steel's Homestead Works. His stepfather and five brothers-in-law worked in the same mill. His fourteen-year-old son expected to follow them. Henry Mikula took home $56 a week and that meant "making out the grocery order with a pruning knife" to make ends meet. When he had recently needed to buy a new furnace and to go to the dentist, both had to be paid for in installments. Yet he owned a small house, entirely paid for, an old washing machine, and a 1935 car. He was clearly better off than he had been when he had started out in the mill at $2.28 a day.[74]

Homestead had been a company town when Henry Mikula started working. As late as 1934, when U.S. Secretary of Labor Frances Perkins came to Homestead to speak at a union-sponsored rally in a public park, the local burgess (mayor) prohibited the meeting, threatening to arrest Secretary Perkins if she spoke. Only after she marched down the street to the U.S. Post Office, climbed the front steps, and ordered local officials off federal property, could she continue. With the coming of SWOC all that changed. In the 1946 strike, local officials declared at the outset their unwillingness to interfere with picket lines, and the burgess and a state senator joined fifteen hundred strikers at the plant gate when it was rumored that supervisory personnel might try to enter the factory to perform maintenance. By 1949 the burgess and ten of sixteen other local officials were steelworkers.[75]

Union power and worker votes guaranteed that companies could no longer break mass strikes by importing strikebreakers or attacking pickets. When local officials occasionally forgot that government labor policies had changed, workers now had the power to remind them. In February 1946, during a strike by the Duquesne Power and Light Independent Employees Association, a local judge jailed the association president for one year after he had defied a judicial injunction against the strike. Although the association was not affiliated with the AFL, CIO, or any other national labor organization, an estimated twenty-five thousand local workers throughout the Pittsburgh region struck in sympathy with the Duquesne Light employees, including six thousand J&L steelworkers and all the streetcar operators and bus drivers—whose walkout paralyzed downtown Pittsburgh. Five thousand chanting workers marched on the jail, driven back by mounted policemen. At midnight that night, the judge summoned the president of the association and released him after he agreed to attend the next strike meeting and urge acceptance of Duquesne Light's most recent contract offer. (He did, but the local rejected the offer by a two-to-one margin and the strike continued.)[76]

Unable to break strikes, companies now had to negotiate, but the range of issues subject to negotiation was essentially confined to the traditional wages, hours, and working conditions plus seniority rights and fringe benefits. Although a few national CIO officials made noises about national economic planning, a broad social-democratic agenda of public programs, and union input into corporate investment decisions, after 1946 such ideas had little relevance to political reality and played no role in labor-management negotiations.[77] This, in effect, is what labor historians have called the New Deal bargain or compromise: mutual recognition of the necessity to negotiate with a trade-off of steadily improving wages and benefits, and a limited range of worker grievance and seniority rights in exchange for labor peace and acceptance of managerial prerogatives in all other areas.[78]

The limited range of negotiable union issues mirrored similar limits to worker and union political power. Public officials were expected to respect union rights to bargain and to organize, but union officials did not, and did not appear to expect to, participate in decisions over the wider range of local public policies. Unions played almost no role in the planning processes which resulted in Pittsburgh's Renaissance, a form, according to Roy Lubove, of corporate-dominated "reverse welfare state" in which taxes generated out of a working-class population went to revitalize corporate properties in the downtown business district.[79]

However, if the broad plot line of this story is clear, almost none of the pieces are. We have virtually no analysis of how each of these pieces happened, how they fit together, or the cause and effect relationships between them.

Conclusion

Even before the Civil War, the rise of a local industrial economy based on large, highly capitalized, heavy industries had led to the formation of a local working class with exceptionally well developed community networks, group loyalties, and trade union organizations. Successive waves of immigration, fundamental technological changes in production, reorganization of industry by large, bureaucratic corporations, and the social changes accompanying the culture of mass consumption continually altered working-class culture and communities. Yet working-class self-identity and solidarity—focused around a bureaucratically organized system of mass production unionism, Democratic political loyalties, and ethnically based community life—seemed as strong in the mid-twentieth century as it had nearly one hundred years before.

Ironically, the continuing viability of these forms of working-class culture in Pittsburgh, long after they had become largely vestigial in many other parts of the country, may also have been a product of the long, slow industrial decline which now threatens to destroy them. Pittsburgh's industries peaked in the 1920s, declining steadily, but very slowly, thereafter. An apparently steady-state economy facilitated neighborhood stability and group loyalties. With limited opportunities for commercial expansion, few older working-class neighborhoods disappeared to make way for urban renewal. With limited new job opportunities, there was far less black in-migration than in other large eastern or midwestern cities (and no Hispanic migration), and far less white flight out of working-class neighborhoods. The community aged as some young people left, but until the late 1970s there were enough mill jobs for sons to follow their fathers into the mills. In most of the postwar era this region had an exceptionally low proportion of high-school seniors who went on to college. Unionization made these good jobs, with high enough wages to maintain a decent and respectable working-class life style.

If Pittsburgh has been a symbol of an earlier industrial culture, it has also been distinctive in the endurance of cultural patterns that grew out of an earlier industrial age. Superficially, some of these older political and cultural loyalties still persist. Unions survive, although they now negotiate terms of surrender. Allegheny County still voted 56.7 percent Democratic in 1984. Many neighborhoods look physically unchanged. There is more than a little nostalgia for old symbols in local popular culture. Some of this nostalgia is certainly misplaced. The Iron City and later the Steel City were gritty

places where generations of immigrants worked themselves
for the magnates of a few giant corporations. But perhaps ⟨
this nostalgia is also born out of some very contemporary fea⟨
will new generations of workers develop new, more effective
of solidarity to deal with the emerging class structure of the ⟨
Reagan?

NOTES

1. Michael Holt, *Forging a Majority: The Transformation of the Republican Party in Pittsburgh, 1848–1860* (New Haven, Conn.: Yale University
Press, 1969), pp. 17–21; Catherine Elizabeth Reiser, *Pittsburgh's Commercial Development, 1800–1850* (Harrisburg: Historical and Museum Commission, 1951), pp. 191–93, 203, 207. As late as 1849 most Pittsburgh iron
was still produced in charcoal furnaces. See Paul F. Paskoff, *Industrial Evolution: Organization, Structure, and Growth of the Pennsylvania Iron Industry, 1750–1860* (Baltimore: Johns Hopkins University Press, 1983), pp. 95,
128–29.

2. Holt, *Forging a Majority*, pp. 23, 29. Blacks made up 4.1 percent of
Pittsburgh's population in 1850.

3. At present the sum total of scholarship on antebellum Pittsburgh's
working class consists of one unpublished graduate seminar paper based
mainly on secondary sources (Barbara Turner, "Organized Labor in Pittsburgh, 1826–1837" [seminar paper, Department of History, University of
Pittsburgh, 1973]), two published essays on local strikes in 1848 and 1850
(Monte Calvert, "The Allegheny City Cotton Mill Riot of 1848," *Western
Pennsylvania Historical Magazine* 46 [April 1963]: 97–133; and James
Linaberger, "The Rolling Mill Riots of 1850," *Western Pennsylvania Historical Magazine* 47 [Jan. 1964]: 1–18); and scattered references to Pittsburgh
labor in local histories, in national labor histories, and in William A. Sulli-
van's *The Industrial Worker in Pennsylvania, 1800–1840* (Harrisburgh:
Pennsylvania Historical and Museum Commission, 1955).

4. Robert Eugene Harper, "The Class Structure of Western Pennsylvania in the Late Eighteenth Century" (Ph.D. diss., University of Pittsburgh,
1969), pp. 32, 55, 63–67, 78–79, 90–101, 103, 192–93; *Return of the Whole
Number of Persons Within the Several Districts of the United States* (Philadelphia, 1791), p. 45; *Return of the Whole Number of Persons Within the
Several Districts of the United States* (Washington, D.C., 1801), p. 20; Lee
Soltow, "Wealth Inequality in the United States in 1798 and 1860," *Review
of Economics and Statistics* 66 (Aug. 1984): 467. Figures in the table below
are from Harper, pp. 193–220. Based on interpolation from censuses I estimate Fayette County's population in 1796 at 17,400, and the number of
white males over age twenty-one at 3,250. There were 1,158 landowners
including absentees and a few women.

5. Assuming a labor force participation rate equal to the national average of 33 percent in 1800 (based on my own calculations using conventional labor force definitions which exclude most white women and children), and the estimated population of 17,400 (see n. 4 above), Fayette County's labor force in 1796 would have been around 5,700. Given a relatively young age structure and few slaves, the labor force participation rate may have been slightly lower than the national average, but the tax list includes only 2,473 people.

6. Harper, "Class Structure," pp. 214–23.

7. Allegheny County tax records from this period are not preserved.

8. Frank C. Harper, *Pittsburgh of Today: Its Resources and People* (New York: American Historical Society, 1931), pp. 154, 561, 602; Reiser, *Pittsburgh's Commerical Development*, p. 14.

9. Reiser, *Pittsburgh's Commerical Development*, p. 7. See note b, table 1.

10. Tench Coxe, *A Statement of the Arts and Manufactures of the United States for the Year 1810* (Philadelpha, 1814), pp. 49–74; Harper, *Pittsburgh of Today*, pp. 206–08.

11. Harper, *Pittsburgh of Today*, pp. 206–08.

12. Ibid., 206–07, 446–47; "Pittsburgh Cordwainers, 1815, Commonwealth v. Morrow," in John R. Commons et al., *A Documentary History of American Industrial Society*, 10 vols. (Cleveland: A. H. Clark Co., 1910–11), vol. 4, pp. 15–87.

13. Holt, *Forging a Majority*, p. 19; James C. Holmberg, "The Industrializing Community: Pittsburgh, 1850–1880" (Ph.D.diss., University of Pittsburgh, 1981), pp. 124, 145. It is not clear from the context which parts of the city are included in this contemporary estimate, or whether the percentages are of manufacturing employees or of the entire workforce, but these estimates are consistent with Holmberg's totals of the census of manufactures for the metropolitan area.

14. For New York, see Sean Wilentz, *Chants Democratic: New York City and the Rise of the American Working Class, 1788–1850* (New York: Oxford University Press, 1984). For Philadelphia, see Bruce Laurie, *Working People of Philadelphia, 1800–1850* (Philadelphia: Temple University Press, 1980). Laurie found a mean firm size in Philadelphia quite comparable to Pittsburgh in 1850—12.9 compared to 11.8 for Pittsburgh (1850 boundaries) and 13.4 for metropolitan Pittsburgh (1907 boundaries)—but a lower proportion of employees in the largest firms. In Allegheny County in 1850, 43.8 percent of manufacturing employees worked for firms of 100 or more workers compared to 43.1 percent in Philadelphia in firms of 51 or more. In New York City, Richard Stott ("British Immigrants and the American 'Work Ethic' in the Mid-Nineteenth Century," *Labor History* 26 [Winter 1985]: 26) found a mean firm size of 16.7 in a sample of the 1850 manufacturing census, but it is not clear from his discussion how many outworkers are included in this figure. Wilentz estimates that 46 percent of all manufacturing employees in New York City were outworkers, and only about a third of manufacturing employees worked *inside* shops of 21 or more employees, a

better indication of the differences between New York and Pittsburgh than the mean firm size.

15. Random sample of two hundred Pittsburgh residents employed in working-class occupations taken from U.S. Census, 1850, manuscript population schedules, Pittsburgh (microfilm edition). Hereafter referred to as 1850 Census sample.

16. 1850 Census sample; *Historical Statistics of the United States: Colonial Times to 1970* (Washington, D.C.: U.S. Dept. of Commerce, 1975), p. 106. Total national immigration for the 1820s was about 39 percent of the total for the single year of 1850.

17. Turner, "Organized Labor in Pittsburgh," pp. 30–31, 33; Calvert, "Cotton Mill Riot," pp. 105–15; Linaberger, "Rolling Mill Riots," pp. 1–18. The population figure for 1840 is from Holt, *Forging a Majority*, p. 318.

18. Sullivan, *Industrial Worker in Pennsylvania*, pp. 180–90.

19. Calvert, "Cotton Mill Riot," pp. 101–103, Holt, *Forging a Majority*, pp. 40–41, 48, 66–70.

20. Holt, *Forging a Majority*, pp. 328–29, 336–37, 340–42, 356–57.

21. Holmberg, "Industrializing Community," p. 75; S. J. Kleinberg, *The Shadow of the Mills: Working-Class Families in Pittsburgh, 1870–1907* (Pittsburgh: University of Pittsburgh Press, 1989), pp. 4–7; sample of 5,466 Pittsburgh and Allegheny City males twenty-one or older taken from the U.S. Census, 1880 manuscript schedules of population for Pittsburgh and Allegheny City (microfilm edition), hereafter referred to as 1880 Census sample; U.S. Department of the Interior, *Report on the Manufactures of the United States of the Tenth Census, June 1, 1880* (Washington, D.C., 1883), p. 426. The 1880 Census sample was originally collected by students in my graduate research seminar at the University of Pittsburgh, Winter 1984. Details of sample construction available on request.

22. Kleinberg, *Shadow of the Mills*, p. 50.

23. Francis G. Couvares, *The Remaking of Pittsburgh: Class and Culture in an Industrializing City, 1877–1919* (Albany: State University of New York Press, 1984); John William Bennett, "Iron Workers in Woods Run and Johnstown: The Union Era, 1865–1895" (Ph.D. diss., University of Pittsburgh, 1977); Shelton Stromquist, "Working-Class Organization and Industrial Change in Pittsburgh, 1860–1890: Some Themes" (seminar paper, Department of History, University of Pittsburgh, 1973): Kleinberg, *Shadow of the Mills*, and "Technology's Stepdaughters: The Impact of Industrialization Upon Working-Class Women, Pittsburgh, 1865–1890" (Ph.D. diss., University of Pittsburgh, 1973); Paul Krause, "The Road to Homestead" (Ph.D. diss., Duke University, 1987). Generalizations in succeeding paragraphs run throughout these works, and I will only cite specific examples where there is data that is especially distinctive.

24. On intraunion conflicts see especially Stromquist, "Working-Class Organization"; Richard O'Connor, "Cinderheads and Iron Lungs: Window Glassworkers and Their Unions, 1865–1920" (Ph.D. diss. University of Pittsburgh, in progress); Bennett, "Iron Workers," pp. 42–56.

25. Carolyn S. Schumacher, "School Attendance in Nineteenth-Century

Pittsburgh: Wealth, Ethnicity, and Occupational Mobility of School Age Children, 1855–1865" (Ph.D. diss., University of Pittsburgh, 1977), esp. pp. 48, 68, 151, 154, 166, 168–69, 172, 174–75, 179, 189; Ileen A. DeVault, "Sons and Daughters of Labor: Class and Clerical Work in Pittsburgh, 1870s–1910s" (Ph.D. diss., Yale University, 1985). The only social mobility study for Western Pennsylvania is Michael P. Weber, *Social Change in an Industrial Town: Patterns of Progress in Warren, Pennsylvania from the Civil War to World War I* (University Park: Pennsylvania State University Press, 1976). Warren is a small industrial town 135 miles north of Pittsburgh.

26. David Montgomery, *Beyond Equality: Labor and the Radical Republicans, 1862–1872* (1967; rpt. Urbana: University of Illinois Press, 1981), pp. 448–52.

27. Ibid., p. ix. Many scholars have analyzed this working-class republicanism, but see especially Linda Schneider, "The Citizen Striker: Workers' Ideology in the Homestead Strike of 1892," *Labor History* 23 (Winter 1982): 47–66.

28. Couvares, *Remaking of Pittsburgh*, p. 29

29. 1850 Census sample; 1880 Census sample.

30. Holt, *Forging a Majority*, pp. 110–15, 138–45, 154–64, 169–70, 338, 342; Laurie, *Working People*, pp. 169–87.

31. For a series of capsule biographies of local Irish labor activists illustrating the connections between class and ethnic activities, see Victor Anthony Walsh, "Across the 'Big Wather': Irish Community Life in Pittsburgh and Allegheny City, 1850–1885" (Ph.D. diss., University of Pittsburgh, 1983), chap. 9; and Bennett, "Iron Workers," chap. 6. For a discussion of the relationship between working-class radicalism and resistance to assimilation among Detroit's Germans in the same era, see Richard Oestreicher, *Solidarity and Fragmentation: Working People and Class Consciousness in Detroit, 1875–1900* (Urbana: University of Illinois Press, 1986), pp. 43–52.

32. Schumacher, "School Attendance," p. 46. These data are from 1860.

33. 1880 Census sample.

34. Kleinberg, "Technology's Stepdaughters," pp. 61–67, 116–21.

35. Ibid., chap. 6, pp. 236–38; Dan Tuden, "The Impact of Class on Marriage Patterns in the 1900s in Pittsburgh," unpublished paper, 1986, on file in Department of History, University of Pittsburgh.

36. S. Y. Albee, "The Railroad War in Pittsburgh," series of forty-two stereographs (in author's possession). For a description of these events see Robert Y. Bruce, *1877: Year of Violence* (Indianapolis: Bobbs Merrill, 1959).

37. They were published by J. F. Jarvis, Washington, D.C., copyrighted and marketed by Underwood and Underwood. The best description of the events (among many available) is Arthur G. Burgoyne, *The Homestead Strike of 1892* (1893; rpt. Pittsburgh: University of Pittsburgh Press, 1979).

38. Burgoyne, *Homestead Strike*, pp. 204–07.

39. Ibid., pp. 247–48; Paul Krause, "Labor Republicanism and 'Za Chlebom': Anglo-American and Slavic Solidarity in Homestead," in *"Struggle a Hard Battle": Essays on Working-Class Immigrants*, ed. Dirk Hoerder (De Kalb: Northern Illinois University Press, 1986), p. 161.

40. Stromquist, "Working-Class Organization," pp. 18–19; Krause, "Labor Republicanism," pp. 146–60; and "Road to Homestead," chap. 3.

41. Stromquist, "Working-Class Organization," pp. 18–19; O'Connor, "Social Change."

42. John French, "'Reaping the Whirlwind': The Origins of the Allegheny County Greenback-Labor Party in 1877," *Western Pennsylvania Historical Magazine* 64 (April 1981): 97–119; Bennett, "Iron Workers," chap. 6; Couvares, *Remaking of Pittsburgh*, pp. 28–29, 62–69; Montgomery, *Beyond Equality*, pp. 211, 389–92; Krause, "Road to Homestead," chap. 3, pp. 33–36, chap. 4, 34–67.

43. Burgoyne, *Homestead Strike*, pp. 105–106, 137. The sole exceptions to the general acquittal of all the Homestead defendants were Alexander Berkman and two confederates, convicted for an unsuccessful assassination attempt on Frick, and two de ndants convicted of poisoning strikebreakers housed inside the mill.

44. David Brody, *Steelworkers in America: The Nonunion Era* (Cambridge: Harvard University Press, 1960), provides essential background to the changes in the industry; these changes are also described in detail in Krause, "Road to Homestead," chaps. 3–4, and Stromquist, "Working-Class Organization."

45. Daniel Nelson, *Managers and Workers: Origins of the New Factory System in the United States, 1880–1920* (Madison: University of Wisconsin Press, 1975), p. 6.

46. Good descriptions of the nature of the second industrial revolution include ibid., and David Montgomery's essays in *Workers' Control in America: Studies in the History of Work, Technology, and Labor Struggles* (Cambridge: Cambridge University Press, 1979). See also Ronald Edsforth, *Class Conflict and Cultural Consensus: The Making of a Mass Consumer Society in Flint, Michigan* (New Brunswick: Rutgers University Press, 1987), esp. chaps. 1–2.

47. *Fifth Industrial Directory of the Commonwealth of Pennsylvania* (Harrisburg: Pennsylvania Department of Internal Affairs, 1925), pp. 20–78; *Report on Manufactures, 1880*, p. 426.

48. Stromquist, "Working-Class Organization," has a particularly insightful description of the social consequences of this spatial transformation. See also "From Patricia to Woodlawn," *Beaver Valley Labor History Journal* 2, no. 2 (May 1980): 1, 5, 6; and "Roots of Beaver Valley Lodge 200, Amalgamated Association of Iron, Steel, and Tin Workers, 1909–1929," ibid., no. 1 (Mar. 1979): 1, 4, 5.

49. 1880 Census sample; John Bodnar, *The Transplanted: A History of Immigrants in Urban America* (Bloomington: Indiana University Press, 1985), chap. 1; David Brody, *Workers in Industrial America: Essays on the Twentieth-Century Struggle* (New York: Oxford University Press, 1980), p. 15.

50. Michael W. Santos, "Laboring on the Periphery: Managers and Workers at the A. M. Byers Company, 1900–1969," unpublished manuscript, 1987 (in author's possession), pp. 49–73; U.S. Bureau of the Census, *Thir-*

teenth Census of the United States Taken in the Year 1910, vol. 4, *Population 1910: Occupation Statistics* (Washington, D.C.: Government Printing Office, 1914), pp. 590–92; Margaret Byington, *Homestead: The Households of a Mill Town* (1910; rpt. Pittsburgh: University of Pittsburgh Press, 1974), p. 40.

51. Santos, "Laboring on the Periphery," p. 70; John Bodnar, Roger Simon, and Michael P. Weber, *Lives of Their Own: Blacks, Italians, and Poles in Pittsburgh, 1900–1960* (Urbana: University of Illinois Press, 1982), p. 121.

52. Leo Wolman, *The Growth of American Trade Unions, 1880–1923* (New York: National Bureau of Economic Research, 1924), p. 95.

53. Frank H. Serene, "Immigrant Steelworkers in the Monongahela Valley: Their Communities and the Development of a Labor Class Consciousness" (Ph.D. diss., University of Pittsburgh, 1979), p. 195.

54. Carl I. Meyerhuber, Jr., "Black Valley: Pennsylvania's Alle-Kiski and the Great Strike of 1919," *Western Pennsylvania Historical Magazine* 62 (July 1979): 249–65, esp. 253–59; Brody, *Steelworkers*, pp. 258–60.

55. On the cigarmakers, see Ida Cohen Selavan, "The Jewish Labor Movement in Pittsburgh," (seminar paper, Department of History, University of Pittsburgh, 1971), pp. 5–12; Patrick Lynch, "Pittsburgh, the I.W.W., and the Stogie Workers," in *At the Point of Production: The Local History of the I.W.W.*, ed. Joseph R. Conlin (Westport, Conn.: Greenwood Press, 1981), pp. 79–94.

56. Tom Price, "The Westinghouse Strikes of 1914 and 1916: Workers' Control in America?" (seminar paper, Department of History, University of Pittsburgh, 1983); Linda Nyden, "Women Electrical Workers at Westinghouse Electric Corporation's East Pittsburgh Plant, 1907–1945" (seminar paper, Department of History, University of Pittsburgh, 1975), esp. pp. 14–22.

57. Andrew A. Marchbin, "Hungarian Activities in Western Pennsylvania," *Western Pennsylvania Historical Magazine* 23 (Sept. 1940): 167; Selevan, "Jewish Labor Movement," p. 4; Edwin M. Moser, "Jewish Labor in Pittsburgh: Its Ideology and Relation to the National Scene, 1905–1914" (seminar paper, Department of History, University of Pittsburgh, 1961), p. 9. On Croatian socialists, see Frieda Truhar Brewster, "A Personal View of the Early Left in Pittsburgh, 1907–1923," *Western Pennsylvania Historical Magazine* 69 (Oct. 1986): 343–65.

58. Bruce M. Stave, *The New Deal and the Last Hurrah: Pittsburgh Machine Politics* (Pittsburgh: University of Pittsburgh Press, 1970), p. 195; Michael Nash, *Conflict and Accommodation: Coal Miners, Steel Workers, and Socialism, 1890–1920* (Westport, Conn.: Greenwood Press, 1982), pp. 116–18, 174; Price, "Westinghouse Strikes," pp. 36–42; James Weinstein, *The Decline of Socialism in America, 1912–1925* (New York: Random House, 1967), pp. 116–18. Compare these figures to Debs's vote in Detroit (4 percent) or New York City (5 percent).

59. David Montgomery, "The 'New Unionism' and the Transformation

of Workers' Consciousness in America, 1909–22," in *Worker's Control in America*, argues the protosocialist position most explicitly. In later essays Montgomery describes a more varied spectrum of consciousness in response to a range of conflicting pressures. See, for example, "Nationalism, American Patriotism, and Class Consciousness Among Immigrant Workers in the United States in the Epoch of World War I," in Hoerder, ed., *"Struggle a Hard Battle."* Bodnar argues the skeptical position in all of his works, but see especially "Immigration, Kinship, and the Rise of Working-Class Realism," *Journal of Social History* 14 (Fall 1980): 45–65.

60. The most important studies of unionization and union policies in the 1930s and 1940s in Pittsburgh are Ronald W. Schatz, *The Electrical Workers: A History of Labor at General Electric and Westinghouse, 1923–1960* (Urbana: University of Illinois Press, 1983), and Mark McColloch, "Consolidating Industrial Citizenship: The U.S.W.A. at War and Peace, 1939–1946," in *Forging a Union of Steel: Phillip Murray, SWOC, and the United Steelworkers,* ed. Paul F. Clark, Peter Gottlieb, and Donald Kennedy (Ithaca: ILR Press, New York State School of Industrial and Labor Relations, Cornell University, 1987), pp. 45–86.

61. See Schumacher, "School Attendance," and DeVault, "Sons and Daughters of Labor."

62. Serene, "Immigrant Steelworkers," pp. 58–61.

63. Santos, "Laboring on the Periphery," pp. 92–111; Bodnar et al., *Lives of Their Own,* pp. 201–03, 211–33, 237–59.

64. This is the key argument of Frank Serene's study, "Immigrant Steelworkers." While highly anecdotal, there are many suggestive examples in the *Beaver Valley Labor History Journal* and in George Powers, *Cradle of Steel Unionism: Monongahela Valley, Pa.* (East Chicago: Figueroa Printers, 1972).

65. Stave, *New Deal,* pp. 35–40, 80–81.

66. As quoated by David Montgomery in "Nationalism, American Patriotism," p. 346.

67. A stimulating case study of the concept of ethnic democracy is Gerd Korman, "Ethnic Democracy and Its Ambiguities: The Case of the Needle Trade Unions," *American Jewish History* 75 (June 1986): 405–26.

68. For general discussions of these phenomena see David Montgomery, *Workers' Control,* esp. chaps. 2, 4, 5, and Daniel Nelson, *Managers and Workers.* For a detailed account of managerial reform in the steel industry, see Gerald G. Eggert, *Steelmakers and Labor Reform, 1886–1923* (Pittsburgh: University of Pittsburgh Press, 1981).

69. A major exception is David Brody, "The Rise of Welfare Capitalism," chap. 2 of *Workers in Industrial America.*

70. Robert Lewis Ruck, "Origins of the Seniority System in Steel" (seminar paper, Department of History, University of Pittsburgh, 1977). Schatz, *Electrical Workers,* pp. 15–24, 37, 40–46, makes a similar argument about the role of managerial reform in the electrical industry.

71. *Historical Statistics,* p. 301; U.S. Department of Labor, Bureau of

Labor Statistics, *Handbook of Labor Statistics,* Bulletin 2175, Dec. 1983, p. 204. The exact percentage of increase varies depending on which of several BLS statistics are compared, and the various time series only partially overlap. For example, using "after tax mean family personal income per consumer unit" (that is, including both households and unattached individuals), real income increased 100.9 percent between 1936 and 1962. Using "spendable average weekly earnings" for a "worker with three dependents," real income increased 45.8 percent between 1947 and 1972. Using the latter statistic for manufacturing wage earners, real income increased 112.5 percent between 1939 and 1972. The last is probably the most meaningful statistic for blue-collar families, but understates improvements in household income since it does not take into account the increasing employment of married women.

72. Joel Sabadasz, "Steelworkers and Local Politics" (Ph.D. diss., University of Pittsburgh, in progress); Karen Steed, "Unionization and the Turn to Politics: Aliquippa and the Jones and Laughlin Steel Works, 1937–1941" (seminar paper, Department of History, University of Pittsburgh, 1982); Powers, *Cradle of Steel Unionism,* pp. 132–42.

73. Bodnar et al., *Lives of Their Own,* pp. 159, 227, 249–52.

74. David Dempsey, "Steelworkers: 'Not Today's Wage, Tomorrow's Security,' " in *American Labor Since the New Deal,* ed. Melvyn Dubofsky (Chicago: Quadrangle, 1971) pp. 192–201.

75. Ibid., p. 197; Edward Levinson, *Labor on the March* (New York: Harpers, 1938), p. 188; McColloch, "Consolidating Industrial Citizenship," p. 54.

76. George Lipsitz, *Class and Culture in Cold War America: "A Rainbow at Midnight"* (New York: Praeger, 1983), pp. 74–81.

77. Brody, *Workers in Industrial America,* chaps. 5–6.

78. Montgomery, *Workers' Control in America,* chap. 7.

79. Roy Lubove, *Twentieth-Century Pittsburgh: Government, Business and Environmental Change* (New York: John Wiley and Sons, 1969), p. 139.

Government, Parties, and Voters in Pittsburgh

PAUL KLEPPNER

THE STUDY of urban history in the United States has experienced both a revival and a transformation since the mid-1960s. This change involved a shift from an older preoccupation with events, institutions, and personalities to a new emphasis on social relationships and changes in the patterns of these relationships over long periods of time.

What sparked this change in basic outlook is not entirely clear, although I suspect that its sources were diverse and lacked any overarching analytic framework. Indeed, the body of new work that initially emerged under the rubric "urban history" did little to show how the process of urbanization, or even the urban locale, played a distinctive role. The early mobility/persistence studies are good examples.[1] The authors of these works could just as readily have explored the processes of geographic and social mobility in rural or small-town contexts.[2] Their choice of an urban context seems to have been driven by some combination of the availability of sources, especially annual city directories, and the location and possibly the personal preferences of the authors. They gave no attention to determining how much the occupational structure of the city limited opportunities for social mobility, or how much of the observed residential movement was occasioned by larger-scale changes resulting from the spatial (and functional) reorganization of parts of the city.[3]

Whatever their limitations, these efforts to explore the patterns of social relationships within the city had the effect of reorienting the subfield. Urban historians began to look for sources of information about all of the people in the city, and they adopted techniques enabling them to analyze empirical evidence.

At the same time, analogous developments occurred among political historians. The shift there was from a preoccupation with presidential elections and the maneuverings of the elite to a study of grass-roots voting behavior. This new emphasis inevitably led historians to explore the ways in which the social and cultural values of ordinary citizens became politicized. And like their urban colleagues, political historians necessarily turned to different sources and adopted techniques that allowed them to analyze evidence concerning large numbers of people.

These developments occurred independently, with little effort being made to relate them to each other. For the most part, and because they were still concerned with explaining state or national election outcomes, political historians concentrated on contexts larger than cities. And urban historians mainly concerned themselves with social processes, rather than with political developments or with the linkage between social and political changes.

This intellectual separatism was understandable, especially since reactions against an older and dominant political history provided a large share of the stimulus for the new research directions that marked urban and social history. Nevertheless, this parallel-track development prevented historians from reconstructing the full range of experiences in which ordinary citizens were involved. It also inhibited them from seeing whether and how public decisions reflected prevailing social relationships, and whether these decisions in turn then operated to reshape or channel ongoing social processes.

Exploring the possibility and character of these types of interrelationships and interactions ought to be a priority item on the historical research agenda. And the city—much better than the state, the region, or the nation—provides a suitable context within which to undertake the required probes.

Fortunately for Pittsburgh, a considerable body of work exists that does explore both the mutual and dynamic relationships among government, parties, and ordinary voters and how these relationships have changed across time. The authors of the separate studies involved did not undertake them as part of a planned agenda, and so the coverage is chronologically incomplete and analytically uneven. Some matters remained underdeveloped, and other questions were not asked at all; so considerable gaps in our understanding remain, gaps which now can be filled only by plausible inference or large leaps of faith. But by reviewing what has been written, by organizing and ordering the evidence, we can derive a clear sense both of what has been accomplished and of what yet needs to be undertaken.

In its nearly two and a half centuries of existence, Pittsburgh has changed from a sparsely populated commercial center serving the Ohio Valley to a densely populated metropolis housing the headquarters of multinational corporations. This transformation has been accompanied by expansion of the city's physical boundaries; by differentiation of its subareas along functional, economic, and even cultural lines; by the turnover and replacement of economic and ethnic groups within its population; by changes in its form of government; and by shifts in its relationship to the larger regional, national, and international economies. Of course, none of these dimensions of change, or others, occurred accidentally or randomly, and all of them were synergistically interrelated.

While we still need detailed studies of the occupational and residential structures of late eighteenth- and early nineteenth-century Pittsburgh, it was surely a simpler, less diverse place than its modern counterpart. Its active work force—3,329 persons in 1840—were mainly engaged in commerce and related trades, with a secondary concentration in manufacturing.[4] And its whole population— slightly more than 21,000—lived in an area bounded by the two rivers and extending less than two miles eastward from the Point. Like Philadelphia, when it was of comparable size, Pittsburgh was still a walking city, a place of fact-to-face relationships, where patrician landowner and mechanic, rich and poor, lived in close proximity.[5]

Despite economic and cultural divisions among its population, the early nineteenth-century city still functioned as a community. Arising initially from a dense network of business and economic relationships, this sense of community was maintained and reinforced by daily commercial and social interactions. Pittsburgh did not yet experience the sense of geographic and social separateness, nor the resulting problems of communicating across groups, that occurred later in the century.[6]

Pittsburgh's governmental and political arrangements mirrored its basically undifferentiated spatial and social order. At least that is the inference that one could reasonably draw from knowledge of the structure of its city government and from studies of the social backgrounds of its public officeholders.

To be sure, there were several changes in the mechanics of city government between 1794, when the legislature initially approved an act organizing the city government, and 1832–1833, when the office of mayor became popularly elected and the ward system was adopted for elections to the bicameral city council. But these changes represented little more than minor adjustments, fine tuning made necessary by the press of population growth. The 1794 act, for example, had provided for the annual election of two burgesses and four assistant burgesses, with the resulting committee of six serving as a collective executive while the town meeting retained legislative authority. Within nine years, however, the impracticality of this arrangement in the face of a growing population prompted the town meeting to petition for new legislation. The act of 1804 provided for the election of a single burgess and conferred legislative authority on a town council of thirteen members. Finally, the act of 1816 vested legislative power in the Select and Common Councils which, meeting in joint session, then selected a mayor from among the twelve aldermen appointed by the governor.

While these changes in form increased the number of public office-holders and gave the city's government some appearance of being modern, the same underlying sense of community and shared values that had driven the city's politics since the 1790s continued to operate. Regardless of the number or titles of public officials, for example, all elections were citywide affairs, with no separate representation for particular areas of the city. And, as Thomas Kelso's analysis shows, through the late 1820s Pittsburgh's officeholders were mainly recruited from the ranks of the city's "economic dominants."[7] James O'Hara may have acquired larger holdings of land than most of the burgesses, but he and men like Nathaniel Irish and Isaac Craig were otherwise typical of this category of public official. They were land-owners who had inherited their holdings or received them as a consequence of their service in the Revolutionary Army. More important, they were men of social standing, patricians whose families had long roots in the colonies.

Pittsburgh's patrician families were mainly Scots-Irish and Presbyterian, and they often intermarried with each other. Thus, in many cases, family, social, and business relationships created overlapping networks of contacts and associations among those at the top of the city's socioeconomic pyramid. The men who were put forward by these patrician networks to stand for public office were well known and widely respected by their social peers.

It also seems likely that they were about equally well regarded by citizens from the middle and lower social ranks of the city's electorate. At least there is no evidence that electoral combat pitted the "respectable" against the subaltern classes. Instead, the dominance of patricians among the popularly elected burgesses strongly suggests the opposite, that the "authority of names" and "the influence of the great" remained powerful among ordinary people.[8]

In his description of the membership of the city councils, Kelso offers additional support for this notion. Through 1828, employees of various sorts made up only 21 percent of all the elected members of the city councils, while merchants and professionals, the two top-ranking occupational categories, accounted for 58 percent.[9] This glaring imbalance indicates that large numbers of ordinary citizens in Pittsburgh routinely voted for men of property and standing as their public officials.

Two final items round out this picture of politics and government in early Pittsburgh. First, the prevailing sense of community was not much disturbed by parties organized for the purpose of mobilizing electoral coalitions. It was not that the "spirit of party" was dead, but that Pittsburgh was a one-party city. It was a bastion of Federalism surrounded, after 1800, by Democratic-Republican strongholds in the rural areas of Allegheny County and Western Pennsylvania. But party ties were assumed rather than stressed in Pittsburgh's electoral campaigns, and prominent Federalist citizens chose to serve in public office more from a sense of civic duty than as a means of developing party or even personal organizations.[10]

Second, Pittsburgh's patrician officeholders used the instruments of city government much like a modern chamber of commerce. That is, they used city government to promote and aid the city's dominant economic interests. For example, they sent representatives to lobby the state legislature and Congress to pass bills providing for roads in Western Pennsylvania; they inventoried the town's resources for purposes of advertising its advantages; and they spent considerable time and energy to put the city in control of its own wharf. It probably was to be expected that men whose wealth derived from landowning, commerce, and banking would work as public officials to advance these interests. This does not suggest that they were avaricious or unethical men. Rather, it offers more evidence of a pervasive sense of community, the existence of a shared set of values which all citizens, prominent and ordinary, expected their officeholders to articulate and promote.

But Pittsburgh did not long remain a walking city, a community marked by face-to-face relationships, shared values, and a politics of deference and consensus. As its economy diversified and its population continued to grow, the city expanded physically, its subareas became more distinct from each other, and its politics came to reflect this developing sense of geographic and social separateness.

This transformation resulted from processes that operated over long periods of time, marked by starts and stops, ups and downs, with some aspects of the involved changes occurring suddenly and others only slowly and incrementally. Moreover, the processes of change were not organized and implemented from the top down; they involved thousands of individual decisions made by thousands of people, some prominent and others ordinary, some leading the change and others responding to its effects. The new forms of social organization and patterns of human relationships emerged slowly, sometimes hesitatingly, often unevenly, and always they evoked resistance from those committed to older ways of doing things.

Central to understanding Pittsburgh's transformation was its shift from a commercial to a manufacturing center. Manufacturing activities in the early city were handmaidens to its commercial interests. Production was severely limited in volume because it was in the hands of craftsmen, the old artisans, who primarily produced consumer goods for the local market. The War of 1812 created an artificial demand and temporarily expanded the market for Pittsburgh's manufactured goods; between 1810 and 1815 the value of leather and textile manufactures increased threefold and iron manufactures eightfold. But with the end of the war, demand contracted sharply, and Pittsburgh's economy, especially its nascent manufacturing enterprises, suffered a severe postwar depression.[11]

However, the depression of the mid-1810s did not affect all businessmen in the same way. For the most part, the wealthy were hard pressed, and men like Isaac Craig and James O'Hara, pillars of the city's patrician leadership, found themselves "financially embarrassed," as the ironworks, wire factories, and textile mills in which they had interests failed. But other businessmen took advantage of these failures and opened new enterprises, sometimes literally on the ruins of old plants. As a result, by 1826 Pittsburgh had more and larger ironworks, foundries, glass works, and textile mills than it had had at the height of the wartime boom. It also had a new group of rising entrepreneurs, men who were relatively recent migrants to the city, not part of its patrician establishment, and whose success depended on the development and growth of manufacturing and not on extensive land ownership and inherited social status.

Pittsburgh's shift from a commercial center to a manufacturing city had enormous impact on its patterns of human relationships. In turn, changes in these relationships had repercussions on the city's politics and government. Some of the effects on politics and government were immediate; others were long term, showing up only later as the direct products of intervening factors in the causal chain.

In the short run, the growth of the city's manufacturing enterprises contributed to dissolving the earlier commercial consensus that united patricians and artisans and served as the linchpin of public policy. The Pittsburgh that was coming to be, the manufacturing city, challenged the Pittsburgh that was. Several pieces of evidence separately point up the effects of this phase of the city's ongoing contest between old and new ways of doing things.

First, the city abandoned at-large elections for the city councils and adopted a ward system of representation. This change, which became effective in 1833, divided the city into four wards and gave each ward its own representation in both the Select and the Common Council. This shift symbolized a breakdown of the older sense of community and the emergence of identifiable, separate, and often conflicting interests that came to acquire territorial distinctiveness as the city's economic and spatial transformation progressed.

Second, as a new group of manufacturing leaders emerged, they challenged the old elite for control over both the city's institutions of government and the general direction of its public policy. John Dankosky's study of the occupational and business affiliations of city councilmen between 1816 and 1850 shows the result of this challenge. The proportion of councilmen holding commerical occupations peaked in the mid-1830s and declined thereafter; but the percentage from manufacturing backgrounds showed an upward trend, doubling from the mid-1810s and registering a net gain of 39 percent of all council seats after 1833.[12]

Dankosky also shows that analogous shifts occurred within private institutions that played key roles in developing the city's economy and exerting political pressures to shape its public policy. For example, in 1837, one year after its founding, 70 percent of the directors of the Board of Trade represented commercial interests; by 1850, however, manufacturing and commercial outlooks were represented equally, with each accounting for 40 percent of the board's directors. Manufacturers also gained seats on the board of the Bank of Pittsburgh after 1833, occupying a total of 28 percent of the directorships between 1834 and 1850; while those from commercial backgrounds lost their absolute control, falling from 52 percent of the pre-1834 seats to 32 percent over the years between 1834 and 1850.

Finally, this tension between the city's patrician elite and its new cadre of industrialists was also played out in the arena of electoral politics during the 1830s. The rise of a political organization dedicated to extinguishing the supposedly baleful effects of the Masonic Order reinvigorated competition and shattered the calm that usually characterized Pittsburgh's elections. Antimasonry in Pittsburgh was more than an expression of the unbridled zeal of evangelical Christians. As Scott Martin's insightful analysis makes clear, the Antimasonic impulse also represented one final effort by the city's patrician elite to demonstrate its political and moral superiority over its opposition.[13]

Ironically, Masonic lodges had been among the most prestigious social organizations in late eighteenth-century Pittsburgh. Then they counted as members socially prominent men with long ties to the city and region, men who were part of the city's patrician and officeholding elite. But as the Revolutionary generation began dying out and migration and immigration swelled the city's population, the social profile of Masonic members inevitably began to change. The lodges recruited a larger number of naturalized citizens, especially Irish and Germans, as well as other migrants who lacked both long roots in the community and the social standing such ties conferred. Although some of its new members became successful industrialists, the prestige of the Masonic Order declined, making it more vulnerable to attack, especially by evangelicals.

It was this connection between Masonry and an emerging industrial leadership that underlay attacks on the order by representatives of the city's patrician families. Many influential Antimasons had kinship ties to the older elite; for example, Harmar Denny, who ran for Congress as an Antimason, and Neville B. Craig, who used his newspaper, the Pittsburgh *Gazette*, as a vehicle for attacking Masonry, were both members of the city's founding families and themselves the sons of prominent Masons. And other builders of the Antimasonic party in Pittsburgh, men like William Hays and William Eichbaum, were contemporaries and close associates of leading patricians of the previous generation. The nature of these connections indicates that Antimasonry represented an attempt by some of the core families of an embattled older elite to reassert their influence. Seeing what they regarded as their rightful dominance over the economic and political life of the city threatened by newcomers, these old families fought back by attacking Masonry.

For a brief time, at least, Antimasonry allowed these descendants to regain the economic and political initiative that their fathers' generation had lost after 1815. Once on the offensive, Pittsburgh's Antimasons chose to establish their own party and bitterly resisted efforts to arrange fusion slates with the Whigs. In part, this tactic reflected their judgment that both existing parties were already controlled by the very entrepreneurial newcomers whose influence they wanted to restrain. More important, organizing a separate party was the best way for its founders to differentiate themselves from other politicians and to establish a basis for claiming moral superiority over all their opponents. Above all else, Antimasons emphasized their moral claims and the need to elevate to public office only the "right sort" of men, that is, loyal supporters of the patrician establishment.

Like other politicomoral movements aimed at restoring or revitalizing a rapidly disappearing social order, Antimasonry enjoyed only marginal success. As a separate movement, it was a flash in the pan, unable to stop or deflect the pace and direction of social change. Its larger significance lay in its impact on the major parties, especially the city's dominant Whigs. For in Pittsburgh, as elsewhere, Antimasonry's "Blessed Spirit" brought an evangelical and moralizing style into political life, imbuing it with a tone of uncompromising righteousness and a sense of revulsion against party discipline.[14] Both of these traits marked Whiggery's battles against its political foes, suggesting that there was strong continuity from evangelical pressure groups, through Antimasonry, to the Whig party. In this sense, while Pittsburgh's patrician elite lost the immediate battle, its evangelical allies won the larger war.

Unfortunately we lack detailed studies of the formation of mass political parties in Pittsburgh during the 1830s and 1840s. We really do not know from what economic, cultural, and social sources Democrats and Whigs recruited their leaders and followers. Nor do we know very much about the rhetorical glue, and the underlying bonds of psychological rapport, that bound leaders and followers together in durable coalitions. And while we can trace changes in the economic backgrounds of elected members of the city councils, we do not yet know how, if at all, this shift from commercial to manufacturing occupations influenced the policies of the city government. For the present, the best we can do is combine a small number of relevant studies with considerable speculation about how continuing population growth, physical expansion, and manufacturing development reshaped the city's politics and public policy during the 1840s and 1850s.

Pittsburgh's population boomed during the 1840s, increasingly by 121 percent, from 21,115 to 46,601, and then grew only by an additional 2,616 (5 percent) during the 1850s. The economic character of the population also changed considerably, and by 1850 the number of unskilled laborers (3,628) was larger than the city's total work force had been ten years earlier.[15] This increase in the number of factory wage workers testifies to the successful development of the city's industrial activities. Initially small-scale enterprises, allied to the city's commercial function and sited within its original core, by the 1840s these had become large-scale, heavy industries, hiring unskilled wage workers and often locating on the periphery of the city, especially along the Allegheny and Monongahela rivers. While the artisans who controlled early manufacturing activities produced mainly consumer goods for a local market, the city's new factory workers produced industrial and heavy goods for regional and national markets.

Physical expansion of the city accompanied these changes. The city pushed beyond its original four wards, all less than two miles from the Point, and by 1860 its boundary was nearly four miles east of the Point along the Allegheny River and just under three miles east along the Monongahela. The area within these boundaries and between the rivers was divided into nine wards. While the distances between the Point and the city's eastern limits were between two and three times greater than they had been in the 1820s, for most inhabitants Pittsburgh was still a walking city in the 1850s, with its original wards registering population densities almost one-and-one half times those of its peripheral wards.[16]

This developing, expanding city experienced another change that greatly affected its politics. Much of its population increase after 1830, especially the growth of its manufacturing work force, resulted from immigration. Germans and Irish, especially, Protestants as well as Catholics, settled in the city and labored in its factories. At best, these recent immigrants lived in a state of uneasy tension with the city's older and predominantly Anglo-Protestant population groups. And at its worst, the relationship was one of open political warfare.

Obviously, these dimensions of change were interrelated, and they worked to shape Pittsburgh's politics during the 1840s and 1850s. The shifting occupational composition of councilmen, which began in the mid-1830s, was one indicator of increased political representation of the city's manufacturing interests. By the end of the 1840s, there were almost twice as many councilmen from manufacturing backgrounds as there were from commercial occupations, a shift which nearly reversed the ratio of the late 1810s. Moreover, most of these manufacturers were elected from the central wards of the old city (wards 1 through 4), while the outlying areas (wards 5 through 9) returned mostly skilled laborers (32 percent) and self-employed artisans (21 percent). Even at this comparatively early point in its process of development, the city's rudimentary spatial differentiation had begun to produce a corresponding specialization of interests and representation.[17]

This shift in control from the old to the new economy, from commercial to manufacturing interests, also occurred among party leaders. Michael Holt's analysis of the economic and occupational status of party candidates and secondary leaders shows the change quite clearly. In choosing candidates, for example, over the years from 1848 through 1851, the Whigs showed a modest preference for merchants, shopkeepers, and clerks (37 percent) over manufacturers (29 percent), while the opposing Democrats were about as likely to select their candidates from one occupational grouping as the other. By the end of the next decade (1858–1860, inclusive), however, the pattern had changed. Then the Republicans tilted strongly toward candidates from manufacturing and professional (51 percent) rather than commercial (28 percent) backgrounds, while the Democrats moved in the opposite direction, preferring merchants and shopkeepers (42 percent) to manufacturers and professional men (34 percent).[18]

The replacement of the dominant Whig party by an emerging Republican party symbolized the change in the city's economic function. The Whigs represented Pittsburgh's old economy, its commercial and patrician past; the Republicans represented its new economy, its industrial and entrepreneurial future. And it was not coincidental that a political change of this sort occurred in the 1850s. It was in that decade that railroads were completed that linked Pittsburgh directly into a larger transportation network, destroying its function as a transshipment center and thus displacing many of its merchants. Replaced as economic dominants, Pittsburgh's merchant group finally lost control over its instruments of politics and government.[19]

At other levels, the political turmoil of the 1840s and especially the 1850s reflected conflicts that arose as the city's population became increasingly diverse. As a jumping off point to the Northwest and a growing industrial center, Pittsburgh attracted a varied population of native-born whites, free blacks, and immigrants, especially Germans, Irish, and British who worked in the factories and mills. By 1850 immigrants made up over a third of Pittsburgh's population and an even larger share of its voting-age population. For the most part, the foreign-born lived in close proximity to their fellow countrymen, eventually establishing a dense network of fraternal, social, and religious organizations that reinforced group identities. Although divided along lines of nationality and religion, the city's growing immigrant population was sometimes regarded as a monolith by its native-born citizens, which aroused their own sense of group consciousness. Seeing their cultural hegemony threatened by these newcomers, native-born inhabitants reacted by forming their own exclusive social organizations.[20] As relations among groups deteriorated, and as each set of groups turned to politics to assure its own control, these social and fraternal organizations became the building blocks of political action.

The nativist uprising of the 1840s, which led to the formation of a Native American party in 1847, differed from the nativism of the following decade. The leaders of the agitation in the 1840s seem mainly to have been workers whose central complaint was the loss of jobs due to competition from cheap immigrant labor. The economic slump of the 1840s, which occurred when the city's popula-

tion was doubling, created enormous problems. People who came to
Pittsburgh expecting jobs and prosperity found instead only layoffs
and wage cuts. The labor strikes that resulted were only one indicator of the ensuing social tension. The anti-immigrant movement
was another. The city's street preachers, disdained by wealthy and
middle-class citizens, played important roles in tapping the anxi-
eties and mobilizing the discontent of their working-class followers
and directing it against the immigrants.[21]

During the 1850s, on the other hand, middle-class workers and
artisans organized in secret societies spearheaded the nativist agitation. Some of these Know-Nothing leaders were among the city's
wealthiest men, and 54 percent of them owned some sort of real
property, in a city in which only 15 percent of the white males
owned any property at all. They lamented the social behavior of the
immigrants, especially their poverty, drunkenness, and habits of
bloc voting.[22]

What in the 1840s had been an uprising against immigrants by
the city's factory workers became in the next decade an economically more inclusive movement aimed primarily against Catholic
immigrants. Thus, native and immigrant Protestants found common cause against their religious enemies. In the process, as the
studies by Holt and Miller Myers show, the electoral coalition that
sustained the city's new majority party, the Republicans, took on a
distinctly anti-Catholic coloration. Temperance supporters, Sabbatarian crusaders, and antislavery zealots were all part of the new
electoral majority that emerged in the mid-1850s, but it was a
shared, powerful, and politically relevant anti-Catholicism that held
together the separate components of this coalition.[23]

The voting shift that occurred among the mass public paralleled
what took place among councilmen and party candidates.[24] The
groups that comprised the Republican voting coalition shared a common outlook—an evangelical or pietistic disposition that collided
with the distinctly ritualistic or liturgical mind-set of the city's
Catholic voters. Pittsburgh's pietists, like their counterparts elsewhere, were imbued with a sense of mission to transform the world
in their image. Catholic voters in Pittsburgh, and other parts of the
North, were equally determined to preserve their own values and
ways of doing things. It was these irreconcilable outlooks, each
deeply rooted in a set of religious beliefs, that underlay the social
and political conflict between these sets of groups.[25]

Specific measures sparked these conflicts, of course; measures
such as prohibition statutes, Sunday closing laws, and efforts to re-
strict the parochial schools or to "Protestantize" the public schools.

But underlying such episodes was a larger conflict of values and interests, pitting pietistic modernizers, as Republicans, against ritualistic traditionalists, as Democrats.

For the most part, pietists shared a modernizing psychology. They were future-oriented, inclined to look outside themselves, and committed to personal and social improvement. Reason, albeit from religious premises, was their guide to decision making; education their remedy for social ills; and efficiency their ideal. Despite their sometimes strident moralizing and their association with causes—like prohibition—that appear anachronistic from the vantage point of the late twentieth century, pietists proved to be friends of science and technology, willing to apply and test it in every line of endeavor, whether farming, education, business, or even child-rearing.[26]

Ritualists, on the other hand, mainly espoused traditionalistic values. Their religions produced inwardly directed believers, members who avoided outsiders, married within the fold, paid close attention to orthodox beliefs, and followed the leadership of their pastors. Believing that theirs were the only true churches, ritualists were intensely loyal both to the institution and to fellow believers. Unlike their modernizing opponents, they valued education primarily as a means to pass on their beliefs to the next generation, rather than as an opportunity to expand human consciousness. Consequently, ritualists were not innovators or experimenters, always trying to push back the frontiers of knowledge; they were content instead with things as they were, accepting authoritative explanations and praying for improvement in a life to come.

Thus, among the general public, as well as among political and governmental leaders, the 1850s witnessed a major shift. Individuals and groups espousing change, dedicated to modernizing economic and social relations, organized themselves and took control of Pittsburgh's political parties and its city government. Native and immigrant Protestants, many of them factory workers, became partners in a Republican governing coalition with leaders of the city's industrial establishments.

Pittsburgh in 1900 was a far different city than it had been fifty years earlier. Its population had grown to 321,616, a nearly sevenfold increase in fifty years, making it the country's eleventh largest city. Most of its male work force (40 percent) was engaged in manufacturing, but large proportions were also employed in trade and transportation (27 percent) and domestic and personal service (28 percent). Immigrants from Germany, Ireland, and the British Isles were still the largest components of the city's foreign-born population, but now there were also sizeable settlements of people from the Austro-Hungarian Empire, Italy, Poland, and Russia.[27]

The city had also expanded geographically, annexing Lawrenceville and the East End in 1867 and South Pittsburgh, Birmingham, and East Birmingham in 1872.[28] Most of the city's land area in 1900 still lay between the two rivers, with its boundary extending eastward along the Allegheny to the western edge of Penn Township and along the Monongahela to Swissvale, both nearly eight miles from the Point. The area that comprised Pittsburgh's nine wards in 1850 was divided into twelve wards by 1900, with the annexed territories between the rivers forming wards 13 through 23. Pittsburgh's remaining political subdivisions, wards 24 through 36, were on the south side of the Monongahela, extending northwest beyond the Point and along the south bank of the Ohio River to the eastern boundary of Chartiers Township.

Growth was accompanied by an enormously significant internal reorganization of the city. As Joel A. Tarr shows, after 1850 a series of innovations in transportation fundamentally altered commercial, industrial, and residential patterns within the city. The introduction of the omnibus and commuter railroad, followed by the horse-drawn streetcar, and finally by the cable car and, especially, the electric streetcar transformed Pittsburgh's appearance and its spatial organization. Whereas the nascent manufacturing city of 1850 had been characterized by a mixture of land uses within its core, the Pittsburgh of 1900 far more specialized patterns of land use. Most salient was the emergence of distinctive residential areas. The development and elaboration of an intracity transportation network made it possible for many people to increase the distance between their place of work and their place of residence. The large office buildings that increasingly dotted the central business district, for example, were mainly staffed by white-collar workers who commuted daily from outlying residential areas. These new transportation possibilities, in turn, generated construction booms as new lines penetrated previously untapped areas.[29]

But not all Pittsburghers were equally able to take advantage of the improvements in transportation. Most of the city's factory workers, for instance, had neither the income nor the time required for commuting and continued to live within walking distance of their places of work. In addition to factory hands whose housing clustered close to the mills, unskilled and semiskilled workers, whose employment was irregular and work locations changeable, lived close to the central business district because it offered a large number of job alternatives.

In other words, innovations in transportation set in motion a sorting-out process. Given the differences in income levels and work-discipline demands of particular social groups, improvements in transportation made possible a pattern of residential separation along economic lines, with distinctive upper, upper-middle, central-middle, lower-middle, and working-class areas emerging. Moreover, within each of these areas, people selected their places of residence on the basis of real and/or symbolic attachments to a network of human relationships. Thus, within each economic band of residences, subareas of distinctive ethnic and religious settlement emerged.[30]

In all of this, there was a clear pattern of residential succession, along with a filling in of formerly open space with new construction to meet the housing needs of an in-moving population. For example, what in the 1870s was a central-middle-class area by 1900 had become lower-middle class, with its former residents having moved further outward.

Each wave of succession increased the area's population density. In 1870 wards 5 through 12, which were within three miles of Pittsburgh's central business district, had a population density of fifty-four persons per acre. By 1900, when they were all served by streetcar lines, their density increased to seventy-five persons per acre. In wards 13 through 33, the outlying wards to the eastern end of the city, population density increased sixfold, from two to twelve persons per acre, over the thirty years.[31] These increases created demands for more city services, especially for sewers, street lighting, police, and fire protection. They also created a need for entrepreneurs to undertake subdivision, development, financing, and construction, and for local retail stores—groceries and druggists, for example—to serve the growing population.

This growth, expansion, and spatial differentiation affected the city's political parties and government. As people sorted themselves out residentially, and as areas of the city became increasingly distinctive, it was nearly inevitable that Pittsburgh's decentralized ward system of government would reflect these changes. Geographic separateness likely produced a heightened sense of social difference, leading people within communities to seek to elect one of their own to represent their interests. Political distinctiveness accompanied residential distinctiveness, in other words, and this tendency was probably accelerated by the developmental demands created by each new wave of population succession.

Over the last fifty years of the nineteenth century, councilmen tended to be quite similar to the people in the communities they represented. Working-class wards elected workingmen, or retailers who were widely known in the community. Bakers, druggists, grocers, and saloon keepers, well known because their places of business were centers of contact and communication within the community, often were elected councilmen. Upper-class and upper-middle-class wards, on the other hand, elected bankers, lawyers, and professionals, persons closely linked with the dominant economic and social groups of those communities. Whatever ward he came from, each council-man functioned as an ambassador linking his community into the larger political life of the city, articulating its interests and allying with councilmen from similar areas to advance them.[32]

Since the city councils were based on ward representation, it was no coincidence that many other governmental institutions—schools, police, and fire services, for example—were also organized along ward lines. The result was a political system and a city government which, as Samuel P. Hays has expressed it, "reflected the dominant community focus in human experience and social organization."[33]

For some groups of Pittsburghers, however, that focus was too narrow and limiting, a cause of inefficiency and even an occasion for corruption. They sought to change the city's governmental system, to bring it into line with their interests and vision of what government ought to be and do. The twin economic and political crisis of the mid-1890s gave them their opportunity to act.

The industrial depression that began in the spring of 1893 shattered the complacency and optimism of Pittsburgh's economic leaders. Their concern probably turned to alarm as unemployment grew, requests for relief multiplied, and the jobless became more assertive in calling for action to relieve the distress. Fear of social upheaval probably permeated the boardrooms of Pittsburgh's banks and corporations, as it did similar institutions in other northern cities.[34] But Pittsburgh's citizens, or at least those who voted, voiced their collective preference for the more familiar and reassuring solutions offered by William McKinley and the Republican party.

Pittsburgh had been a Republican city since the 1860s, although usually somewhat less so than the suburban areas in Allegheny County, and Republican support increased sharply when the depression hit. While Pittsburgh's mean Republican percentage in presidential and gubernatorial elections between 1876 and 1890 had been 53 percent, the city returned 67 percent for the Republican candidate for state treasurer in 1893; 76 percent for the Republican gubernatorial candidate in 1894; and 70 percent for McKinley in 1896. Once the excitement of the 1890s waned, Republican voting declined somewhat, but the party's margin over the Democrats still remained larger and more secure than it had been in the 1870s and 1880s.[35]

As Carmen DiCiccio suggests, this newly forged Republican hegemony was largely due to shifts in working-class wards that previously had voted Democratic. Even before the depression, the Democrats received the votes of only a minority of the city's working-class voters. The depression and the sectional candidacy of William Jennings Bryan virtually extinguished even that level of support. As a result, instead of polling 30 to 40 percent of the citywide vote, as they had in the 1870s and 1880s, Democrats were cut to 20 to 25 percent, and they were sometimes reduced to third place when factional bolts split the dominant Republicans.[36]

DiCiccio's paper is the only one that examines voting results, and it does not describe the grass-roots patterns or link them with social characteristics as thoroughly as one might wish. Given the composition of Pittsburgh's work force, it is not likely that the Republicans could have forged so large a majority without solid working-class support. But we still do not know whether this swing to the Republicans cut across all of the ethnic and religious components of the laboring force. And we do not know whether the voting changes were accompanied by shifts in the composition of each party's leadership. Better understanding of what happened to party politics in Pittsburgh during the 1890s must await studies that fill these gaps.

Another change occurred in politics in Pittsburgh, and elsewhere, around the turn of the century. And it, too, was set in motion by the events of the 1890s.

New ways of expressing and organizing interests and bringing them into politics challenged the older ways of doing things. During the late nineteenth century, electoral politics provided the best opportunities for articulating interests and converting them into public policy. And electoral politics was preeminently partisan politics, with parties organizing and administering the entire process. But since parties are organized geographically, as institutions they are only capable of representing the shared concerns of inclusive spatial groups. They are uniquely unsuited to mobilize and represent interests that lack any geographic base of organization. Thus, the political concerns of exclusive membership groups—for example, associations of professionals whose members do not share any common residential area—are more appropriately organized and represented through functionally focused interest groups.[37]

After the turn of the century, exclusive membership groups came to play an increasingly influential role in shaping Pittsburgh's public policy. Two sets of factors converged to produce this shift. First, as a result of the political upheaval of the 1890s, electoral competition between the city's major parties was virtually eliminated. Thus, grass-roots interests and values could not be organized and represented meaningfully through the mechanism of party competition for elected office. True, in some other places these were expressed through factional conflict within the hegemonic Republican party. But even where this occurred, as it did in Wisconsin, for example, this channel of expression proved unsatisfactory because it was more difficult to arouse and organize a public following along factional lines than under party labels. And in Pittsburgh it is not even clear that factionalism reflected this sort of underlying conflict. At least Joshua Chasan's analysis of the factional groupings that battled each other in 1912 failed to detect any revealing social distinctions. Both Taft and Progressive Republicans seemed generally to come from the same high-status occupational and social backgrounds, and those active in the municipal reform movement were about as likely to enlist in Taft's camp as in Theodore Roosevelt's. Only the Socialists recruited leaders from the city's laboring-class groups and among those who had fought against the shift from ward-based to citywide government.[38]

Second, and more significant, was the fact that many of the concerns that drove people to action around the turn of the century simply lacked any geographic base of organization. For example, public health professionals were concerned with the spread of communicable diseases, a problem that cut across geographic subdivisions and called for citywide—indeed regionwide—solutions. They saw the city's ward-based system of government as an obstacle preventing efficient and effective action. So did associations of teachers and other professional and business groups whose interests and perspectives were not bounded by geography.

As Samuel P. Hays has shown in his influential study "The Politics of Reform in Municipal Government," it was these types of people who were responsible for political change in Pittsburgh in the early twentieth century. The city's two major reform organizations, the Civic Club and the Voters' League, recruited their members mainly from the upper class: 65 percent were listed in upper-class directories that contained the names of only 2 percent of the city's families. And these reformers were not the descendants of the city's old patrician and commercial elite. For the most part, they were a new upper class, persons whose wealth came from the city's industrial establishments, especially its iron and steel mills.[39] Just over half (52 percent) were bankers, corporate officials, or their wives.

This group included the presidents of the city's fourteen largest banks and officials of its major industrial concerns—Westinghouse, Pittsburgh Plate Glass, the H. J. Heinz Company, the Pittsburgh Coal Company, U.S. Steel, Jones & Laughlin, as well as smaller steel companies like Crucible, Pittsburgh, and Superior. These were not the city's small businessmen; they were the directors of its most powerful banking and industrial organizations.

The city's small businessmen, especially those engaged in retail activities, typically opposed the municipal reform movement. Wayne Lammie shows that publications issued by the associations of Pittsburgh and Western Pennsylvania druggists favored enactment of pure food and drug laws by the state, principally to eliminate dealers who offered lower prices because they sold adulturated drugs, and they were also concerned with price competition from chain stores. But apart from these areas of direct, tangible interest, they were indifferent to other reform efforts. And the same pattern characterized the reactions by coal dealers, grocers, and other retail merchants. Unlike the larger business organizations represented by the Pittsburgh Chamber of Commerce and Pittsburgh Board of Trade, both of which supported the drive for municipal reform, the city's small-scale merchants were satisfied with a decentralized system of government.[40] After all, as notables within their communities, they frequently were elected to represent their wards in the larger political life of the city.

The reformers who were not associated with banks and corporations (the remaining 48 percent) were professionals—doctors, lawyers, architects, engineers, and ministers. Some of them were upper class, but their involvement in reform activities reflected their professional rather than their class outlooks. They were among the leaders in professional movements aimed at acquiring new knowledge and applying it more widely to public affairs.[41]

What the groups that dominated the reform movement shared in common was an involvement "in the rationalization and systematization of modern life." They wanted a form of government that was consistent with that belief, not a ward-based system that, as they perceived it, represented narrow, particularistic interests. In Pittsburgh, as in other northern cities, the struggle for municipal reform was really a battle between groups with different visions of the city's future. It pitted the social elite, professionals, and officials of large corporations, persons with citywide and cosmopolitan outlooks, against lower- and middle-class groups—small businessmen, white-collar workers, skilled artisans, and unskilled factory workers— which had dominated the city's ward-based government. These contending groups "came from entirely different urban worlds, and the political system fashioned by one was no longer acceptable to the other."[42]

Reformers fought their battle for the city's future on at least two important fronts. First, they attacked political parties and their role in the electoral process. For example, they supported the movement for the Australian ballot, believing that this reform would reduce party influence over voting. But the centerpiece of their antiparty efforts, as Loomis Mayfield shows, was a successful attempt to impose personal registration as a requirement for voting. Enacted in 1906, with support from such Pittsburgh organizations as the Chamber of Commerce, the Municipal League, and the Civic Club, Pennsylvania's personal registration law applied only to its cities, leaving the old nonpersonal system in effect in small towns and rural areas.[43] Second, reformers attacked the city's ward system of government, and in 1911 they succeeded in replacing it with a City Council whose members were elected by the city as a whole. At the same time, they abolished ward-elected school boards and shifted control to a central board whose members were to be appointed by the judges of the court of common pleas.

These changes in the form of city and school government had the effect of centralizing decision making, shifting it upward and divorcing it from public opinion organized at the community level. Groups deployed geographically and dominating particular wards were no longer assured of representation on the City Council or of control over public institutions that operated in their neighborhoods. Citywide election or appointment to the school board changed the social character of representation. Members of the upper class, representatives from the professions and large businesses dominated the new centralized agencies of government, while the city's middle- and lower-class elements were underrepresented.[44]

Reformers did not content themselves with changing the form of government, and two other studies convey a sense of the range of their involvements. Janet Daly shows that the movement for a comprehensive zoning program, finally enacted in 1923, was supported by the city's business and professional reform groups, those with a communitywide perspective, while it was opposed by the Pittsburgh Board of Trade, the Pittsburgh Real Estate Board, and the city's savings and loan institutions, groups whose collective outlook was narrower and whose short-term interests called for maximizing building and real estate transactions. To be sure, some of the groups supporting zoning also did so for particularistic reasons. The Civic Club, for example, saw it as a way to protect the integrity of high-status neighborhoods and to maintain property values. More generally, however, zoning neatly complemented reformers' attempts to realize their larger vision. It was citywide in its coverage and provided for a central decision-making mechanism in its board of appeals. Thus, a zoning ordinance offered centralized control over individual decisions and promised to prevent the seeming chaos that resulted when individuals acted alone without any vision of the larger community.[45]

Pittsburgh's business and professional elite also used voluntary civic associations to inaugurate a flood control movement. Roland Smith's analysis of this movement from its inception in 1908 through 1960 shows how its dynamics changed over time. During its first phase, through about 1936, the Chamber of Commerce, then dominated by Pittsburgh's larger industrial and financial institutions, led the way, establishing a Flood Commission in 1908 to investiage the causes of floods and to develop a plan to control them. This commission, which eventually included public officials and received financing from the city and county governments, operated as a quasi-governmental agency. Because its report, issued in 1912, called for a comprehensive mixture of preventive and protective measures, the commission's work correctly became identified with the multiple-purpose approach to developing inland waterways. That approach and one of the key measures recommended by the commission to implement it, the construction of a series of dams and reservoirs in the Upper Ohio Valley, brought Pittsburgh's flood control proponents into conflict with the Corps of Engineers and its allies in Congress. The corps was authorized only to deal with navigation and consequently saw congressional approval of a comprehensive water development program as a threat to its independence. It rejected the proposal to build dams and reservoirs, arguing instead that only a system of levees and flood walls provided an economically feasible solution to the problem of controlling floods. Pittsburgh's local elite, which became allied at the national level with those advocating rational and systematic water resource management, thus found itself in conflict with groups favoring a decentralized approach. Until the corps's levee system failed to contain the floods that struck the Mississippi Valley in the 1920s, and the depression of the 1930s created a demand for work relief projects, those favoring centralized decision making and bureaucratic rationalization in the management of the country's inland waterways were on the losing side of this struggle.[46]

By June 1936, when Congress inaugurated the country's first nationwide flood control program and approved a comprehensive plan for Pittsburgh, local political conditions had changed considerably.

The interlocking network of upper-class individuals acting through voluntary associations, which had driven the flood control movement earlier, disintegrated in the 1930s. So did the Republican party's popular appeal, and with it the hegemonic control that William Larimer Mellon and the upper class had exercised over city government during the 1920s. Even the Chamber of Commerce, the initiator of the flood control movement, had changed. It lost two-thirds of its members in the early 1930s and was beset by internal bickering, as downtown merchants and small businessmen grew in influence at the expense of the large corporations. As a result, the chamber could no longer be used by the business and professional elite to forge consensus on issues of public policy. Finally, this fragmentation of authority and decentralization of the decision-making process mirrored a "changing of the guard" from one generation of leaders to another. William Larimer Mellon's retirement from business and political affairs and the subsequent rise of Richard King Mellon in the corporate world and David L. Lawrence in politics represented this transfer. It also represented a shattering of the fusion of business and political control that William Larimer Mellon's leadership had embodied.[47]

Bruce Stave has thoroughly chronicled the transition from Republican to Democratic electoral control of the city, showing both its social base and the roles of the LaFollette and Smith presidential campaigns as its forerunners. And he has perceptively delineated its central feature—the transfer of power from a once entrenched Republican machine to a newly emerging Democratic machine. Far from undermining the economic basis for machine politics, New Deal social programs served as a catalyst for building a new political machine. Pittsburgh's Democrats turned the public payroll, and especially the work-relief programs of the Works Progress Adminstration (WPA), into a refuge for party workers. Democratic committeemen won appointment to coveted WPA jobs, such as foremen, supervisers, and timekeepers; and for many of them, work relief was only the first rung on the ladder of public employment.[48]

But in Pittsburgh, and elsewhere, the 1930s represented not so much a turning point as a transition. The Depression shattered public confidence in old ways of doing things, in old political and economic leaders, while Franklin Roosevelt and the New Deal represented the ascendency of new leaders in government and new directions in public policy. But while the old order was dying during the 1930s, new arrangements were still struggling to be born; and these would not become viable until the early 1940s, when the historic relationship between Pittsburgh's government and its voluntary associations changed.

The historic formula, as Roy Lubove has shown, delegated constructive responsibility to voluntary institutions, such as the Civic Club, the Chamber of Commerce, the Voters' League, the Civic Commission, for active intervention to centralize decision making. Government's role was limited to a comparatively negative, regulatory function.[49] It was this relationship that drove the reform movement of the early twentieth century, including the efforts to change the form of city and school government, to enact a zoning ordinance, to improve housing conditions, to control floods, and to reduce smoke pollution.[50] However, the formula never worked as effectively as its originators intended, or as its critics charged, principally because Pittsburgh's business community was not a monolith. There was ongoing conflict between business leaders representing large metropolitan and nationally oriented corporate enterprises and the city's small entrepreneurs and tradesmen. In many areas of public policy, decisive action was impossible, and the result was only stalemate or incremental change. In any case, this historic relationship dissolved under the impact of the Depression and the ensuing fragmentation of sociopolitical authority; when the city's economic and political leaders faced another crisis after World War II, they had to develop new ways to centralize decison making.[51]

The Pittsburgh Renaissance, a massive physical renewal program that emerged from the city's postwar economic crisis, represented this shift in the historic allocation of public and private responsibilities. Under the leadership of Richard King Mellon, a newly emergent group of business and professional leaders organized the Allegheny Conference on Community Development (ACCD) and initiated a program calling for an increase in the use of public power and a dramatic expansion of public enterprise and investment. City governnment was no longer to be confined to a negative, regulatory role, although its enhanced functions were still to be aimed at serving corporate needs, especially to revitalize Pittsburgh's decaying central business district and its regional economy. David Lawrence's decision in 1945 to give the endorsement of his Democratic machine to the ACCD program and to ally with the leaders of the conference restored centralized decision making among the leaders of the city's sociopolitical power structure.[52]

New leaders, new organizations, and a new relationship between public and voluntary institutions underlay Pittsburgh's Renaissance. In the early days of this renewal program, flood control, smoke control, and Point Park were the key projects, and each involved the use of public power or investment to promote private economic ends. Each also involved drawing upon the resources of other levels of government. State legislation was required for smoke control, and state funds were needed for Point Park and for related highway and bridge construction. The civic coalition of leading businessmen and Democratic machine politicians was diverse enough to reach out and recruit the support it needed from all quarters. In pursuing later Renaissance projects—for example, the development of Gateway Center and unified ownership and management of mass transit in Allegheny County—the civic coalition exhibited the same capacity along with sufficient flexibility to adopt whatever administrative expedient was necessary. And in all of these projects and others, the coalition demonstrated that decisive intervention required the fusion of economic and political power, the centralization of decision making.[53]

Two overarching themes run through this fragmentary account of the evolution of politics and government in Pittsburgh. First, there was recurring conflict between old and new ways of doing things, with the old invariably succumbing, although often only after waging a rear-guard resistance. Relationships between economic groups, commercial versus industrial, for example, and even between established and rising generations within the city's elite can be cast in this context. So can the differences between core and developing areas of the city, and between earlier and later immigrant groups. Moreover, in looking at developments from this perspective, one uncovers a built-in dynamic, a persisting sequence of relationships, for what is once new ages and soon becomes the older way of doing things challenged by yet another new approach.

The second overarching theme is one of persisting tension between decentralizing and centralizing impulses. This has involved an interplay between the tendency to disperse and fragment, on the one hand, and the desire to aggregate power and control, on the other. Technology unleashed both of these contending impulses. New products and manufacturing processes pushed in the direction of decentralizing urban life, while changes in communications opened the way to greater coordination and central control. Thus, the tendencies toward centralization and bureaucratization have technological roots as well as administrative and managerial ones.

These two themes, or contexts for analysis, are overlapping rather than mutually exclusive. For example, the application of new technology to urban transportation had clearly decentralizing consequences, by allowing larger and larger numbers of people to increase the distance between where they lived and where they worked. The range of choices always remained limited by income and job demands, but every innovation increased options and broadened the area of settlement open to each economic group.[54] In turn, the patterns of movement unleashed at each stage led to the development of economically distinctive communities, a decentralization of residential life that had significant implications for city politics and government. The city's leaders, some of whom had been involved in applying the innovations in the first place, then had to develop new forms of public organization and management to control the consequences of technologically induced fragmentation. In other words, innovation that affected human relationships in one way often unleashed reverberating effects, leading particular groups to seek change in other patterns of interaction. Analyses of Pittsburgh's political development must remain sensitive to such ostensibly nonpolitical sources of change in its government and politics.

Future analyses of political change in Pittsburgh might well shed light on the following question. First, who got what from city government, and when did they get it? As the city grew in population and physical size, the City Council probably became a forum for resolving contention over the distribution of resources. It may be a fair and reasonable guess that the better-off wards got larger shares of the benefits, and faster, than the working-class wards.[55] But our knowledge of this allocation process is at best fragmentary and limited to a single period. We need to know more about how the very definition of allocable resources changed across time, and whether shifts in the form of government and in party control significantly changed the patterns of allocation.

Second, how did the limited, almost casual government of the eighteenth and early nineteenth centuries become the specialized, bureaucratic institution of the late twentieth century? In other words, what was involved in the process of centralizing and systematizing city government? Why did the city take control and centralize some activities and services fairly early, and others not until much later? Which groups pushed for these changes, which resisted them, and why? What were the consequences of developing a city bureaucracy? Once formed, for example, did city departments themselves become sources of impulses for further change, or did they resist subsequent innovation? And did an increasingly specialized bureaucracy itself become a career alternative for some individuals, producing a corps of professional civil servants whose values, outlooks, and loyalties differed significantly from those in the general populace with comparable levels of education and occupational status?

Finally, how did the grass-roots community life of the city's social groups change across time, and how did these changes affect the ways in which they related themselves to the larger political life of the city? For example, in the 1850s, closely knit and church-centered networks of human relationships were the vital building blocks of an intensely contested ethnocultural politics. Individuals related themselves to the world of politics through their involvement in ethnic and religious associations. But these networks themselves did not remain static over time. As residential life became increasingly decentralized, contact across economic levels within groups eroded, and so did the group's sense of cohesiveness. As some group members moved away and began to think of themselves as different from those who remained behind, did they also redefine their relations to politics and parties? What new groups or networks emerged to serve as intermediaries between those who moved and the broader political life of the city? And what of those who stayed? How did they react politically to this erosion of the old bases of the group's solidarity?

More generally, we need to know a great deal more about how the relationship between community life and political involvement changed over time. Through what processes and organizations did spatially defined groups accommodate to the shift from ward-based to citywide politics? And as political life has become increasingly complex, technically focused, and remote from the experiences of daily life, how have the city's citizens managed to make sense of this "great, booming, buzzing confusion" in the world about them?

To enhance understanding of how the relationships among government, parties, and voters in Pittsburgh changed over the past two centuries, these new lines of inquiry must be integrated with and guided by the results of earlier work. Future studies, for example, should show how the overarching conflicts that underlay the city's development, whether between old and new ways of doing things or between centralizing and decentralizing tensions, influenced its public decision-making process; how efforts to cope with the effects of these contending impulses led groups to involve government, to reallocate resources, and, in the long run, to encourage tendencies toward specialization and bureaucratization in city administration. And especially, future studies should show how groups sought to bring the institutional and physical environments of the city into line with their own perceptions, and how the alterations they wrought then further worked to reshape their perceptual worlds. For it is this focus that ultimately allows us to see how technological, social, and political developments became linked in human consciousness. And it is this focus, at the same time, that compels us to realize that linkages once established do not remain forever frozen and that, since urban development is a continuous process, understanding it requires an equally dynamic and cumulative approach.

NOTES

1. Stephan Thernstrom, *Poverty and Progress: Social Mobility in a Nineteenth-Century City* (Cambridge, Mass.: Harvard University Press, 1964), stimulated a series of similar studies. An obvious and important exception to the generalization is Sam Bass Warner, *Streetcar Suburbs: The Process of Growth in Boston, 1870–1900* (Cambridge, Mass.: Harvard University Press, 1962). Although it was more distinctly urban in conception and design, and published two years prior to Thernstrom's book, Warner's study did not set the trend.

2. An earlier mobility study used Trempealeau County, a rural area in Wisconsin, as its locale; see Merle Curti, *The Making of an American Community: A Case Study of Democracy in a Frontier County* (Stanford, Cal.: Stanford University Press, 1959).

3. For an attempt to do this, see Lawrence E. Hazelrigg, "Occupational Mobility in Nineteenth-Century U.S. Cities," *Social Forces* 53 (Sept. 1974): 21–32.

4. U.S. Department of State, *Compendium of the Enumeration of the Inhabitants and Statistics of the United States as Obtained at the Department of State, from the Returns of the Sixth Census* (Washington: Thomas Allen, 1841), pp. 26–27.

5. Sam Bass Warner, *The Private City: Philadelphia in Three Periods of Its Growth* (Philadelphia: University of Pennsylvania Press, 1968), pp. 11–14, for the low levels of residential separation (or segregation) in the 1770s, when Philadelphia's population was about 23,700.

6. This is a paraphrase of Warner's description of late-eighteenth century Philadelphia; see *Private City*, pp. 10–11.

7. Thomas Kelso, "Pittsburgh's Mayors and City Councils, 1794–1844: Who Governed?" (seminar paper, Department of History, University of Pittsburgh, 1963).

8. The quoted expressions are from Gordon S. Wood, *The Creation of the American Republic, 1776–1787* (Chapel Hill: University of North Carolina Press, 1969), pp. 489–90. Also see Ronald P. Formisano, *The Transformation of Political Culture: Massachusetts Parties, 1790s–1840s* (New York: Oxford University Press, 1983), pp. 128–39.

9. The observation comes from my reanalysis of the raw data reported

in Kelso, "Pittsburgh's Mayors and City Councils," p. 12, for the 1815–1820 and 1821–1828 councils.

10. Kelso, "Pittsburgh's Mayors and Councils," pp. 7–8.

11. See the data for 1810, 1815, and 1819 in Catherine Elizabeth Reiser, *Pittsburgh's Commercial Development, 1800–1850* (Harrisburg: Pennsylvania Historical and Museum Commission, 1951), pp. 19 and 23.

12. John Dankosky, "Pittsburgh City Government, 1816–1850" (seminar paper, Department of History, University of Pittsburgh, 1971).

13. Scott C. Martin, "Fathers Against Sons, Sons Against Fathers: Antimasonry in Pittsburgh" (seminar paper, Department of History, University of Pittsburgh, n.d.). Duane E. Campbell, "Anti-Masonry in Pittsburgh" (seminar paper, Department of History, Carnegie Mellon University, n.d.), shows that Antimasonic voting was not rooted in socioeconomic differences.

14. Ronald P. Formisano, *The Birth of Mass Political Parties: Michigan, 1827–1861* (Princeton, N.J.: Princeton University Press, 1971), pp. 60–67.

15. See the data reported in Michael Fitzgibbon Holt, *Forging a Majority: The Formation of the Republican Party in Pittsburgh, 1848–1860* (New Haven, Conn.: Yale University Press, 1969), pp. 318 and 322.

16. Ibid., data on p. 321. The distances were measured on maps of Pittsburgh published in 1855 by J. H. Colton & Co., and in 1860 by Geo. H. Thurston.

17. Dankosky, "Pittsburgh City Government," data in tables 4, 19, and 20.

18. Holt, *Forging a Majority*, data on pp. 323–24, 361–64, and see his discussion of the city's upper class on pp. 32–35.

19. Samuel P. Hays, "The Development of Pittsburgh as a Social Order," *Western Pennsylvania Historical Magazine* 57 (Oct. 1974): 438, notes this significance of the replacement of the Whigs by the Republicans. And see Holt, *Forging a Majority*, pp. 228–30, for the impact of the railroads.

20. Holt, *Forging a Majority*, pp. 25–28.

21. Ibid., p. 25; and Robert Kaplan, "The Know-Nothings in Pittsburgh" (seminar paper, Department of History, University of Pittsburgh, 1977).

22. Kaplan, "Know-Nothings in Pittsburgh," pp. 6–7.

23. Holt, *Forging a Majority*, pp. 123–74 and 263–303; Miller Myers, "An Analysis of Voting Behavior in Pittsburgh, 1848–1856" (seminar paper, Department of History, University of Pittsburgh, 1963); and see Paul Kleppner, "Lincoln and the Immigrant Vote: A Case of Religious Polarization," *Mid-America* 48 (July 1966): 176–95.

24. Holt, *Forging a Majority*, data on pp. 326 and 354, shows the religious backgrounds of candidates and secondary leaders.

25. For the general formulation, see Paul Kleppner, *The Third Electoral System, 1853–1892: Parties, Voters, and Political Cultures* (Chapel Hill: University of North Carolina Press, 1979), pp. 143–97.

26. Richard Jensen, *Illinois: A History* (New York: W. W. Norton, 1978), pp. 34–60.

27. Each of these four groups comprised over 5 percent of Pittsburgh's

foreign-born population. For the data used in this paragraph, see U.S. Census Office, *Abstract of the Twelfth Census of the United States 1900* (Washington, D.C.: Government Printing Office, 1902), pp. 100–01.

28. In 1907 Pittsburgh annexed Allegheny City, which in 1900 had a population of 129,896, ranking it twenty-seventh in the nation in size.

29. Joel A. Tarr, *Transportation Innovation and Changing Spatial Patterns in Pittsburgh, 1850–1934,* Essays in Public Works History, no. 6 (Chicago: Public Works Historical Society, 1978), pp. 4–24, for the information in this and the following paragraph.

30. The process in Pittsburgh was comparable to that which was described for Boston in Warner, *Streetcar Suburbs.*

31. Tarr, *Transportation Innovation,* pp. 10–11 and the data on p. 41, appendix B.

32. Hays, "Development of Pittsburgh," pp. 443–44. And for a more modern example, see Herbert J. Gans, *The Urban Villagers: Group and Class in the Life of Italian-Americans* (New York: The Free Press, 1962), pp. 3–41 and 142–226.

33. Hays, "Development of Pittsburgh," p. 445.

34. I discuss the impact of the depression on the thinking of elites in *Continuity and Change in Electoral Politics, 1893–1928* (Westport, Conn.: Greenwood Press, 1987), chap. 4.

35. These observations are based on percentages calculated from the raw data reported in *Smull's Legislative Handbook and Manual of the State of Pennsylvania* (Harrisburg, Pa., 1890–1898).

36. Carmen DiCiccio, "The 1890s Political Realignment and Its Impact on Pittsburgh's Political Structure" (seminar paper, Department of History, University of Pittsburgh, 1983). Other observations are based on my own analysis of city-level voting data.

37. This distinction between types of groups follows Scott Greer, *The Emerging City: Myth and Reality* (New York: The Free Press, 1962), pp. 36–39.

38. Joshua Chasan, "The Election of 1912 in Allegheny County: A Comparative Study of Progressive, Standpat, and Socialist Leadership" (seminar paper, Department of History, University of Pittsburgh, 1967).

39. Samuel P. Hays, "The Politics of Reform in Municipal Government in the Progressive Era," *Pacific Northwest Quarterly* 55 (Oct. 1964):160.

40. Wayne Lammie, "Political Attitudes of the Small Pittsburgh Merchants in the Progressive Era" (seminar paper, Department of History, University of Pittsburgh, n.d.).

41. Hays, "Politics of Reform," p. 160.

42. Ibid., pp. 161–62.

43. Loomis Mayfield, "Voter Corruption and Reform in Early Twentieth-Century Pittsburgh: Evidence and Analysis," paper presented at the History Forum, Duquesne University, Pittsburgh, Oct. 1986.

44. Hays, "Politics of Reform," p. 165.

45. Janet R. Daly, "The Political Context of Zoning in Pittsburgh, 1900–

1923" (seminar paper, Department of History, University of Pittsburgh, 1984); Roy Lubove, *Twentieth-Century Pittsburgh: Government, Business and Environmental Change* (New York: John Wiley, 1969), pp. 94–95.

46. Roland M. Smith, "The Politics of Flood Control, 1908–1936," *Pennsylvania History* 42 (Jan. 1975): 5–24.

47. Roland M. Smith, "The Politics of Flood Control, 1936–1960," *Pennsylvania History* 44 (Jan. 1977): 4–7.

48. Bruce M. Stave, *The New Deal and the Last Hurrah: Pittsburgh Machine Politics* (Pittsburgh: University of Pittsburgh Press, 1970), pp. 36–39 and 162–82.

49. Lubove, *Twentieth-Century Pittsburgh*, pp. 20–40.

50. On the latter, see Robert Dale Grinder, "From Insurgency to Efficiency: The Smoke Abatement Campaign in Pittsburgh Before World War I," *Western Pennsylvania Historical Magazine* 61 (July 1978): 187–202.

51. Lubove, *Twentieth Century Pittsburgh*, pp. 23 and 106.

52. Smith, "Flood Control Politics, 1936–1960," pp. 16–17; Lubove, *Twentieth-Century Pittsburgh*, pp. 108–11.

53. Lubove, *Twentieth-Century Pittsburgh*, pp. 120–24.

54. Tarr, *Transportation Innovation*, pp. 24–39, on the impact of the automobile.

55. Hays, "Development of Pittsburgh," pp. 435–36, makes this claim.

Metropolis and Region: A Framework for Enquiry Into Western Pennsylvania

EDWARD K. MULLER

THE NAME PITTSBURGH still evokes the image of the nation's preeminent steelmaking city. Pittsburgh meant steel, coal, metal fabricating, and a host of other heavy industries like aluminum, electrical machinery, and glass. However, Pittsburgh, that is the city of Pittsburgh, was the relatively smallish urban center of a larger industrial region. Although factories lined the rivers and the railroad corridors of the city and its contiguous industrial suburbs, Pittsburgh's commanding position in the country's metal manufacturing sector derived from an extensive network of railroads, coal mines, steel mills, foundries, machine shops, metal fabricators, glass works, and other miscellaneous plants located throughout the hilly topography of the Allegheny Plateau of western Pennsylvania, southeastern Ohio, and northeastern West Virginia. Numerous farms, isolated hollows, and wooded hills belied the region's industrial character in the twentieth century. The residents of small settlements and towns frequently derived their livelihoods from the mills, factories, and mines of the region's major river valleys—the Ohio, Monongahela, Allegheny, Beaver, Youghiogheny, and Kiskiminetas—and their many tributary creeks and runs.

Even though the traditional industries declined severely in the 1980s, shutting down scores of mines and mills, the region has retained a sense of identity and functional integrity. Changing economic activities, competing communities, and new leadership groups have altered the structure of the region in ways not yet understood. As the center of emerging regionwide banking corporations, the hub of USAir's regional commuter service, and the home of major educational, cultural, and sports institutions, Pittsburgh has continued to function as the region's integrating focus. The city may have become more dependent on such region-serving functions than at any time in the twentieth century. The Pirate Caravan, a group of

Pittsburgh Pirate baseball players and broadcasters, tours the region each winter to promote the upcoming summer season in Pittsburgh. The Carnegie Institute, Buhl Science Center, and Station Square retail complex, to name a few examples, promote increased patronage from the region for continued successful and ambitious programming and for justification of state financial support.

Despite these interdependencies and others of long standing, Pittsburgh and the region's communities often function independently of, or at odds with, each other. Mistrust, antagonism, and jealousy between the metropolis and hinterland persist. Pittsburgh's concentration of economic and political power has provided the foundation for numerous examinations of its economic restructuring and the proliferation of programs for coping with change. The reports invariably claim to represent the broader region; but communities around the region feel that such planning efforts are ignoring them, and they look to state government for assistance.

In much the same manner, historians have examined Pittsburgh's past, especially the rise of an urban industrial society, and neglected the industrial transformation of the region's communities and institutions. No general work attempts to describe the development and structure of the region, much less its recent restructuring and possible demise. Yet, city regions provide compelling spatial units for the examination of social change by virtue of the functional relationships that define the region. The purpose of this essay is to outline a historical and geographical framework for understanding Pittsburgh and its region, drawing particularly upon unpublished seminar papers and theses for evidence and insight.

The Western Pennsylvania Region

The term *region* is elusive and confusing. Since the late nineteenth century, scholars have recognized that cities function interdependently with the surrounding rural countryside in a complex web of economic and social relationships.[1] The spatial extent and character of these complementary relationships change over time, but they define a city region such that change in either city or countryside profoundly affects the other. In this essay, *region* refers to the city region, Pittsburgh and its hinterland, rather than to either large areas of the country often called regions such as the South, or to physiographic areas such as the Ohio Valley. The former areas might be termed sections that include several city regions.[2] Despite their appealing spatial logic, physiographic areas typically cut across the man-made functional areas of city regions, thereby confusing the analysis of social change.

It is easier to conceive of a city region than to delimit its spatial extent, especially for past eras. What constellation of relationships captures the full extent of the city region, in which there exists a functional integrity and sense of regional identity? The geographical extent of various city and hinterland relations varies considerably from the short range of daily commuting to the extensive reach of wholesale trade and electronic communication. At the same time, major cities compete for influence and profit at the margins of their regions, so that regional boundaries not only change but also overlap. Finally, the character, or perhaps intensity, of city influence varies within the region, often diminishing with increasing distance from the city center. Clearly, the definition and measurement of a particular city region present formidable problems and vary with the issue and period under examination.

Since there are no studies explicitly concerned with Pittsburgh's city region, some working definitions must be accepted. While portions of southeastern Ohio and northeastern West Virginia clearly functioned as part of Pittsburgh's hinterland, *region* in this essay will be confined to Western Pennsylvania, stopping at the state's boundaries. This convenient definition ignores important, linked industries in nearby cities such as Wheeling and Steubenville. It also overlooks Cleveland and Buffalo interests that compete with Pittsburghers for business in portions of Western Pennsylvania. The eastern boundary of the region within Pennsylvania has been assumed to be the Allegheny Mountains, though research is required to gauge the extent of Philadelphia's influence. The region is called both *Western Pennsylvania* and the *Pittsburgh region*. Although the latter name may more aptly fit the concept of a city region, people residing outside the metropolitan area do not readily accept the city's name because of Pittsburgh's geographical remoteness, resentment at its neglect or exploitation of their interests, or alienation from its urban life-style.[3]

A region can be subdivided into six zones or areas that vary geographically with different historical periods. This generally follows the principles of economic patterning established by Von Thunen, but unfortunately the historical and social correlates of hinterland differentiation are not well understood, particularly beyond the most urban zones.[4] At the center is the densely built-up urban core, the city. An area of urban expansion surrounds the city and in time becomes part of the urban core, whether politically annexed or not. This area of expansion includes both developing industrial and residential suburbs and a rural-urban fringe of mixed land uses, rapidly changing activities, and land speculation. Beyond the rural-urban fringe in the city's shadow lies an area which is rural in appearance and contains both farm and nonfarm activities. The city affects this third area directly through its economic and social influences on land values, agriculture, and demography. The three remaining areas exist beyond the direct impact of the city, but they experience urbanization through functional relationships with the city. These three areas are the rural settlements and associated villages, the small cities of the hinterland, and the border zones of the region that function within the orbit of more than one metropolis.

This discussion of the Pittsburgh city region also draws on the literature of North American urban growth, regional economic development, and industrial transformation. Three concepts of the city's role in the region organize the essay. In N.S.B. Gras's classic statement, a metropolitan city links the region to other regions and the nation. This economic and social connectivity through commerce, communication, and personal networks distinguishes the metropolitan city from the region's other urban centers. Second, the metropolis is the functional focus for the region. As the center of information, capital, and commodity flows *within* the region, the city integrates intraregional activity and provides the greatest diversity of goods and services.[5] Third, the intense concentration of people and activity in the city places developmental pressures on adjacent areas, transforming them eventually into integral parts of the built-up city.

These three concepts—external connectivity, regional integration, and urban development—assume that city residents and institutions are the agents of economic and social change in the region. Scholars have questioned the appropriateness of this formulation, especially the implication that the hinterland's people and institutions play passive roles. Some also have challenged its spatial accuracy for the second half of the twentieth century, when new technologies of transportation and communication diminished traditional spatial hierarchies of flow and movement.[6] It is not hard to imagine nonurban sources of change in a region as complex as Western Pennsylvania (and its tristate extensions). But in the absence of research on capital investment, social and kin networks, migration, and economic links, which might lead to an alternative formulation, the city and its region provide a useful framework for addressing the issues of social and economic change in Western Pennsylvania.

External Connectivity: The Metropolitan Function

The point of land formed by the junction of the Allegheny and Monongahela rivers seemed destined to become the site of a metropolitan settlement. Not only did these two major rivers and their tributaries provide a means for movement in Western Pennsylvania, but they also formed the source of the Ohio River, a primary route to the western frontier. Thus, as long as travel depended on water and crude roads, the site at the source of the Ohio lay at the focus of both inter- and intraregional movement. It commanded strategic military importance during the French, Indian, and intercolonial struggles of the eighteenth century. The two major military roads that crossed the mountains and terminated at the forks of the Ohio enhanced the advantages of the site. Traders found the military settlement a convenient place to assemble furs for their conveyance to eastern markets and to resupply the trappers and Indian agents for subsequent western ventures. Colonial and later national officials concerned mainly with military and Indian matters also operated from this settlement.[7] In this manner, during the decades prior to substantial agricultural settlement, Pittsburgh and its military predecessors functioned primarily as a locus for long distance, interregional contacts between the settled East and the amorphous trans-Appalachian frontier.

The tide of settlers that crossed the Allegheny Mountains into Western Pennsylvania after 1770 and journeyed on to Ohio Valley frontiers in the first half of the nineteenth century magnified opportunities for land speculators and merchants who congregated at the Ohio River's source. At the junction of the major overland routes from the East and the frontier's riverine main street, Pittsburgh merchants endeavored to dominate the frontier trade, manufacturing items for western markets which were too costly to freight over the mountains, and supplying migrants for their journey downriver. Busy wharves, merchant warehouses, boat yards, artisan shops, teamster services, and manufactories blossomed with the growth of the trans-Appalachian frontier. The Pittsburgh business community supported canal schemes, turnpike companies, and river navigation improvements to enhance its commercial position and ward off competitors for the Ohio Valley trade. Nevertheless, Pittsburgh's far-flung trade aspirations could not survive the growth of rival cities down the Ohio River which were closer to each succeeding frontier, the superior Great Lakes interregional route, and eventually the competition of railroads. By the 1850s, local merchants operated increasingly within a more regional (albeit tristate) hinterland.[8]

For most Western Pennsylvania settlements during these early decades, Pittsburgh was a distant market with limited economic potential. It was, to be sure, the seat of interregional communication and, accordingly, the symbol of eastern commercial interests and urban power. Along with the needs of its own residents, Pittsburgh's access to frontier armies, migrants, and Ohio Valley settlements created the region's first major market, but the difficulty of overland travel to the region's navigable rivers restricted commercial opportunities for many settlers. After 1820 steamboats, canals, and road improvements extended Pittsburgh's economic reach in Western Pennsylvania. Multiple points of overland entry into the region from the East and a few alternative river locations encouraged some competition from Washington, Uniontown, and Wheeling, but Pittsburgh's early advantages and activities established a regionally powerful commercial community that protected and multiplied its interests. Postal and banking information, indicating the volume of commercial activity, demonstrates Pittsburgh's unrivaled position in antebellum Western Pennsylvania. According to Ann Marie Dykstra, between 1826 and 1855 no other communities west of the mountains compared with the city in the amount of either postal receipts or discounted bills of incorporated banks.[9]

Catherine Reiser's able study of Pittsburgh's commercial activity sheds some light on relations between city and region from the perspective of Pittsburgh. Scattered reports of commodity flows via steamboats, the Pennsylvania Main Line Canal, and the Monongahela Navigation Improvement Company reflect the marketing of regional forest and agricultural products and the increasing commercialization of rural Western Pennsylvania.[10] Pittsburgh's access to western markets, which generated some demand for agricultural products, stimulated a rural iron industry. Pittsburgh merchants found a vigorous market for "merchant" bar iron and various ironwares in the Ohio Valley frontier. The development of rolling mills, nail factories, and other ironworks in Pittsburgh created an expanding appetite for the pig iron of country furnaces.[11] Unfortunately, little research exists on the extent, intensity, and diversity of the city's economic relationships with Western Pennsylvania communities in the antebellum period.

The broad outlines of economic development in rural Western Pennsylvania are presented in several studies. Descriptions of early frontier life in the late eighteenth century portray a perilous, subsistence-level existence with considerable impoverishment and vulnerability to life-threatening dangers. Robert E. Harper estimates that in the 1790s only one in six farmers produced a commercial surplus.[12] The precarious existence and limited commercial opportunities created a number of landless residents, tenant farmers, and absentee landowners. Wealth was concentrated in the hands of a relatively few rural families. Nevertheless, the accessibility to markets provided by major rivers like the Monongahela and Youghiogheny encouraged the commercialization of adjacent townships in Allegheny, Washington, Westmoreland, and Fayette counties. Harper noted the development in these areas of agricultural processing, cottage manufactures, boatworks, tanneries, iron furnaces and forges, and small towns. Accordingly, there was an increase of people employed in nonfarm occupations as well as farm tenants, farm laborers, and landless inhabitants. Not surprisingly, this commercialization produced an increasingly unequal distribution of landownership and wealth.

After 1820, the growing Ohio Valley markets for agricultural and manufactured products, as well as regional urban market demands, spread commercial opportunities more widely throughout the region. Using a variety of agricultural census measures at the county level, Ann Marie Dykstra charts the slow increase in commercial agriculture between 1820 and 1860, and Linda K. Pritchard's analysis of development after 1840 supports her picture. Counties in southwestern Pennsylvania of long settlement duration and with better access to Pittsburgh were more commercially developed in the 1820s and 1830s than the northwestern counties. These farmers grew grain and tended livestock for marketable products. Commercial agriculture increased along similar lines in the northwest after 1840, while the older farm areas to the south diversified into dairy and vegetable products. The eastern tier of countries along the Allegheny Front remained less commercially developed throughout the period. Overall, however, the agricultural prosperity of the antebellum decades failed to raise regional farm values and per capita crop production to levels comparable to those of the older fertile counties of southeastern Pennsylvania or the contemporarily settled areas of western New York and eastern Ohio.[13]

Dykstra and Pritchard also identify the spread of rural, nonfarm extractive and manufacturing industries. Agricultural processing industries were widely spread throughout Western Pennsylvania from individual milling or distilling sites to rural complexes of several small manufactories. The latter is illustrated by the settlement at West Overton in Westmoreland County, where the Overholt family milled flour and manufactured whiskey, textiles, wooden barrels, and eventually coke. They also built tenant quarters and a store for workers. Iron furnaces, coal mines, and lumbering generated some development in particular portions of the region. The sparsely populated northeast counties along the front ranges of the Allegheny mountains depended almost exclusively on lumbering, using the Allegheny River for access to markets. Coal mines and iron furnaces, producing fuel and blooms for Pittsburgh manufacturers, were located in counties south and east of Pittsburgh and after 1840 in those immediately to the north. Most of these enterprises were small in this period, although a few, such as the iron plantation at Brady's Bend, represented substantial capital investment.[14] Besides these rural extractive industries and the city's considerable industrial sector, there was little manufacturing in the region. Agriculture remained the primary economic activity.[15]

In the early frontier years when settlement was predominantly south and east of Pittsburgh, the city not only dominated interregional communication, but also symbolized the different world of the settled East. Westerners resented eastern policies concerning land acquisition, Indian defense, access to markets, and, of course, excise taxes. When violence associated with the Whiskey Rebellion against the tax on stills erupted in 1794, rural folk initially directed their wrath against the tax collectors; but soon they turned their anger toward the commercial elite of the countryside and Pittsburgh, which they identified with federal authority and eastern economic interests. In her study of the Whiskey Rebellion, Dorothy E. Fennell places most rebellious activities and participants in the river townships that were undergoing commercialization. Here, she argues, rebellion should be interpreted as action by disadvantaged farm tenants and landless nonagricultural workers against the concentrated wealth and privilege of the commercial elite who powered economic development. A mob of several thousand, for example, assembled at Braddock's Field near Pittsburgh intent upon plundering the city, though the assault never took place. Ironically, the rural participants in this movement sometimes used the Pittsburgh *Gazette* for announcements and communication, further demonstrating Pittsburgh's central position in the region.[16] Divisions between the traditional rural society and the more cosmopolitan world of commercial, often urban, groups persisted into the antebellum decades. In his study of the Antimasonry movement in Somerset County during the 1820s and 1830s, for example, John W. Brant found that the Antimasons' strength lay among farmers, especially Germans, in the most rural townships, in contrast to the Democratic strongholds where the county's more powerful merchants, manufacturers, and professionals resided.[17]

While we know that Pittsburgh's access to interregional markets spurred economic development in the region, the city's importance for social change is largely inferred. Pittsburgh experienced considerable social change as the recipient of a large foreign immigration after 1820, the location of substantial industrial growth, and the imitator of eastern urban municipal and social institutions.[18] However, most of the region's communities had little regular contact with Pittsburgh. Before the railroad arrived in the 1850s, the rivers and roads supported only sporadic movement of goods and people over the long distances that separated the city from hinterland communities. The hinterland remained largely a rural world. Economic activity and innovations, it is generally theorized, filtered between city and region through the urban hierarchy.[19] The participants in economic interactions—transporters, businessmen, and professionals—presumably became potential agents of social change in the region. Pittsburgh's contacts and conflicts with regional institutions and peoples, especially the elite, have not been examined, as Edward J. Davies and Burton W. Folsom have done for eastern Pennsylvania.[20] Accordingly, we do not know, nor can we gauge, the extent or character of Pittsburgh's social impact.

Social change in the hinterland did, however, accompany the opportunities and impacts of economic development. The processes of social and economic differentiation that Harper reported for the frontier years were extended more widely with the increase in commercial agriculture during the nineteenth century. In a study of Peters Township in Washington County between 1800 and 1850, J. E. Davidson captures the impact of commercialization. Better roads, including a locally sponsored turnpike built sometime after 1840, connected the township to Pittsburgh's markets. Steam-powered mills began to replace the ubiquitous small stills and grist mills that had processed local grain. With flour now marketable, land values presumably rose, and the number of landowners declined. Land and wealth became more concentrated among the township's few wealthier residents, while the number of tenant farmers and farm laborers increased. Nonfarm occupations showed some increase with the rising demand for goods and services. Nevertheless, Davidson argues that economic opportunities among Peters Township residents generally declined, because little fertile farm land was available at this late stage of settlement. Economic gains were registered largely among extant landowning families.[21]

In contrast, Cross Creek Township, also in Washington County, remained isolated from Pittsburgh, Wheeling, and Washington (Pennsylvania) markets as late as 1860. With the aid of the agricultural census and a surviving farmer's dairy, Edward H. Hahn describes Cross Creek as a settled rural community with a self-sufficient economy reminiscent of earlier eras. Farms were small and production diversified. Trips to local towns were infrequent, and business was still conducted with cash and barter. Although quantitative comparisons of these studies are difficult, wealth among Cross Creek property holders apparently was not as unequally distributed or concentrated as it was in more commercialized rural communities.[22]

Narrowing economic opportunities in Western Pennsylvania's agricultural communities despite their increased commercial character probably explains the greatest social change that occurred, the loss of young adults and families from migration. In an examination of Pennsylvania's rural social structure in 1860s, including the western portion of the state, Reginald P. Baker finds the rural farm population to be strikingly deficient in men and women between twenty and forty-four years old and correspondingly burdened by high dependency rates of young children and the elderly. It was a largely native-born, indeed Pennsylvanian, population with a high fertility rate. Confirming Davidson's picture of Peters Township, Baker reported some occupational diversity in the heavily agricultural areas and a markedly unequal distribution of wealth. The young adults who stayed in the communities generally worked as farm laborers or at other unskilled tasks, obtaining farm property and some economic mobility by inheritance in late middle age. There were even fewer opportunities for young women besides working as servants in other rural households. Many women left for towns and cities, which in Baker's analysis developed an excess of females in their populations. Migration, therefore, balanced the diminishing ability of the region's agricultural economy to provide for all the children of farm families.[23]

Pittsburgh's allure as the fastest growing city within the region's transportation network must have been great, but again we are hamstrung by little research. We do know, for example, that a young Thomas Mellon left the family farm in Westmoreland County during the 1820s for Pittsburgh in order to obtain an education and become a lawyer. Of course, he also succeeded in establishing the city's foremost banking house, that in turn greatly influenced lives throughout the region.[24] While many who left the farm moved out of the region altogether, others sought nonagricultural jobs in the new mines and small manufacturing operations of the Western Pennsylvania countryside, often near Pittsburgh. Baker reports that in comparison to Pennsylvania's agricultural communities the state's rural nonfarm population displayed more foreign immigrants, greater occupational diversity, more young adults, and an even more unequal distribution of wealth. These social characteristics, Baker argues, reflected the nonfarm population's increasing integration with the emerging urban industrial economy.

As the main center of external contact for the region, Pittsburgh was both a focus and generator of economic development and atten-

dant processes of social change. This development coursed unevenly throughout the region, affecting differently small cities and
towns, rural nonfarm areas, and commercialized agricultural communities, while leaving some areas largely untouched. The work of
Harper, Fennell, Davidson, Hahn, and Baker provide important insights, but the processes and patterns of development and change
remain unexamined.

Regional Integration: Metropolitan Dominance

Pittsburgh continued to function as the main point of interregional contact long past the mid–nineteenth century. However,
after the 1850s, when the railroads diminished the horizons of Pittsburgh's commercial aspirations, the city's growing dependence on
the region's natural resources and its increasing investment in regional extractive and manufacturing industries emphasized its role
as the center of the industrializing communities of Western Pennsylvania. Social scientists have traditionally viewed a city's growing
influence in its region in terms of metropolitan dominance.[25] The
improved physical accessibility created by railroads and later automobiles, the centralizing forces of industrial capitalism, and the
spatial extension of urban social and cultural institutions generally
expanded the sphere of metropolitan influence to large portions of a
region in a manner quite unlike the interactions of the earlier era.
Once again, the relationships between Pittsburgh and its region are
only sketchily researched because Pittsburgh and a few of the proximate industrial towns have attracted most of the attention of scholars interested in industrialization.

Recent interpretations of urban industrial growth emphasize hinterland development and demand for the initial expansion of metropolitan manufacturing.[26] In this view, a city's merchant community,
with support from rural groups, pressed for improved transportation
into the hinterland. The subsequent stimulus to commercial agriculture not only spurred agricultural processing industries but also
raised purchasing power in the hinterland. Merchants met the growing rural demand, in part, by investing in the local manufacture of
goods or in regional sites where water power was available. Moreover,
a city's merchants and manufacturers looked to the region for raw
materials. In this manner, vigorous manufacturing sectors developed
in metropolitan economies with close relationships to hinterland
markets, resources, and production sites. The primacy of the city-
region context for this development supported an almost ubiquitous
mix of manufacturing products among major cities, except where

local factors of natural resources or entrepreneurship encouraged some early industrial specialization.

As transportation improvements, especially the spread of railroads after 1850, encouraged freight hauling over long distances, local manufactures sought interregional markets for the expansion of production. By the late nineteenth century, a national market for manufactures had evolved and induced increasing specialization. The industrial profiles of metropolitan centers no longer resembled each other as they had earlier in the century.[27] Cities became synonymous with specific industries.

In contrast to this regional demand interpretation, Pittsburgh's industrial expansion before the Civil War may well have depended primarily on interregional markets. From their inception at the beginning of the century, the city's boat building, glassware, and iron manufacturers relied on western markets beyond Pennsylvania. Rural iron furnaces and lumber mills did provide raw materials, and the manufacturers did satisfy regional demand, but it is doubtful that Pittsburgh's manufacturing base would have reached its impressive pre–Civil War dimensions without the Ohio River's link to western markets. The demand for armaments and munitions during the Civil War, the railroad's appetite for bridges, rails, and cars, and generally the nation's vast industrial expansion provided the opportunity for Pittsburgh to shift from limited production for the frontier to a diverse array of integrated iron and steel mills, foundries, fabricators, and machine shops producing for national markets and industries. Similarly, the development of the city's important electrical, aluminum, chemical, plate glass, and other manufactures depended on national demand in the early twentieth century.[28]

While the city's hinterland may not have been the primary market for Pittsburgh's industrial growth, Western Pennsylvania clearly became fully integrated into the city's industrial expansion. The shift from scattered rural iron furnaces to centralized, integrated Bessemer works along the rivers stimulated the growth of bituminous coal mines and coking ovens, initially in the southwestern counties and eventually in areas east and north of the city. More than nine hundred mines operated in Western Pennsylvania in 1903; over thirty thousand beehive coke ovens sent dirty smoke and noxious gases into the air of a narrow rectangular area between Connellsville and Greensburg.[29] A booming oil industry in northwestern Pennsylvania sent barrels of crude oil down the Allegheny River to Pittsburgh for refining and marketing. The exploitation of natural gas in the 1880s added to the region's superior energy resources that not only supported manufacturing, but also contributed to the amassing of enormous pools of investment capital. Small cities such as Connellsville (coal and coke) and Warren (oil) experienced rapid growth as a result of these industries. Railroads fanned out in all directions to tap the resources, and soon the region was honeycombed with the lines of major railroads and dozens of smaller ones. Entrepreneurs throughout the region took advantage of the excellent transportation and natural resources to develop manufactures, while the emerging corporations of Pittsburgh, Philadelphia, New York, and Cleveland established branch plants and purchased subsidiaries. By 1941, half of the region's four hundred thousand manufacturing employees worked in Pittsburgh and Allegheny County. The three adjacent counties of the city's metropolitan area contained 23 percent of the region's employment, and the remaining 27 percent was spread throughout Western Pennsylvania.[30]

With its growing concentration of investment capital and corporate ownership, Pittsburgh exercised enormous economic power throughout the region. Pittsburgh speculators built handsome fortunes with investments in the region's energy resources, and many plowed their profits into manufacturing ventures. The volume of financial transactions in coal, timber, oil, gas, transportation, and manufacturing led to the development of a local stock exchange and proximate concentration of financial institutions known as Pittsburgh's Wall Street. In 1903, for example, Pittsburgh firms, and three in particular, owned 26 percent of the region's bituminous coal mines, principally those located in the southwestern coking coal district. Judge Thomas Mellon's investment in young Henry Clay Frick's Westmoreland County coke enterprises in the 1870s illustrates the metropolitan relationship and concentration process. With sustained Mellon backing, Frick expanded his operations throughout the coking district until he had a dominant position. He then moved to Pittsburgh where wider financial horizons could satisfy his ambitions. He eventually merged his coal interests with Andrew Carnegie's steel firm to form the powerful Carnegie Steel Company, and he participated in many other capital ventures with his friend Andrew W. Mellon. The financial activities of the Mellon brothers, Frick, Carnegie, Benedum, and other wealthy Pittsburghers indicate that the city was a center of investment or venture capital to which inventors and entrepreneurs from the region, indeed from the nation, were attracted. The early histories of area corporations, especially Carnegie Steel, Gulf Oil, Koppers, Carborundum, Alcoa, and Pittsburgh Plate Glass, to mention some of the largest Mellon enterprises, show the dependence on local venture capital. These corporations, like many others, invested heavily in new plants around the region, centralizing tremendous decision-making power in Pittsburgh boardrooms over the lives of regional residents and communities. Information for 1941 indicates that Pittsburgh banks had more than 50 percent of the region's banking capital and that city-owned manufacturing and mining firms employed thousands of regional workers, especially in the industrialized counties adjacent to Allegheny County.[31]

Pittsburgh did not exploit these opportunities without competition from other metropolitan-based corporations. The data for 1941 point to the strong presence of Philadelphia and New York corporations in Western Pennsylvania. While these manufacturing corporations apparently invested throughout the region, the coal companies of Pittsburgh and its metropolitan competitors operated in discrete subregional areas. The prevalence and importance of such investment patterns are unknown.[32] We need a comprehensive study of both the industrial development of Western Pennsylvania and the capital, corporate, and elite connections between Pittsburgh and the hinterland counties which accompanied the process of regional economic integration.

Industrial development and economic integration produced a complex pattern of social change in the region. Despite the absence of a comprehensive overview of development, it is possible to identify a range of settings, from agricultural communities and coal patches to small industrial towns, within which social change took place. A few remote pockets in this northern extension of Appalachia escaped urban industrial influences, and time-worn, self-sufficient ways endured. More generally, however, regional farmers turned to dairying, truck gardening, poultry, and orchard crops.[33] At the same time, the search for energy resources and raw materials led to the establishment of mines, quarries, gas and oil wells, and small manufacturing plants throughout the agricultural countryside. Hahn's study of Cross Creek Township between 1860 and 1880 presents the familiar picture of increasing social stratification and occupational diversification, which resulted from better access to Pittsburgh markets. As the number of farms declined and became larger, the fewer farm owners gained a larger share of the township's wealth. More men in the township found employment as farm laborers and unskilled workers at nearby mines, town businesses, and railroads. However, it is likely that the new nonfarm opportunities failed to stem the outflow of young adults, especially women, which Baker found for Pennsylvania at mid-century. Although Hahn does not address the migration issue, he reports that Cross Creek remained almost entirely native born and family centered.[34]

Despite commercial growth and diversification, the traditional rhythms, values, and daily patterns of rural life were probably not greatly affected until electricity and automobiles altered life-styles and truly broke down rural isolation. Indeed, in her study of the socialization of children in Washington and Greene counties between 1840 and 1900, Anna-Mary Caffee portrays the persistence of a traditional rural society through the concerted efforts of schools, churches, community institutions, and the family. Younger adults, who may have felt confined by the traditional outlook of this world, contributed to its persistence by migrating to urban areas, leaving behind those content with the old ways.[35] Studies of migration patterns and their social implications for Western Pennsylvanian communities are necessary to resolve several issues concerning the impact of regional industrialization on the countryside.

The new settlements that formed around rural mines and manufacturing plants stood in marked contrast to the surrounding farm communities. Life in the coal patch towns was harsh and cut off from the outside world. Families lived under the arbitrary authority of company officials. Wages were low, employment insecure, and working conditions unhealthy and dangerous. Homes were minimally furnished, and patches were without adequate utility services and political autonomy. Company stores provided retail goods and eliminated shopkeeping opportunities for enterprising families. Foreign immigrants with diverse national origins and sometimes black workers mixed uneasily in this tense environment; violence erupted frequently during leisure hours or in periods of labor unrest.[36]

Isolated industrial towns, sometimes company-owned, displayed social environments similar to those of the mining patches. Ray Burkett argues that the model industrial town of Vandergrift was planned in response to the perceived social problems of nearby Apollo, Armstrong County. Blaming a long, debilitating strike that began in 1893 on foreign workers and the union environment of Apollo, millowner George McMurtry decided to develop a separate plant and adjacent town that would foster a more stable, loyal, and home-grown work force. Drawing on environmental theory fashionable in that era, McMurtry planned to control his work force and thwart union organization by creating a community of homeowners with an efficient infrastructure, aesthetically pleasing design, and alcohol-free sales district. The renowned landscape architectural firm of Olmsted, Olmsted, and Eliot planned Vandergrift for McMurtry; and like its more famous predecessor in Pullman, Illinois, the Pennsylvania town received laudatory national publicity. However, corporate paternalism and social engineering failed to al-

ter industrial realities. While Vandergrift remained loyal to the company and resisted union organization until 1936, communities with
speculative housing, segregated immigrant populations, and traditional vice attractions arose adjacent to McMurtry's experiment. As
a whole, the string of towns from Apollo to Vandergrift, including
the unplanned siblings, became differentiated by class and ethnicity
in a manner resembling the larger industrial cities of the region.[37]

In a description of the 1893 strike, Burkett indicates that Apollo
workers resented the company's use of strikebreakers from the surrounding rural countryside.[38] Most accounts of strikebreaking in
Western Pennsylvania portray unskilled foreign or black workers
from Pittsburgh or areas outside the region as unwittingly recruited
for the disagreeable task and sometimes imprisoned at the plant.
Burkett's reference raises the intriguing, though unexplored, issues
of what were the interrelationships between the agricultural populations and the new mining and manufacturing communities that
shared the countryside.

Rapidly growing, small cities of ten thousand or more residents,
some near Pittsburgh and others spread about the region, present a
third setting of industrialization. Specific industries, often the large
plants of a particular corporation, distinguished these cities, for example, oil refining and services in Warren, steel in McKeesport and
other Monongahela and Ohio river towns, glass in Jeannette, and
electrical equipment in East Pittsburgh. Here, as in the smaller
industrial communities, unskilled foreign workers soon dominated
the labor force and transformed the towns. Mary C. Huey finds that
within a few years of the National Tube Works' opening in 1872,
Irish, British, and German workers comprised 44 percent of the
male work force, and the proportion of foreign-born residents in
McKeesport doubled to nearly one-third of the population. This
process was repeated in industrial cities throughout the river valleys, with Southern and Eastern European immigrants becoming
prevalent after 1890. Although large manufacturing plants often
dominated employment, these cities offered a variety of job opportunities in small businesses, nonprofit institutions, and local government, which were unavailable in smaller towns. American-born
residents and members of older immigrant groups disproportionately filled the entrepreneurial, professional, and skilled positions.
Unskilled foreign-born steelworkers toiled under trying conditions,
constantly vulnerable to the catastrophes of disease, injury, or
work stoppages. Coarse patterns of residential segregation reflected
the socioeconomic stratification. From 1892 to the Great Depression, when the large steel corporations enjoyed nearly uninhibited

power, the unskilled workers of these small steel cities obtained little improvement in their precarious existence and exercised little power in local government.[39]

In the oil service city of Warren, Michael P. Weber reports that workers frequently achieved some economic mobility if they remained in town for several years. Persistence resulted in tangible rewards. Despite fear of foreign radicalism and industrial social change, the leaders of Warren worked to integrate industrial workers, including immigrants, into the community through educational and governmental programs. Moreover, homeownership was more accessible in this small city than it was in many other industrial towns. As a result, Weber concludes, industrial conflict was diminished and a sense of community maintained.[40]

Warren's newer immigrant workers failed, not surprisingly, to gain access to political leadership prior to World War I. However, if the small industrial city of McKees Rocks provides a model, Warren's workers would be expected to have achieved meaningful participation in the following decades as their position in the community improved. Joseph P. Fadgen examines political change between 1900 and 1940 in McKees Rocks, a small industrial city of eighteen thousand in 1930, three miles down the Ohio River from Pittsburgh. Several large iron and steel companies entered McKees Rocks after 1880, employing more than eleven thousand workers in the early twentieth century. Familiar patterns emerged: the ethnic division of labor, social stratification, and residential differentiation. As in Warren, after several years of rapid industrial expansion the older political elite of local merchants and businessmen gave way to the progressive politics of new middle-class professionals, white-collar workers, and independent businessmen. Factions of the Republican establishment vied for power during the years following World War I, while factory workers organized politically under the Democratic party. They increasingly participated in the city's politics in the 1920s, making inroads in local municipal offices. This experience built the foundation for claiming power during the Depression.[41] Although the studies of McKeesport, McKees Rocks, and Warren cannot support direct comparisons or generalizations, they remind us that industrialization of the region's small cities occurred under different conditions, such as the domination of a large corporation or the diverse economy of several employers, and created distinctly different social milieus and opportunities for workers and their families.[42] Moreover, studies of other small cities such as Warren which are located far from Pittsburgh should balance the current emphasis on nearby communities of the steel valleys.

While Pittsburgh industrialists and financiers created new industries in a variety of settings throughout Western Pennsylvania, the cosmopolitan ambitions and life-styles of the city's elite affected the region in many other ways. Living and working in a city of rising national stature, Pittsburghers had access to the latest metropolitan trends in material consumption, entertainment, recreation, education, and culture. Consciously imitating other cities, Pittsburgh entrepreneurs, boosters, and philanthropists established the full range of urban delights from department stores and professional baseball to the more refined culture of the Carnegie Music Hall. Western Pennsylvanians learned about Pittsburgh's attractions through business contacts, personal networks, and the city's newspapers. Frequent railroad service made occasional travel to the big city possible. By the early twentieth century passenger service to Pittsburgh extended throughout the region, and commuter trains provided daily communication with towns within a forty-mile radius of the city. Although railroad commuting peaked in the 1920s and declined thereafter, buses and automobiles maintained and extended the city's reach into the hinterland. Railroads not only offered day, overnight, weekend, and longer visits to Pittsburgh, but they also brought the city to the countryside through the stories and advertisements of daily newspapers. We can only speculate as to what impact these interactions had: to what extent students from the region were attracted to the city's colleges and universities, regional towns imitated Pittsburgh businesses and services, and everyday behavior and custom were altered.[43] It would be important to know, in the manner of Edward Davies's study of the Wilkes Barre region, how the elite of the Pittsburgh region interacted through business, social clubs, and marriages, and how these relationships changed during the course of economic integration. In view of the differences between the Pittsburgh and Wilkes Barre regions, one must ask whether evolving social interactions served to weaken or strengthen Western Pennsylvania communities. Many wealthy Pittsburghers, or later their children, left the city for more fashionable national arenas, but did the regional elite gravitate to Pittsburgh in the manner that their counterparts in eastern Pennsylvania moved to Philadelphia and New York?[44]

The counterpart of Pittsburgh's magnetism for the region was the hinterland's attraction for city residents. A century before today's year-round recreational use of Western Pennsylvania, wealthy Pittsburghers established summer homes, hunting and fishing clubs, and children's camps in the region's forests, mountains, river valleys, and lake shores. The Western Pennsylvania Conservancy's efforts to pre serve natural habitats, beginning in the 1940s, probably had distant roots in such late-nineteenth-century activities as the elaborate, though exclusive summer club and community of wealthy Pittsburghers along South Fork Creek in the mountains above Johnstown. Club members traveled by railroad from Pittsburgh, built summer homes, repaired an old canal dam to create a lake, and employed local residents. Although the rupture of the dam in 1889 contributed to the devasting Johnstown flood and the dissolution of the summer community, strong attachments for the splendor of Penn's woods were imprinted in the hearts of Pittsburgh's first families. Other recreational sites established more permanent, though often uneasy links between city and region. The Mellon families' selection of the Ligonier Valley for both leisure pursuits and permanent residences set into motion a series of economic, social, and political changes which intimately bound together the city's wealthy establishment and that area's rural folk.[45] By the 1920s, the rapidly expanding state forest and park system, containing dozens of campgrounds, appealed to a broad spectrum of urban residents, who were looking for recreation in a decade of increasing automotive mobility. The impact of these activities on the region's rural communities and landscapes has yet to receive adequate scholarly attention.

Urban Development: Metropolitan Expansion

During periods of sustained urban growth, cities inevitably expanded outwardly onto adjacent lands that had not yet been intensively developed. At such times active decentralization from the congested city core exerted tremendous developmental pressure on surrounding communities. But even when development was most concentrated on the city's center, some people or businesses found it desirable or economically advantageous to locate on the urban periphery.

Measured in time and cost of movement, access to the diverse job opportunities, markets, and personal and business services of the central city broadly established the boundaries of potential expansion. Each new form of transportation and elaboration of the transport network extended these boundaries farther into the surrounding countryside. Consequently, expectation of either development within the limits of feasible access or extension of these limits caused an assessment of the periphery's potential use and value and initiated the process that converted rural land into urban uses. While the actual timing and character of development depended upon the building cycle, antecedent land uses and ownership patterns, location in the metropolitan area, and the developer's needs and predilections, the developmental process commonly involved speculative real estate ownership, subdivision and other planning preparations, construction, eventual intensification of use, and possibly annexation to the city.[46]

At any given time, the different phases of the developmental process can be traced in the patterns of land values, population densities, and land uses which extended out from the city. Older residential and industrial suburbs near the city frequently experienced further intensive development, while newer areas simultaneously underwent residential subdivision or initial industrial development. An urban fringe area formed at the edge of this suburbanization, where the limits of daily commuting or feasible economic attachment to the city's business markets and services were reached. In the rural-urban fringe, land held as a speculative investment mixed with other diverse uses, including wealthy estates, small farms, clusters of poor residents, scattered small industries forced out of the city by high land costs (brickyards) or by the production's noxious qualities (slaughterhouses), and other rejected city functions like prisons, cemeteries, and dumps. Beyond this fringe, but still within the city's economic shadow, lay rural agricultural and nonfarm communities that also directly felt urban market pressures and opportunities.[47] The histories of many Western Pennsylvanian communities reflect these various developmental phases and patterns.

During the first half of the nineteenth century, development was restricted to the small area accessible by foot or wagon from the city's commercial center at the Monongahela Wharf. Steamboat, canal, and turnpike improvements after 1820 primarily enhanced access to the hinterland, while circulation within the urban areas remained pedestrian. Thus, growth before the Civil War placed substantial pressure on the small triangular point of land between the rivers, forcing some businesses to seek flat land along the interregional routes beyond the crowded center. New settlements across the rivers created autonomous communities separated by the water barriers to easy daily access. Despite close economic relationships, these vigorous little towns on the south shore of the Monongahela River and north shore of the Allegheny River maintained their independence from Pittsburgh until after mid-century when numerous bridges eased communications and coercive annexations incorporated many of them into the larger urban fabric.

Development on the urban periphery occurred east of the city along both rivers and on the adjacent hill. The floodplain of the Allegheny's south shore, with the Pennsylvania Main Line Canal

Basin at one end and the Allegheny Arsenal at the other, attracted many industrial enterprises. Business and residential blocks spread northeast toward the Arsenal in Lawrenceville, and an omnibus route provided passenger service in the 1840s. While the communities across the rivers grew as autonomous urban centers, the contiguous eastern townships experienced the developmental patterns more typical of urban expansion, culminating in their annexation to the city by mid-century. Scattered settlements, businesses, farms, and the expansive Allegheny Cemetery occupied the complex, hilly topography of the eastern townships between the rivers. In his study of Minersville, Bradley W. Hall traces the conversion of land from farms and country estates to urban residences. Located two miles east of the city, Minersville was the site of prosperous farms until one landowner, James Herron, opened a coal mine in the area and built tenements for his workers. Some wealthy Pittsburghers, including Herron, moved their year-round residences to these country lands in the 1830s and 1840s, co-existing with the nearby boisterous industrial village. By using the Will Books of the Probate Court and Orphans Court Proceedings for Allegheny County, Hall finds that the second- and third-generation inheritors of the estates, eschewing older paternalistic visions of the elite gentry, either subdivided them into urban lots for speculative profits or sold them to investors in order to cover estate debts. With the increasing density of the original city and new omnibus lines reaching Minersville, subdivision attracted middle-class buyers.[48] Thus, the passing of the older generation combined with the developmental pressures of burgeoning Pittsburgh to transform adjacent peripheral areas such as Minersville into city neighborhoods.

The belated opening of railroads in Pittsburgh in the 1850s dramatically extended the area of potential urban development. William K. Schusler reports that a dozen railroads provided commuter service to communities within a thirty to forty mile radius of the city by 1900. Service peaked in the 1920s with more than three hundred fifty commuter trains entering and departing each day. Over thirty thousand people traveled daily to downtown Pittsburgh by train.[49] At the same time, the growth of the iron and steel industry pushed the search for large, suitable manufacturing sites outward along the major river and smaller tributary valleys. New factories brought rail service to these areas.

Development occurred at these industrial sites and around the stations of the railroad, which were strung out along the rail lines every four or five miles. During the first thirty years of rail service, communities up to a dozen miles from Pittsburgh attracted growth—

places such as Sewickley and Wilkinsburg. The adoption of street
railways, or horsecars, during the second half of the century spurred
the more intensive filling in of areas contiguous to the city and between the railroad lines. The construction of new commuter routes
and continued industrial growth after 1890 generated another round
of urban development at rail stops and river sites in the outer townships of Allegheny County and in neighboring counties, while the
rapid adoption of electrified street railways, or trolleys, spread intensive residential construction into the older railroad suburbs. Highway
construction in the mid-twentieth century initiated a third phase of
development. One can envision waves of spatial expansion, followed
by intensification.[50]

Together, industrial growth, commuter railroad service, and
street railways transformed countless small towns and rural townships into pieces of the sprawling metropolitan mosaic. The social
and economic character of these new urban communities ranged
from industrial mill towns to elite residential enclaves. Farms and
country estates, often owned by wealthy Pittsburghers, preceded the
opening of the railroads that initiated urban development. Farms, for
example, covered Homestead until 1871 when a local bank purchased them and began a suburban residential development in antici-
pation of the Pittsburgh, Virginia, and Charleston Railroad's arrival.[51] Farther up the river, a few wealthy owners of farm tracts and
a coal patch town, mostly inhabited by Germans, controlled the site
that became the small steel city of Duquesne.[52] Similarly, Hazelwood on the Monongahela, Shadyside in an eastern township near
Pittsburgh, and Sewickley twelve miles down the Ohio River entered the 1850s as rural settlements. In each place, prosperous farmers and prominent Pittsburghers owned land, usually residing on it
at the time improved access to Pittsburgh was obtained in the 1850s.
Thus, each area had a prestigious appeal, when landowners subdivided their tracts and sold off the new lots. However, the resulting
urban communities contrasted markedly.

Joan Miller identifies the beginning of Hazelwood's urban development with the building of Braddock Field Plank Road around mid-
century. Then in 1861 the Pittsburgh and Connellsville Railroad
established a short commute into the city. Wealthy Pittsburgh families built permanent homes, weekend estates, and even a private
club for a summer retreat. However, Hazelwood's river frontage,
proximity to Pittsburgh, and location across the river from South
Side iron mills also attracted the attention of industrial investors.
The building of coke ovens, blast furnaces, rolling mills, Baltimore
and Ohio Railroad shops, and several other smaller factories during

the ensuing twenty years altered the upper-class direction of early development. By 1880 businessmen, professionals, and other white-collar workers constituted only one-quarter of Hazelwood's employed residents, while unskilled, often foreign, workers increased rapidly in number. Horsecars and later trolleys also encouraged middle-class families to move to Hazelwood. Housing developments for industrial and white-collar families engulfed the former estates, and in the 1880s wealthy residents left the community. With fourteen thousand residents in 1900 and twenty-eight thousand in 1920, Hazelwood no longer resembled the weekend retreat that it had been at mid-century. As Pittsburgh's Twenty-third Ward, its transformation to a big city industrial neighborhood was complete.[53]

Some of Hazelwood's wealthy families may have sought refuge in Shadyside, a few miles to the north. The eastern townships between the rivers had accommodated the estates and farms of wealthy Pittsburghers for decades. The relatively flat lands of Shadyside were no exception; but when the Pennsylvania Railroad laid tracks along its northern edge in 1852, some wealthy local residents and prominent city investors (notably Judge Thomas Mellon) began the speculative subdivision of their large tracts. Robert J. Jucha's study captures the social distinctions that resulted from the different practices and goals of the area's developers. In the western end of Shadyside the Aikens sold off only large lots and helped establish a Presbyterian church, thereby ensuring that upper-middle-class buyers would be attracted. Prominent private schools and social organizations completed the affluent character of the community. Judge Mellon, Alfred Harrison, and a few others, in contrast, planned modest rectangular subdivisions in the area's eastern section, carving up their properties into less expensive, smaller, city-sized lots. Detached single-family houses on these narrow lots created a dense middle-class suburb. The desire to capitalize on suburban development persisted past the subdivision phase and engulfed middle-income residents. One-third of Shadyside's homeowners built more than one house. Although the absence of heavy industry spared Shadyside the immense pressure of accommodating large numbers of working-class families, trolley transportation and a prestigious reputation intensified developmental forces after 1890 in both the affluent and middling sections. In the wealthy western end, dead-end streets penetrated the larger properties in order to create several smaller, but still expensive lots. To the east, terraces of row houses and duplexes and twenty-six apartment buildings increased residential densities. Moreover, nearly half of Shadyside's residents were now renters. By the time the automobile offered alternative means of commuting, Shadyside had become a fashionable urban neighborhood; and suburban development had moved several miles farther eastward.[54]

The construction of the Ohio and Pennsylvania Railroad along the north shore of the Ohio River in 1851 triggered similar developmental pressures in the Sewickley valley. However, Fred Wallhausser argues that Sewickley's remote position from Pittsburgh (in comparison to Hazelwood or Shadyside), its established upperclass rural society, and snobbish determination to preserve its social character prevented the second round of intense development that overtook most older industrial and residential suburbs at the close of the century. Nevertheless, development produced internal tensions and conflicts. With Pittsburgh and Allegheny City only thirty minutes away, professionals and businessmen found Sewickley's well-heeled Scots-Irish and English rural residents, established Presbyterian church, and private academies attractive. Rural isolation disappeared with the invasion of the newcomers; soon city newspapers were regularly distributed, and borough status was obtained. Despite their upper-class origins and manners, the newcomers' liberal urban attitudes ruffled the older gentry. As leadership in the church shifted, conflict between old and new Sewickley families arose over a variety of issues, ultimately leading to an irreparable division and the establishment of a new church. Eventually, the older generation passed away, and their descendants joined the newcomers in creating a typical late-nineteenth-century elite suburb, complete with exclusive recreational clubs, arts associations, and private schools. Wealthy Sewickleyans remained active in local affairs in order to shape and control the bucolic, affluent character of the community. In 1904 they rejected a proposed trolley line to Sewickley over the objections of local workingmen who desired inexpensive access to broader employment opportunities. Whereas the trolley intensified development in Hazelwood and Shadyside, its rejection in Sewickley helped to preserve the suburb's low density, upper-class character.[55]

A new era of metropolitan expansion unfolded in the 1920s with the automobile. A precipitous decline in railroad commuter ridership between the 1920s and 1950s mirrored the acceptance of the automobile by middle-class families. The automobile primarily intensified development in the interstitial hilly townships away from older river and railroad towns until the 1950s, when the construction of high-speed highways initiated another significant outward expansion of suburban growth.

From modest residential communities in townships like Shaler or West Mifflin to affluent developments like Mt. Lebanon and Fox

Chapel, automobile-oriented suburbs continued and accentuated the economic and social differentiation of previous suburban eras. In contrast to the fine studies of Pittsburgh's early suburbanizing communities, we know very little about the development of these post–World War II communities. For all the eras of suburban growth, we need to learn a great deal more about who settled them, the investors and developers who created them, and the many relationships between them and the city's population and institutions. To date, scholarly attention has emphasized the industrial mill towns and the conversion of rural land to suburban uses at the expense of understanding the socioeconomic composition, social processes, and significance of urban expansion. Further, we need a temporal and spatial framework within which to place these community case studies. Even a casual knowledge of family ties among residents in particular urban neighborhoods and suburban townships today (for example, between Lawrenceville and Shaler) suggests the sorting processes that bring order and specificity to the metropolitan area's complex historical social geography and social structure.

The Pittsburgh urban area has always been one of the nation's most politically fragmented metropolises. The city of Pittsburgh's proportion of the total metropolitan population remained throughout the twentieth century well below the average for other large cities. Even in the 1980s, suburban communities jealously protected their political autonomy and local identities despite successful school consolidations, metropolitan service authorities, and increasingly inadequate revenue bases.[56] Moreover, suburbanites steadfastly resisted efforts to establish revenue sharing with the city. Nevertheless, they appropriately recognized that their welfare depended on the successful economic health of the entire metropolitan area, and they identified enthusiastically with Pittsburgh sports teams. The ambivalent relationships between the city and suburban communities have deep historical roots that bear directly on the difficulty of orchestrating metropolitanwide programs and exercising leadership. Large subsections of the metropolitan area such as the Monongahela and Beaver valleys feel neglected by Pittsburgh's self-centered private and public leadership that emphasizes the city, the corporations, and the interests of its own residential communities.

The problems of mistrust within the metropolitan area are magnified at the regional level, where long-standing relationships have divided the city and hinterland despite (or partially because of) economic interdependence. A few centralizing institutions such as the universities, sports teams, major banking corporations, and the media keep alive a regional identity, but a regional voice has yet to be

found for major political and economic issues. The original settlements and subsequent industrialization of the Western Pennsylvania region with Pittsburgh at its center involved a complex, dynamic
set of social and economic relationships between diverse hinterland
communities and the urban center. The paucity of historical scholarship on the emergence and evolution of this city region leaves an
inadequate understanding of the region as a whole, the diversity of
its many component parts, and the important relations between
Pittsburgh and the parts.

NOTES

1. Adna F. Weber, *The Growth of Cities in the Nineteenth Century: A
Study in Statistics* (1899; rpt. Ithaca, N.Y.: Cornell University Press, 1963).

2. In making this distinction between region and section, I am following the lead of Diane Lindstrom in *Economic Development in the Philadelphia Region, 1810–1850* (New York: Columbia University Press, 1978).

3. The names commonly used by residents to identify their region often reveal important insights into the region. It would be interesting to
ascertain what is the appropriate historical vernacular name for Western
Pennsylvania as a whole or for specific subregional areas. See Wilbur
Zelinsky, "North America's Vernacular Regions," *Annals of the Association of American Geographers* 70 (March 1980): 1–16.

4. Peter Hall, ed., *Von Thunen's Isolated State: An English Version of
"Der Isolierte Staat,"* trans. C. M. Wartenberg (New York: Pergamon Press,
1966). Reginald P. Baker assessed social patterns within a geographical
framework that was based on the economic differences of various rural
settings in "Economic Development and Rural Social Structure: Pennsylvania in 1860" (Ph.D. diss., University of Pittsburgh, 1976).

5. N.S.B. Gras, *An Introduction to Economic Growth* (New York:
Harper and Brothers, 1922); James Vance, Jr., *The Merchant's World: The
Geography of Wholesaling* (Englewood Cliffs, N.J.: Prentice-Hall, 1970); and
Allan R. Pred, *The Spatial Dynamics of U.S. Urban and Industrial Growth*
(Cambridge, Mass.: MIT Press, 1967).

6. Donald F. Davis, "The 'Metropolitan Thesis' and the Writing of Ca-
nadian Urban History," *Urban History Review/Revue d'histoire urbaine* 14
(Oct. 1985): 95–114.

7. Leland D. Baldwin, *Pittsburgh: The Story of a City, 1750–1865* (Pittsburgh: University of Pittsburgh Press, 1937).

8. Richard C. Wade, *The Urban Frontier: Pioneer Life in Early Pittsburgh, Cincinnati, Lexington, Louisville, and St. Louis* (Chicago: University of Chicago Press, 1959); and Catherine E. Reiser, *Pittsburgh's Commercial Development, 1800–1850* (Harrisburg: Pennsylvania Historical and Museum Commission, 1985).

9. Ann Marie Dykstra, "Region, Economy, and Party: The Roots of Policy Formation in Pennsylvania, 1820–1860" (Ph.D. diss., University of Pittsburgh, 1986), chap. three; and R. Eugene Harper, "Town Development in Early Western Pennsylvania," *Western Pennsylvania Historical Magazine* 71 (Jan. 1988): 3–26.

10. Reiser, *Pittsburgh's Commercial Development.*

11. Arthur C. Bining, *Pennsylvania Iron Manufacturing in the Eighteenth Century* (Harrisburg: Pennsylvania Historical and Museum Commission, 1938); and Louis C. Hunter, "Influence of the Market Upon Technique in the Iron Industry in Western Pennsylvania up to 1860," *Journal of Economic and Business History* 1 (Feb. 1929): 239–81, and "Financial Problems of the Early Pittsburgh Iron Manufacturers," ibid. 2 (May 1930): 520–45.

12. Robert Eugene Harper, "The Class Structure of Western Pennsylvania in the Late Eighteenth Century" (Ph.D. diss., University of Pittsburgh, 1969); Thomas P. Slaughter, *The Whiskey Rebellion: Frontier Epilogue to the American Revolution* (New York: Oxford University Press, 1986), pp. 62–74; and Hal Kimmins, "Westmoreland County, 1783–1790: A Study of the Economic Base and Local Government" (seminar paper, Department of History, University of Pittsburgh, 1964).

13. Dykstra, "Region, Economy, and Party"; and Linda K. Pritchard, "Religious Change in a Developing Region: The Social Contexts of Evangelicalism in Western New York and the Upper Ohio Valley During the Mid-Nineteenth Century" (Ph.D. diss., University of Pittsburgh, 1980).

14. James E. Fell, Jr., "Iron from 'The Bend': The Great Western and Brady's Bend Iron Companies," *Western Pennsylvania Historical Magazine* 67 (Oct. 1984): 323–45.

15. Baker, "Economic Development," pp. 83–103.

16. Slaughter, *Whiskey Rebellion;* and Dorothy E. Fennell, "From Rebelliousness to Insurrection: A Social History of the Whiskey Rebellion, 1765–1802", (Ph.D. diss., University of Pittsburgh, 1981).

17. John W. Brant, " 'Those Damn Ignorant Somerset Dutch': An Analysis of Antimasonic and Whig Voting in Somerset County, Pa., from 1828 to 1840" (seminar paper, Department of History, University of Pittsburgh, 1969).

18. Michael F. Holt, *Forging a Majority: The Formation of the Republican Party in Pittsburgh, 1848–1860* (New Haven, Conn.: Yale University Press, 1969); and Wade, *Urban Frontier.*

19. Allan R. Pred, *Urban Growth and the Circulation of Information: The United States System of Cities, 1790–1840* (Cambridge, Mass.: Harvard University Press, 1973).

20. Edward J. Davies II, *The Anthracite Aristocracy: Leadership and Social Change in the Hard Coal Regions of Northeastern Pennsylvania, 1800–1930* (DeKalb, Ill.: Northern Illinois University Press, 1985); and Burton W. Folsom, Jr., *Urban Capitalists: Entrepreneurs and City Growth in Pennsylvania's Lackawanna and Lehigh Regions, 1800–1920* (Baltimore, Md.: The Johns Hopkins University Press, 1981). Harper suggests the role of towns as leaders of change in "Town Development," pp. 18–26.

21. J. E. Davidson, " 'God Speed the Plough': A View of Agricultural Society" (seminar paper, Department of History, University of Pittsburgh, 1969).

22. Edward H. Hahn, "Social Changes in a Small Community: 1860 to 1880" (seminar paper, Department of History, University of Pittsburgh, 1974).

23. Baker "Economic Development," pp. 104–216.

24. Burton Hersh, *The Mellon Family: A Fortune in History* (New York: William Morrow, 1978).

25. R. D. McKenzie, *The Metropolitan Community* (New York: McGraw-Hill, 1933); Donald J. Bogue, *The Structure of the Metropolitan Community: A Study of Dominance and Sub-Dominance* (Ann Arbor: University of Michigan, 1949); Michael P. Conzen, *Frontier Farming in an Urban Shadow* (Madison, Wis.: The State Historical Society of Wisconsin, 1971); and Roberta Balstad Miller, *City and Hinterland: A Case Study of Urban Growth and Regional Development* (Westport, Conn.: Greenwood Press, 1979).

26. Lindstrom, *Economic Development*, pp. 8–21; and Allan R. Pred, *Urban Growth and City Systems in the United States, 1840–1860* (Cambridge, Mass.: Harvard University Press, 1980).

27. David R. Meyer, "Emergence of the American Manufacturing Belt: An Interpretation," *Journal of Historical Geography* 9 (April 1983): 145–74; and Beverly Duncan and Stanley Lieberson, *Metropolis and Region in Transition* (Beverly Hills, Cal.: Sage Publications, 1970).

28. James C. Holmberg, "The Industrializing Community: Pittsburgh, 1850–1880" (Ph.D. diss., University of Pittsburgh, 1981), chap. 2; and George Littleton Davis, "Greater Pittsburgh's Commercial and Industrial Development, 1850–1900" (Ph.D. diss., University of Pittsburgh, 1951).

29. Muriel E. Sheppard, *Cloud by Day: The Story of Coal and Coke and People* (Chapel Hill: University of North Carolina Press, 1947); and David P. Demarest, Jr., and Eugene D. Levy, "A Relict Industrial Landscape: Pittsburgh's Coke Region," *Landscape* 29, no. 2 (1986): 29–36.

30. *Tenth Industrial Directory of the Commonwealth of Pennsylvania* (Harrisburg: Department of Internal Affairs, 1941); *Second Industrial Directory of Pennsylvania, 1916* (Harrisburg: Department of Labor and Industry, 1916); and Roger B. Saylor, *The Railroads of Pennsylvania*, Industrial Research Report No. 4 (University Park: Pennsylvania State University, 1964).

31. "General Map of the Bituminous Coal Fields of Pennsylvania," (Pottsville, Pa.: Baird Holberstaat, 1903); *Tenth Industrial Directory*; Hersh, *Mellon Family*; and Harold C. Livesay, *Andrew Carnegie and the Rise of Big Business* (Boston: Little Brown, 1975).

32. "General Map of the Bituminous Coal Fields"; and *Tenth Industrial Directory*.

33. Ronald Abler et. al., *The Atlas of Pennsylvania* (Philadelphia: Temple University Press, forthcoming), pp. 93–95.

34. Hahn, "Social Changes"; and Baker, "Economic Development."

35. Hahn, "Social Changes"; and Anna-Mary Caffee, "The Socialization of Rural Children in Southwestern Pennsylvania, 1840–1900" (seminar paper, Department of History, University of Pittsburgh, n.d.).

36. Linda Nyden, "Black Miners in Western Pennsylvania, 1925–1931: The National Miners Union and the United Mine Workers of America" (seminar paper, Department of History, University of Pittsburgh, 1974).

37. Ray Burkett, "Vandergrift: Model Workers' Community" (seminar paper, Department of History, University of Pittsburgh, 1972); and Anne Mosher Sheridan, "Scribing the Outlines of a Model Company Town: The McMurtry/Olmsted and Eliot Plan for Vandergrift, Pennsylvania," paper presented at the Eastern Historical Geography Association Meeting, University Park, Pa., October 1987 (part of a forthcoming Ph.D. diss., Department of Geography, Pennsylvania State University).

38. Burkett, "Vandergrift."

39. Mary C. Huey, "Occupational and Nationality Structure of Mc-Keesport, 1880" (seminar paper, Department of History, University of Pittsburgh, n.d.); Frank Huff Serene, "Immigrant Steelworkers in the Monongahela Valley: Their Communities and the Development of a Labor Class Consciousness" (Ph.D. diss., University of Pittsburgh, 1979); Margaret Byington, *Homestead: The Households of a Mill Town* (1910; rpt. Pittsburgh: University of Pittsburgh Press, 1974); Francis G. Couvares, *The Remaking of Pittsburgh: Class and Culture in an Industrializing City, 1877–1919* (Albany: State University of New York Press, 1984); and Matthew S. Magda, *Monessen: Industrial Boomtown and Steel Community, 1898–1980* (Harrisburg: Pennsylvania Historical and Museum Commission, 1985).

40. Michael P. Weber, *Social Change in an Industrial Town: Patterns of Progress in Warren, Pennsylvania, from the Civil War to World War I* (University Park: Pennsylvania State University Press, 1976).

41. Joseph P. Fadgen, "McKees Rocks: Study in Political Change" (seminar paper, Department of History, University of Pittsburgh, 1971). Steve Agoratus et. al. found similar political patterns in Donora; see "Donora: An Historical Perspective" (seminar paper, Department of History, Carnegie Mellon University, 1982). Elaine P. Sloan addressed the issue of different political behavior occurring in different types of communities in "The Political Behavior of Mining Communities" (seminar paper, Department of History, University of Pittsburgh, 1962).

42. This point is made clear in the comparison of Johnstown and Pittsburgh in Michael P. Weber and Ewa Morawska, "East Europeans in Steel Towns: A Comparative Analysis," *Journal of Urban History* 11 (May 1985): 280–313.

43. Seymour Martin Lipset's brief article on experiences in California indicates that city and country relationships held important implications for evolving social patterns. See "Social Mobility and Urbanization," *Rural Sociology* 20 (Sept. 1955): 220–28.

44. Davies, *Anthracite Aristocracy*. Although fictionalized, John O'Hara's voluminous writings on the Pottsville region of the anthracite

fields describe the social behavior and values of the upper class during the
initial half of the twentieth century in a way that should whet the appetite
of the historian of Western Pennsylvania. See Matthew J. Bruccoli, *The
O'Hara Concern: A Biography of John O'Hara* (New York: Random House,
1975).

45. Hersh, *Mellon Family*; M. Graham Netting, *Fifty Years of the Western Pennsylvania Conservancy: The Early Years* (Pittsburgh: Western Pennsylvania Conservancy, 1982); David G. McCullough, *The Johnstown Flood*
(New York: Simon and Schuster, 1968), pp. 39–78; and William C. Forrey,
History of Pennsylvania's State Parks (Harrisburg: Department of Environmental Resources, 1984).

46. Kenneth T. Jackson, *Crabgrass Frontier: The Suburbanization of the
United States* (New York: Oxford University Press, 1985); Sam Bass Warner,
Jr., *Streetcar Suburbs: The Process of Growth in Boston, 1870–1900* (Cam-
bridge, Mass.: Harvard University Press, 1962); and David Ward, "A Compara-
tive Historical Geography of Streetcar Suburbs in Boston, Massachusetts, and
Leeds, England," *Annals of the Association of American Geographers* 54
(Dec. 1964): 477–89.

47. Henry C. Binford, *The First Suburbs: Residential Communities on
the Boston Periphery, 1815–1860* (Chicago: University of Chicago Press,
1985); and John Kellogg, "Negro Urban Clusters in the Postbellum South,"
Geographical Review 67 (July 1977): 310–21.

48. Bradley W. Hall, "Elites and Spatial Change in Pittsburgh: Minersville as a Case Study," *Pennsylvania History* 48 (Oct. 1981): 311–34.
There has been little other research on the developmental and conversion
processes associated with the adjacent communities of these early years. For
a general description, see Baldwin, *Pittsburgh*, pp. 231–47.

49. William Kenneth Schusler, "The Economic Position of Railroad
Commuter Service in the Pittsburgh District: Its History, Present, and Future" (Ph.D. diss., University of Pittsburgh, 1958).

50. Joel A. Tarr, *Transportation Innovation and Spatial Change in Pittsburgh, 1850–1934*, Essays in Public Works History, no. 6 (Chicago: Public
Works Historical Society, 1978).

51. Byington, *Homestead*, pp. 4–5.

52. Karen Cowles, "The Industrialization of Duquesne and the Circulation of Elites, 1891–1933," *Western Pennsylvania Historical Magazine* 62
(Jan. 1979): 1–17.

53. Joan Miller, "The Early Historical Development of Hazelwood"
(seminar paper, University Department of History, of Pittsburgh, n.d.); and
Joel A. Tarr and Denise DiPasquale, "The Mill Town in the Industrial City:
Pittsburgh's Hazlewood," *Urbanism Past and Present* 7 (Winter/Spring
1982): 1–14.

54. Robert J. Jucha, "The Anatomy of a Streetcar Suburb: A Development History of Shadyside, 1852–1916," *Western Pennsylvania Historical
Magazine* 62 (Oct. 1979): 301–19; Renee Reitman, "The Elite Community
in Shadyside, 1880 to 1920" (seminar paper, Department of History, Univer-

sity of Pittsburgh, 1964); and Ethel Spencer, *The Spencers of Amberson Avenue: A Turn-of-the-Century Memoir* (Pittsburgh: University of Pittsburgh Press, 1983).

55. Fred Wallhausser, "The Upper-Class Society of Sewickley Valley, 1830–1910" (seminar paper, Department of History, University of Pittsburgh, 1964); and Stephen J. Schuchman, "The Elite at Sewickley Heights, 1900–1940" (seminar paper, Department of History, University of Pittsburgh, 1964).

56. See Jon C. Teaford, *City and Suburb: The Political Fragmentation of Metropolitan America, 1850–1970* (Baltimore, Md.: The Johns Hopkins University Press, 1979).

Infrastructure and City-Building in the Nineteenth and Twentieth Centuries

JOEL A. TARR

L IKE OTHER AMERICAN cities in the late twentieth century, Pittsburgh depends for its operation upon pipes, wires, streets, tracks, and other technological systems collectively known as the urban infrastructure.[1] This infrastructure has evolved gradually over time as a vital part of the city-building process. It accommodates flows of people, vehicles, messages, energy, and water and sewage, and includes structures such as bridges, public buildings and parks. The process by which it has been constructed and operated has varied greatly: sometimes it has been publicly created, sometimes privately, and occasionally, by combined public and private efforts. Over time, there have been important shifts from one form of ownership to the other, and fierce political battles have been fought over questions of public or private ownership as well as the terms of franchises to private corporations.

Private infrastructure has been built either in response to a perceived market demand or in an attempt to create a market, while publicly constructed infrastructure ("public works") has usually been the product of public demand, concern over private monopoly control of public goods, or calculations of political benefits. Public works projects have served the purpose of providing political patronage, unemployment relief, or aiding politically sensitive or powerful neighborhoods in addition to enriching politically well-connected contractors. More than in other areas of technological development, infrastructure decisions have been marked by clashes between those who supposedly value expertise, efficiency, and cost-effectiveness, those who consider equity factors, and those who are concerned with political benefits.

Until the 1970s, in spite of its importance, historians of American cities in general, and of Pittsburgh in particular, largely neglected the development and effects of infrastructure and its relation to the his-

213

tory of city-building. As a result, although earlier studies made contributions in specific areas, they seldom attempted to relate their findings to the larger urban context.[2] In the 1970s, when interest in both the history of technology and social change in its urban context began to increase, so did research into infrastructure technologies and their relation to city development.[3] More specifically for Pittsburgh, students and faculty at Carnegie Mellon University and at the University of Pittsburgh, as well as scholars outside the city, began to develop a body of research on infrastructure and, more generally, on the city-building process. The material exists in many forms, including seminar and class research papers, senior and master's theses, doctoral dissertations, and published articles and books. This chapter will attempt to use this literature to construct an integrated essay as well as to identify those areas still in need of exploration.

Samuel P. Hays has suggested that Pittsburgh's physical growth is the "overriding context" for the history of the city. That is, growth from the perspective of "people undertaking physical development, and especially of the sequences of development."[4] This concept, therefore, links the construction of infrastructure to the territorial growth of the city, a position adopted in this essay. Six questions seem especially pertinent: What was built and where? Did the private or the public sector or some combination of the two provide the service? If public, how would these services be paid for? If private, what were the terms of the franchises? How did they relate to urban development and expansion and on what basis were they distributed throughout the city? And, what effects did they have on the urban fabric and the quality of life in the city? Some of these questions can be answered but others cannot, either because the data is unavailable or because it has not yet been examined in a systematic fashion by scholars. A great deal more work therefore remains, as will be indicated below.

It is clear that physical development was unevenly distributed throughout the city's neighborhoods. Construction depended upon a range of factors including city or private willingness to provide infrastructure, the ability of individual homeowners to pay for improvements, voter approval of bond issues, and the availability of options given the Pittsburgh environment. As Christine M. Rosen has shown in her study, *The Limits of Power: Great Fires and the Process of City Growth in America*, there were barriers to urban physical change and improvements that impeded the ability of the city to grow on what might be called a rational and equitable basis. Some of these barriers were economic, but others included institutional factors such as the lack of an appropriate decision-making structure as well as the rigidity of property divisions, a reluctance of both the elite and ordinary citizens to embrace change, the resistance of governmental organizations to create debt, and political stalemate. Power to cause change, and especially rapid change, was often sharply constrained.[5]

Pittsburgh had both similarities to other cities in regard to the city-building process and important differences in the timing of innovation and change. The similarities were the result of the city's involvement in a larger national economy and membership in a network of cities whose members communicated about ideas and technology.[6] The common historical sequences went in the following order: mercantile cities experienced economic development, population growth, and political changes in the first half of the nineteenth century that accelerated pressures for city-building. City governments underwent structural changes that reoriented them from mercantile, regulative activities toward the provision of services. Transportation innovation stimulated urban territorial expansion, and annexations of contiguous communities enlarged the city's political boundaries. Political machines with a neighborhood base developed in the latter part of the century and often facilitated the expansion of infrastructure and services as a means to solidify their position. Toward the end of the century, engineering and public health professionals assumed larger roles in government. In the early part of the twentieth century cities went through a period of reform and institutional change that resulted in centralization of decision making and heavy investments in city-building, although comprehensive planning had little success. In the first third of the twentieth century the automobile emerged as the key shaping element in the city-building process as cities increasingly decentralized. In general, physical improvements were slow in accomplishment and uneven in allocation, because various socioeconomic interests, different neighborhoods, and political parties and factions fought over their desirability or acquisition.

While this sequence of events occurred in Pittsburgh, the city also had important local differences that complicated the process and affected the timing of development. One constraining feature that raised costs and impeded communications between neighborhoods and communities was the topography. Commenting in 1910 on the city's topographical characteristics, Frederick Law Olmsted, Jr., observed that "no city of equal size in America or perhaps the world, is compelled to adapt its growth to such difficult complications of high ridges, deep valleys and precipitous slopes as Pittsburgh."[7] Initially located at the point of land formed where the

Allegheny and Monongahela rivers converge to form the Ohio River, Pittsburgh eventually grew to encompass land on both sides of the rivers. Flat land was in limited supply, and the sharp contours, valleys, and river breaks affected the pace and pattern of development.

In comparison with other commercial cities, Pittsburgh industralized early, developing a base in glass and iron products as well as in textiles. Cheap energy in the form of coal, access to raw materials, and its location, were primarily responsible for Pittsburgh's industrial prowess. Industrialism brought great wealth and changing social patterns as well as high environmental costs—severe air and river pollution and the scarring of the landscape. The shift from a mercantile to an industrial city produced a range of changes, including shifting elites, new occupational groups, additional forms of community activities and conflicts, and altered residential patterns.[8] Throughout the nineteenth and twentieth centuries, the Pittsburgh business community and substantial members of the upper class supported expansion of the municipal boundaries through a process of annexation. Finally, compared to other cities, Pittsburgh made an early and energetic attempt to cope with its various physical problems following World War II. Driven by the direct involvement of corporate leaders in civic affairs and a powerful prodevelopment public-private coalition, the Pittsburgh Renaissance became the model for renewal attempts by a number of other cities.

This essay will divide the history of city-building in Pittsburgh into four periods: the walking or pedestrian commercial city, 1794–1867; the development of the networked or wired, piped, and tracked industrial city, 1868–1899; centralization, decentralization, and the impact of the automobile, 1900–1944; and the rebuilding of Pittsburgh in the decades of the Pittsburgh Renaissances, 1945–1988. These divisions are based upon a combination of factors relating primarily to changing technologies, altered roles for government, and spatial changes in the city's area. They are not meant to be inclusive, and many elements in the infrastructure persisted for long periods of time, continuing to serve their original purposes or undergoing adaptation or retrofitting to fill other needs or even serving as a physical barrier to needed change. Each period captures, however, enough linked technological and political developments to provide unity.

The Pedestrian City, 1794–1867

Pittsburgh thrived first as a mercantile and then as an industrial center in these years. Its location on the three rivers provided it with access to both raw materials and to markets. Because of the city's strategic location, it developed as a transfer station for goods coming from the east over the Allegheny Mountains. State construction of the Pennsylvania Main Line Canal, completed in 1834, and the development of numerous railroad lines after 1851, reinforced its commercial links to other cities and regions. Pittsburgh, as well as surrounding towns, industralized somewhat earlier than many other nineteenth-century cities. The rich beds of bituminous coal that underlay the region supplied cheap fuel for numerous manufactories of iron, glass, and textiles as well as for copper and brass foundries. By 1850 ironmaking was the city's chief industry, and rolling mills, foundries, machine shops, and boiler yards employed a substantial part of the region's work force.[9]

From 1794 to 1867, the city's population grew from approximately twelve hundred to over fifty thousand. Like other American cities, it was primarily a pedestrian city. No public transportation existed until the 1840s, although innovations followed fairly rapidly after this decade. Work and residence were often closely linked, and the city's elite—bankers, merchants, industrialists, and professionals—lived largely in the urban core close to governmental, commercial, and mercantile activities while the working class resided in nearby alleys and streets or in the outlying wards.[10] A number of towns grew up around the city, the most important of which were Allegheny City on the north side of the Allegheny River, the industrial boroughs of Birmingham and East Birmingham on the south bank of the Monongahela River, and the Northern Liberties, Pitt Township, Oakland and Lawrenceville boroughs, and Peebles Township toward the east.

Between 1804 and 1846, the city sought political integration by annexing some of the surrounding townships and boroughs, increasing its land area to 1,130 acres. Pittsburgh's commercial elite took the lead in the drive for territorial expansion. The original city was divided into four wards in 1833 with a fifth added in 1837 (Northern Liberties); four more wards were created in 1845 and 1846 as a result of the annexation of Pitt Township. Some of the city's annexations were welcomed, but others were carried out with difficulty or even prevented. The 1837 annexation of Northern Liberties Borough, for instance, was unopposed and required no popular vote, but in 1854, when the Pittsburgh Board of Trade led a campaign to consolidate Pittsburgh with Allegheny City and several surrounding boroughs, the targeted communities blocked the legislation.[11] James C. Holmberg has suggested that the opposition can be explained by Pittsburgh's particular form of growth. Rather than undergoing sequential development, separate urban communities grew simultaneously.

Each cherished and strove to protect its independence.[12] In other cities, outlying communities often sought consolidation with the central city because it provided them with access to urban services. It is not clear how important this factor was in Pittsburgh, although once an annexation occurred the older and newer sections of the city often competed for services.[13]

Territorial and population growth created a demand for various types of services, many of which involved governmental action. Richard L. McCormick has argued that the main function of government in the nineteenth century was to promote development by "distributing resources and privileges to individuals and groups," rather than to administer or regulate.[14] The mercantile elite was the chief advocate of these policies although industrialists played important roles in Pittsburgh. One level of aid involved the state government. The Commonwealth of Pennsylvania, like a number of other states, followed a policy from the 1820s through the 1840s of investing in transportation projects that were either under state control or were "mixed enterprises," combining public and private operations. The motives for these policies included promotional goals, a desire for public profit, concern over the limitations of private corporations, and even a belief in the employment opportunities offered. The most important example of state public works was the Pennsylvania Main Line Canal, connecting Pittsburgh with Philadelphia. The Main Line, which involved a mixture of canal, railroad, and portage railway, reached Pittsburgh in 1834. It improved the city's trade position and also stimulated the development of clusters of businesses, warehouses, and hotels around the canal basins. Eventually, however, the cumbersome system, with its alternative modes and break-of-bulk features, proved an economic failure, and by 1859 all sections of the works had been sold to private interests.[15]

The state was also involved as an investor in various transportation enterprises and bridges in the Pittsburgh area as part of the mixed enterprise strategy. Bridges were important for commercial purposes and helped provide social as well as economic integration across river barriers. Companies to construct bridges over the Monongahela and Allegheny rivers were chartered in 1810, with the Monongahela Bridge completed in 1818 and the Allegheny Bridge in 1819. The state subscribed to approximately one-third of the stock of the companies with Pittsburgh merchants holding the remainder.[16] Both were covered wooden toll bridges designed by Louis Wernwag who had built the famous bridge Colossus in Philadelphia. Private investors hoped to profit from the tolls but they were also interested in land speculation. A majority of the stockholders in the Monongahela Bridge Company, for instance, owned property located near the bridge approaches on both sides of the river. Public funds thus helped integrate the city and neighboring communities (to be annexed at a later date) but also benefited private capitalists.[17] Other bridges were constructed in subsequent years (three by the famous engineer John Roebling), and by the end of the 1850s there were five bridges crossing the Allegheny and two over the Monongahela River. All but the Allegheny River aqueduct for the Main Line Canal were privately owned and were operated as toll bridges with no state investment.[18] State aid to various transportation-related projects and state-owned public works reached its peak in the 1820s and 1830s, with rapid disinvestment after 1843 due to citizen dissatisfaction with overinvestment, taxes, and corruption.

As state investment in public works declined, municipal involvement in service provision expanded, both through outright ownership and through shared investments. Much of this expansion resulted from the growing demand for services from mercantile leaders and industrialists, although broader public support was available for some projects. The role of municipal government as service provider, however, represented a shift from the eighteenth-century pattern. While eighteenth-century municipalities provided limited infrastructure, their chief concern was with the regulation and protection of commercial activities. In addition, some services that later became a municipal responsibility were handled by volunteer groups or remained an individual responsibility.[19] These patterns changed gradually in the nineteenth century for most American cities, including Pittsburgh. During the first half of the century, Pittsburgh government still regulated local trade and checked food quality and measurements. By the middle of the century, however, the municipality was largely concerned with supplying city services such as streets, water, and sewers.[20] Structural changes in city government were necessary preconditions to this reorientation and depended on state constitutional provisions and legislation.

Pittsburgh was first incorporated as a borough by act of the legislature in 1794. Government was limited under the original charter. Freeholders and "other inhabitants" elected two burgesses, and a town meeting served as the legislative body. During the borough period, the town meeting made provision for limited city services and elected street regulators and supervisors; the burgesses appointed a clerk of the market and the first night watch. Revisions of the borough charter in 1804 provided for an annually elected but unpaid thirteen-member council as a substitute for the town meeting. Among the council's powers was the right to enact ordinances and to assess, apportion, and appropriate taxes. The council was also empowered to regulate the market, to supervise the streets, and to ascertain the depths of vaults, sinks, and privy pits.[21]

As the city grew in population, the existing governmental framework became inadequate to meet municipal needs. In 1816, when the population had grown to over six thousand, the state legislature approved a statute incorporating Pittsburgh as a city, creating a government with a bicameral governing body and a mayor. The voters at large elected the councils, which then chose a mayor who possessed little executive power. In 1834 new state statutes divided the city into four wards and provided for election of the councils on a ward basis. The mayor was elected to serve a one-year term.[22] The councils still controlled the government, having the power to provide the ordinances and regulations "necessary for the government and welfare" of the city. They exercised their executive and administrative powers through joint standing committees dealing with areas such as streets and water, gas lights, fire engines, and markets, and could enter into contracts and disburse public moneys.

As the city's business expanded, the number of joint committees grew from six in 1816 to eighteen in 1866. Finance was the single most powerful committee, handling appropriations, taxes, and loans. Methods of fiscal administration were cumbersome, requiring the cooperation of the councils, the treasurer, the mayor, and several committees. City funds were expended through warrants, drawn by the mayor from the city treasurer at the request of a council committee. This system continued until 1857, when state legislation created the office of city controller with authority over municipal fiscal affairs.[23]

The state constitution and statutes limited municipal taxation and bonding powers. The city depended primarily on a property tax until 1846, when the state legislature authorized a municipal sales tax and taxes on various businesses. Not until 1857 and 1858, however, did the legislature allow the city to levy special assessments on property owners for street and sewer improvements. The 1857 law also permitted the city to assess property owners for past improvements that had been previously paid for out of general taxes, a procedure that resulted in considerable litigation.[24] Even in the face of demand, these fiscal limitations provided an institutional restriction on governmental city-building activities.

The budgetary process was tightly linked to the political process, but unfortunately the relationship between politics and fiscal policy and between politics and developmental policy in Pittsburgh has not been fully studied.[25] The major projects for which the city spent public funds can be traced, however, in an attempt to understand their evolution and to arrive at answers to the question of who got what. Funds were allocated for a variety of services, ranging from investment in private transportation facilities to the provision of streets, water, and sewers. The latter investments involved intracity infrastructure needs, while those in transportation were made in the late 1840s and 1850s and related primarily to external commercial connections.

The councils spent more of their time on matters relating to streets, such as financing, openings, maintenance, lighting, and cleaning, than on any other. The first charter had given the councils the responsibility for "improving, repairing and keeping in order the streets, alleys and highways."[26] They responded by appointing a street commissioner and by making streets the concern of their first standing committee. Over the years a variety of methods were used to pay for street improvements. General tax funds and loans were used initially, but consistent attempts were made to shift the burden to abutters. An 1807 ordinance provided that two-thirds (changed to three-quarters in 1816) of the property owners on any street could petition the councils for paving at their own expense under the direction of the street commissioners. The property owners received a credit on their city taxes, but few streets were paved under this approach.[27] In 1835 the city attempted to stimulate new street construction by promising to compensate citizens who incurred damages by assessing those who benefited.[28] After the annexation of Pitt Township in 1837, many new streets were paid for by small loan certificates called "script," which was accepted by the city as legal tender for payment of taxes. The script fluctuated considerably in price until the councils redeemed it at full price in 1850, thereby stimulating a heated controversy between the losers and the gainers.[29]

In 1850 the legislature permitted the city to levy a special tax to create a fund for street and sewer improvements, and finally in 1857 the councils were permitted to assess abutting property owners for street improvements as well as for past improvements. Considerable litigation ensued over the question of assessments on property owners who had not petitioned for improvements, with some final settlements not being reached for almost a decade. In 1864 special assessment procedures were revised to permit appeals and to allow for different approaches in cases of street improvements as compared with street openings.[30] By the end of the 1860s, paving was largely confined to streets in the city's commercial sections and in the older and wealthier residential areas; in total, less than half the city streets were paved.

In addition to paving, the city also gradually illuminated the streets for reasons of public safety. Whale oil lamps were briefly tried to 1816 and again in 1830, but with limited success. In 1837, belatedly following the lead of cities like Philadelphia and Baltimore, the city began using manufactured gas. The financial arrangements for supplying the gas again reflected the willingness of government in this period to experiment with various forms of enterprise and with "mixed" public-private projects. In 1835 the city organized the Pittsburgh Gas Works as a joint-stock company to be managed by a board of trustees appointed by the councils. All revenues were to be paid into the city treasury, and stockholders were guaranteed a return of 6 percent per year for fifteen years. The gas company sold gas to private residents as well as providing street lights. In 1848, for reasons that are unclear, the gas works was incorporated as the Pittsburgh Gas Company, managed by a board of trustees half appointed by the councils and half by stockholders, with the city given lower prices for gas consumed for lighting.[31]

Water supply was the second priority for city governments after provision of a basic street network. Like other urbanites at the beginning of the nineteenth century, Pittsburghers drew their water from local sources such as rivers, ponds, and cisterns. Water was provided by mixed private and public suppliers almost from the city's beginnings. Private ownership, however, provoked continual citizen's complaints. Private vendors peddled water in the streets and the 1815 city directory listed five water carters. An 1802 ordinance provided for borough construction of four public wells and for the purchase of private wells "in useful and necessary parts of the Borough."[32] Wells and ponds, however, were inadequate for the needs of the growing city, and increasingly Pittsburghers demanded improved supplies. Citizens debated whether private firms or the government should provide the service. In 1818 the councils refused to approve an attempt by private interests to obtain a municipal water franchise, and in 1821 sixty-one prominent citizens successfully petitioned the councils to provide new wells and to make all existing pumps public. In 1822, citizens again petitioned the councils, requesting that the municipality build a waterworks to supply Allegheny River water to the city.[33] The petitions maintained that municipal ownership was required to guarantee improved fire protection and to secure lower fire insurance rates; to serve domestic and manufacturing needs; and to meet public health needs. Represented on the petitions were prominent members of the business community who were concerned about the threat of fire, and industrialists and craftsmen who needed clean water in their production processes.[34]

In response to these demands, in 1826 the Pittsburgh Select and Common Councils approved the construction of a waterworks that would, the council presidents boasted, provide protection against fire and "beneficial effects to every manufactory and . . . family in the city."[35] The waterworks was completed in 1828. The system utilized a steam pump to draw water from the Allegheny River and raise it to a million-gallon reservoir located on Grant Hill for gravity distribution throughout the city. The councils appointed a joint standing committee to supervise the system through three appointed water commissioners and a superintendent.[36] The councils expanded the system in 1844 and again in 1848, probably in response to the great fire of 1845 and the needs of the territory annexed in 1845 and 1846. By the end of 1850, over twenty-one miles of water pipe had been laid and 6,630 dwellings, stores, and shops were served.[37]

The funding of the waterworks was the single largest expenditure made by the city during its first fifty years. The initial cost of construction constituted 40 percent of all municipal spending from 1827 to 1833. The expansion in the 1840s increased the size of expenditures, and in 1854 the Water Committee estimated the total cost of the water system as $677,709. Of this amount, $243,240 was paid out of annual appropriations and $377,069 was borrowed at 6 percent interest.[38] Pittsburgh was not unusual in the extent to which waterworks costs constituted a substantial part of the total municipal budget. The building of New York's Croton Aqueduct in 1842, for instance, increased the city's debt from $500,000 to over $9 million and caused many citizens to predict financial disaster.[39] Waterworks were ordinarily the most expensive capital project undertaken by nineteenth-century American cities. The willingness of municipalities to make such large expenditures for a public good can be explained by the joining of a variety of interest groups— merchants and industrialists, homeowners, fire insurance companies, and those concerned with the public health—to demand the construction of an adequate waterworks.[40]

While a supply of potable water was a major achievement, it formed only one part of the city's metabolic system. Storm water, domestic used water, and human and animal wastes, as well as garbage, needed to be removed from the city to avoid nuisance and dangers to human health. This required additions to the city's infrastructure as well as improved administrative and financial procedures. In Pittsburgh, as in other cities in this period, sanitary conditions were dismal.[41] Until the 1840s, all sewers were above ground and made of wood or brick.[42] An 1816 statute specified that gutters be located twelve inches from the curbstone, but they were frequently constructed in the middle of streets as well as along their sides. Their purpose was to remove water from the streets and to eliminate pools breeding miasmas, but they often became receptacles for decaying wastes; in 1819, for instance, letters in the Pittsburgh *Gazette* complained of the "filthiness of the gutters and sewers," the "greenish hue" of their contents, and their "noxious exhalations" which created a hazard to health.[43]

Household wastes and waste water were usually disposed of in cesspools and privy vaults, not in the sewers.[44] The borough charter of 1804 gave the municipality the right to regulate these receptacles, but no statutes were enacted. Increasing citizen's complaints about overflowing filth and smells from privy vaults caused the councils in 1816 to approve the levying of fines in the case of nuisances. Many problems were caused by private scavengers who, under city contract, were responsible for cleaning privy vaults and removing garbage, but who continually fouled the streets and polluted the rivers with the wastes. In order to try to regulate disposal, in 1844 the city designated a barge located on the Allegheny River as the official city dump.[45]

Public health concerns played a limited role in stimulating spending for improved sewerage systems in this period. The cholera epidemic that beset Pittsburgh and most other American cities in 1831–1834, for instance, caused only a temporary improvement in waste collection methods. Local officials and the medical profession argued about rival theories of disease etiology, especially contagionism and anticontagionism, although some physicians and ministers still maintained that personal failings were responsible for sickness. Contagionists believed that specific contagia, probably animate and originating primarily from outside the community, caused disease. Anticontagionists, on the other hand, held that disease resulted from miasmas eminating from the decay of organic material. A belief in the first hypothesis led to a demand for a quarantine policy, while adherence to the second led to an emphasis on improving sanitary conditions.[46]

In June 1832, as the city awaited the visitation of cholera, the councils established a Sanitary Board to "direct all such measures as they think necessary for averting the introduction of the frightful epidemics." The board had the power to "cause the streets, lanes, alleys, buildings, lots and shores of the rivers to be explored, cleansed and purified in an efficient manner."[47] It proceeded to organize the city into sanitary districts, attempted to clean the streets, and sought to control unsanitary privy vaults and cellars. The councils also, in these crisis years, enacted ordinances to improve waste collection and to extend the water system. The response to the public health threat, however, remained limited and temporary, and conditions soon reverted to their usual unsanitary state.[48]

Fear of epidemic disease alone could not persuade the councils to make the large expenditures necessary to build a sewerage system. Confusion over disease etiology as well as uncertainty about the technical and design requirements for an efficient system had a discouraging effect. The city constructed its first underground sewers in 1848 and 1849 in the commercial district, and by 1866 there was a "fairly adequate" system of main sewers in this section. The motivation for construction, however, was the removal of storm water from the streets to avoid nuisance, not the protection of public health. Sewer-building at this time related primarily to the facilitation of commerce rather than to health.[49]

The pattern that emerges from an examination of the provision of streets, water, and sewers in Pittsburgh shows that the needs of the city's downtown or older sections received first priority. The spatial distribution of services often reflected a division between older and newer sections, with territorial consolidation frequently heightening political conflict. The 1846 annexation, for instance, created a "line of demarkation" in the city councils between the older wards and the annexed sections. Representatives of the new wards wanted the city to pay for the grading and paving of streets, the installation of gas lights, and the extension of water systems to their neighborhoods, but the old wards resisted. Such resistance may have had a class and ethnic as well as a territorial and fiscal basis. The new wards were inhabited by many working-class Irish Catholic and German immigrants and were represented on the councils by self-employed men and skilled workers. The old wards held concentrations of wealthy, Protestant, native Americans and Scots-Irish Presbyterians, and were generally represented by men with business affiliations.[50] The largely native-American Whig party, which dominated the councils, strongly opposed providing the new wards with city services and blocked their access to the water supply system for several years.[51] Whether class, ethnicity, or spatial competition alone accounted for such opposition, however, is still largely undetermined. While we do know that skilled workers or artisans formed the largest single group on the councils from 1816 to 1850, with manufacturers and merchants following, no systematic analysis has been conducted of their voting patterns.[52]

Municipal policy in this period involved not only allocations for public works but also investment in private companies for developmental purposes. By the 1840s property owners throughout the state had largely rejected the concept of state ownership and operation of public works as well as the mixed enterprise strategy. Local mercantile groups and industrialists, however, still clamored for improvements, and in 1848 the state legislature authorized local subscriptions to various railroad lines. Pittsburgh, Allegheny City, and Allegheny County subscribed heavily in railroad bonds, apparently with wide public support. Pittsburgh's railroad debt reached $1.8 million by 1855, and that of the county was over $3.3 million. Whether because of the public subsidies or economic prosperity, railroad development in the 1850s was rapid, and by the middle of the decade the city councils had approved ordinances that allowed four more railroads to enter the city.[53] Increased county taxes to pay for these investments, however, led to political protests, and by the end of the decade the county railroad debt had been partially repudiated.[54]

The railroads were constructed primarily to encourage interurban and interregional trade and traffic, but they also had the unforeseen effect of facilitating suburbanization. Until the 1850s, Pittsburgh's rugged topography and a lack of transportation had limited the housing choices of the city's upper class and growing middle class to the urban core. The omnibus or horse-drawn bus was the city's first public transportation system, appearing in the city in the 1840s; by the 1850s four lines were operating.[55] These were privately operated with municipal charters and charged twelve cents for a ride.[56] The speed and capacity of the omnibuses, however, were limited. The railroads did not suffer from this disadvantage, and they rapidly developed a commuter traffic. Pittsburgh merchants and industrialists bought houses in outlying towns and used the train to commute to work. By the 1860s regular commuter or accommodation trains connected the city with towns along their routes, offering season tickets to commuters and excursion trains to nearby towns.[57]

The greatest boast to urban transportation, however, was provided by the streetcar, introduced into Pittsburgh in 1859. This was eight years after it appeared in New York and Philadelphia. Privately owned, the streetcar lines operated with city charters (usually of a twenty-year duration) that required them to pay a percentage of their earnings to the city and to undertake some obligations in regard to street maintenance. The granting of the privilege of using the public streets to a private corporation was a form of development policy that was contingent on urban politics, but this relationship has yet to be explored. First powered by horses and mules, then by cable, and ultimately by electricity, the streetcar dominated urban transport in

Pittsburgh from the Civil War through the 1920s. Horsecars were initially little more than omnibuses operated on rails laid on city streets. Pittsburgh's hilly topography necessitated small cars and forced the lines to follow essentially the same routes that had been used earlier by the turnpikes and the omnibuses—those of least travel resistance. By 1869, five horsecar lines were operating approximately twenty-three miles of track. Passenger traffic was about 8 million or approximately fifty rides per inhabitant per year, with ridership still limited generally to the middle and upper classes.[58]

Transportation technology did offer a form of integration that had not yet been politically accomplished. These innovations gradually altered basic patterns of work and residence from those of the pedestrian city. One study shows that in the 1850s nearly 60 percent of the Pittsburgh primary elite, comprised mostly of merchants and manufacturers, lived and worked downtown; another 30 percent lived in Allegheny City and probably also worked in Pittsburgh. By the 1860s, 70 percent still worked downtown, but less than 40 percent lived there.[59] The residential locations outside the downtown predominantly chosen by the elite were in Allegheny City and the East End.

The residential shift of the Pittsburgh elite from the old core of the city toward its periphery signified the movement of the city into a different stage of development. During the first two-thirds of the century Pittsburgh changed from a mercantile to an industrial city, grew in population and territory, and began to supply its citizens with important services. The structure of municipal government became more elaborate as it assumed responsibilities for streets, water, and sewers as well as the public safety and fire prevention.

These developments, however, had proceeded in an incremental and often halting manner and seldom had a systematic character. Methods of financing evolved slowly, with much recourse to the courts and the state legislature. Private-sector investments in the built environment, aside from housing, about which relatively little is known, focused on transportation and bridges and were often linked to public investment. What should be public and what should be private was a matter of dispute. The city, for instance, rejected private ownership of the water and gas works and constructed them itself, operating the gas works as a joint-stock company. The municipality provided infrastructure unevenly, centering on the downtown and the city's older areas. These allocations were apparently influenced by territorial and ethnocultural factors as well as by class. Thus, a plurality of interests had determined the development of

Pittsburgh's built environment when it emerged as a major urban center. The nature of these interests still needs to be investigated, however, as do such basic issues as how the needs of the industrial city differed from those of the commercial city, and the changing nature of the political structure and its relationship to fiscal policy.

Infrastructure and Services for the Expanding City, 1868–1899

During the last quarter of the nineteenth century Pittsburgh experienced great expansion and prosperity, becoming one of the world's major industrial centers. Iron and steel production was Pittsburgh's dominant industry, and the mills, the blast furnaces, and the coke ovens occupied the river floodplains, determined the location of workers' housing, and polluted the environment. Increasingly, workers labored in large industrial establishments rather than small shops and factories. The managerial and residential needs of Pittsburgh industrialists were reflected in the development of a central business district with a high-rise profile, elite neighborhoods with magnificent mansions, miles of substantial housing for the growing middle class, and the construction of extended infrastructure networks. Thus the city developed both a new infrastructure for production and a new infrastructure for consumption.[60] Both were contingent on the city's evolving political structure and organizations as well as relating to factors of production and consumption.

From 1868 to 1900, the city grew in population from about fifty-five thousand to over three hundred thousand and increased its land area from 1.77 to over 28 square miles. Population increase resulted primarily from the arrival of thousands of European immigrants and Pittsburgh's continued annexation of contiguous boroughs and townships. The two most significant annexations, in 1868 and 1872, increased the city's land area by over twenty-five square miles and its population by approximately sixty-five thousand inhabitants. The 1868 consolidation involved a number of townships and boroughs to the east of the city having a population of about thirty thousand and occupying an area of 21.3 square miles, while that of 1872 included a group of industrial communities on the south side of the Mononga-hela River having a population of about thirty-six thousand and occupying 4.2 square miles.

While the annexations were an important step toward a "Greater Pittsburgh," they also raised important questions about methods of consolidation and payment for existing debt. Advocates of a Greater Pittsburgh argued that consolidation would improve services for the annexed areas and provide for more integrated economic development and planning; those opposed were concerned about the integrity of their communities and the size of the Pittsburgh municipal debt. Small businessmen were conspicuous in opposing annexation, but the elite of the East End, were divided.[61] No uniform policy existed in regard to taxation and debt. After the East End annexation in 1868, for instance, almost $500,000 of indebtedness was charged to the "old city." The consolidation of the South Side, however, resulted in increased taxes for most of the annexed boroughs, some of which paid special taxes over and above their regular Pittsburgh taxes into the twentieth century.[62]

The development of this newly annexed territory and the supply of services to the older areas of the city and the expanding central business district (CBD) depended not only on public policy and/or private initiatives but also on various combinations of the two. Government provided some services while private companies operating under the terms of government franchises and contracts supplied others. Transportation was provided by a number of traction companies under franchises granted by the city councils. The Magee-Flinn political machine largely controlled the councils in the 1880s and 1890s, and strong links existed between the traction companies and the bosses. These links as well as the terms of the franchises have not been fully examined, but the actual process of traction development and its impact on the city can be traced.

During the 1870s and the 1880s, the city's four original horsecar lines were extended and ten additional companies chartered and developed. Existing lines were extended to distances about five miles from the city; feeder lines and competitive parallel lines were developed; and, where topography permitted, crosstown lines were constructed. By 1888 there were 172 horsecars running on about 56 miles of track; they carried over 23 million passengers for the year or 68.2 rides per Pittsburgh inhabitant. In the years from 1888 to 1890, three horsecar lines operating to the east shifted from horse to cable, and by 1890 the city had 15.5 miles of cable line with cars that averaged twice the speed of horsecars. In addition, commuter rail service was extended and by 1890, 235 commuter trains a day entered the city; for the year, they carried 2,698,633 passengers. Finally, beginning in 1870, several inclined-plane passenger railways were put into operation and helped open the hilly areas of the city, especially on the South Side, to settlement. The transportation system, however, was badly integrated. Many separate companies held franchises and no single streetcar line followed a route through the downtown area and across the bridges to another part of the region.

Transfers existed between some lines but the privilege was not universal, resulting in passenger delays and inconveniences.[63]

Two major developments occurred in the 1890s: electrification of the horsecar lines and consolidation of street railway companies. Pittsburgh was an early innovator in electric traction, and by 1896 all major Pittsburgh traction companies not using cable had converted to electricity. Pittsburgh entrepreneurs not only electrified existing lines but also invested in new routes. The region's total street length of track increased from 113.3 miles in 1890 to 337 miles in 1896 and to 469.5 by 1902. Passenger traffic rose 264 percent from 46,299,227 passengers in 1890 to 168,632,339 in 1902, while rides per inhabitant jumped from 108 to 263.[64] The capital needs of electrification also drove consolidation, and by 1902 almost all Pittsburgh streetcar lines had been absorbed by the Pittsburgh Traction Company, controlled by George Westinghouse's holding company, the Philadelphia Company.

Private firms holding municipal franchises developed other infrastructure systems. These related to the transmission of messages, the provision of electricity, and the supply of natural gas. In 1851 Pittsburgh became linked via the telegraph with a network of other cities. By 1874, a business telegraphic network and a district messenger service had been established to facilitate communication within the city. When the telephone became available in the late 1870s, the business community quickly adopted it, and by the turn of the century the telephone had supplanted the telegraph for intraurban business communications and had begun to make inroads into the residential markets. By 1900, 30 percent of the telephones in service were residential.[65]

Operators of electric power companies also focused service initially in the downtown. The Allegheny County Light Company was founded in 1880, just six months after Thomas Edison had demonstrated the first practical electrical incandescent light bulb. The initial pattern in electrical development was similar to that of traction: many small companies, each with a municipal franchise for a limited territory, divided the market, supplying direct current to users; many firms also generated their own power. In 1887, however, George Westinghouse's Allegheny County Electric Company began supplying alternating current to downtown users and by 1900 his Philadelphia Company controlled most of the county's electrical system, providing power to the CBD, to industry, and to residences. Natural gas was also an important energy source in Pittsburgh in the 1880s and the 1890s. Based on the exploitation of wells close to the city, it was piped into many glass and steel firms as well as into several residential areas, providing some relief from the smoky skies produced by local coal.[66]

The expansion and electrification of the Pittsburgh traction system caused large changes in space utilization and in the city's settlement patterns. Beginning in the 1880s, three important city-building trends accelerated. These were the displacement of residential population from the downtown wards and their development into a commercial, office, and shopping center or CBD; the industrial movement from downtown to the urban periphery; and residential development of the city's outlying areas.[67] High-rise skyscrapers in the downtown, integrated steel mills along the river flood plains, and housing developments formed from the subdivision of farms and estates provided visual evidence of these trends.

The transformation of the downtown into a CBD occurred as former residents moved elsewhere. From 1870 to 1890, residential density in the four downtown wards (the oldest area of the city) dropped slightly from 47.9 to 45 persons per acre, and from 1890 to 1900 it slid to only 29.1 persons per acre. What had been an area devoted to mixed industrial, commercial, and residential uses now lost almost half its population as well as the institutions, such as churches, that served it.

Manufacturing also moved out, pressured by rising land costs and the need for space to expand. Those warehouses and factories that remained were largely confined to land near the rivers and in the floodplains.[68] But many large firms that moved their manufacturing facilities continued to maintain their central administrative offices in the CBD, reflecting the separation of manufacturing from the central office that was an important step in the evolution of the modern corporation. The downtown transformation began in the 1880s, when hundreds of new "business blocks," took the place of "old soot covered structures." In 1893 Pittsburgh's first steel-frame skyscraper, the thirteen-story Carnegie Building, was completed, and the profile of the CBD began to change dramatically. A number of skyscrapers were built in the first decade of the twentieth century, several of which were nineteen stories high. Bankers and industrialists constructed many of these buildings, pointing to the growth of mass production industries in Pittsburgh that required specialized and differentiated administrative tasks.[69]

While the downtown was losing residential population, other formerly sparsely populated areas of the city were gaining. In 1870, after the East End annexation (1868), Pittsburgh had a density of 5.8 persons per acre. Densities, however, varied greatly throughout the city. The first four wards had a density of about 48 persons per acre, while wards five through twelve, covering the belt of land annexed in 1846, had a density of 54 persons per acre. The enormous East End, much of its land in large estates, had a density of only 2.25 persons per acre. The South Side communities annexed in 1872 had a density of 11.7 persons per acre in 1870.[70] Thirty years later, in 1900, the city's density had increased to 17.9 and that of the East End wards to 12.4. An extensive residential building boom occurred in the outlying wards. Between 1870 and 1900, for example, the number of dwellings in the land annexed in 1868 increased from 5,350 to 28,278. The new housing included middle-class, suburban style detached homes, workers' row houses, and mansions for the wealthy.[71]

Government played an increasingly important role in the city-building process as the city expanded, although in a somewhat different manner than in the antebellum period. During these earlier years, the municipality as well as the state had engaged in a number of mixed-enterprise projects, such as the gas works, or projects in which public capital was invested in private corporations, such as the railroads. After the Civil War, however, such activities by municipal or state government were either constitutionally prohibited or in political disfavor. Other methods of linking government and development, often extralegal, were resorted to.[72] Little is known about the process of municipal politics in the post–Civil War years, but such information that is available suggests that it was a period of considerable turmoil and a search for organizational forms.[73]

Several important governmental changes occurred when the Republican political machine of Christopher L. Magee and William Flinn, formed in 1879, was in power. The political machine organized the political domain and formed links with private interests involved in city-building, which politicians regarded as a source of patronage and party funds. One of their primary functions in various cities was to centralize political power that had previously been badly fragmented, hampering development.[74] The machine's consolidation of political power was especially useful to businessmen anxious to secure franchises and contracts from municipal government. Magee and Flinn appear to have supported changes in the structure of Pittsburgh city government in order to better control administrative policies and the city-building process. The changes they sponsored also enabled them to channel contracts and patronage to their own companies and those of their followers.[75]

Magee was a member of the Pittsburgh elite who served as city treasurer, fire commissioner, and state senator at various times during the last three decades of the century. A student of politics, he supposedly visited New York City after the collapse of the Tweed ring to study its strengths and weaknesses.[76] In 1870 his fledgling organization led the way in the reorganization of the fire department, a move strongly supported by the business community. Unlike the pattern in other cities, this change was accomplished without conflict because volunteers were absorbed into the new department. They also provided Magee with a strong base of political support. Magee also increased his own personal economic assets. By 1892 he owned the Pittsburgh *Times* and served as president of two transit companies and director of five others. Flinn was chairman of the city and county Republican organizations and state senator in the same period. He was a partner in the construction firm of Booth & Flinn, the company that handled most of the city's construction during the last few decades of the century.

The Magee-Flinn machine controlled the city councils for most of the 1880s and 1890s. It had strong links with streetcar companies and other public utilities as well as with banks that held public deposits. Small businessmen such as saloonkeepers, contractors, building supply proprietors, and real estate and insurance agents, many of whom could benefit directly from city contracts, made up much of the machine's membership. Few blue-collar workers or immigrants were involved. The machine's representatives on the common councils were drawn from the same social groups as its general membership, but its candidates for citywide office were wealthy and socially prominent individuals.[77]

A major reorganization modernizing the structure of Pittsburgh government occurred in 1887, when the state legislature approved a charter act reorganizing the city government and creating a set of executive departments. The old council arrangement of standing committees, which dated back to 1816, was inadequate to meet the city's growing needs and suffered from serious problems in coordination, implementation, and financial management.[78] The 1887 act formed the departments of Public Works, Public Safety, and Charity (heads appointed by the councils) and gave the councils the authority to form others. The councils later provided for a Board of Assessors and Awards, composed of the mayor and three departmental heads with the authority to let public contracts. The new departments replaced twenty of the city councils' standing committees, although control remained in the hands of the councils rather than the office of the mayor. Included in the Department of Public Works were the bureaus of Engineering and Surveys; Highways and Sewers; City Property; Public Lighting; Water Supply and Distribution; the Water Assessor; Re-paving; Viewers; Public Parks; and Bridges.[79]

Although they have not been closely examined, these reforms appear to have strengthened the ability of the Magee-Flinn machine to control Pittsburgh government.

The Department of Public Works was the most significant governmental unit in the city-building process, and Magee and Flinn pushed through the councils the appointment of Edward M. Bigelow, a cousin of Magee's, as director. Bigelow was a young civil engineer who had been appointed city engineer in 1880, a post that had been created in 1872 in response to the developmental challenges presented by the newly annexed territory. The appointment of trained professionals to municipal posts was a major advance, representing the growing bureaucratization of city government regardless of the group in power, but we know relatively little about the use of trained professionals in Pittsburgh during these decades of governmental change.[80] Bigelow ran the department for over fifteen years, directing a vast program of activities. His most visible achievement was the creation of a park system, but he was also responsible, along with Magee and Flinn, for equipping large areas of the city with streets, water pipes, and sewers. As Barbara Judd notes in her study of Bigelow and the parks, his engineering background and familiarity with European and American planning provided him with a vision of Pittsburgh's urban possibilities.[81] In return for a relatively free hand in directing his department, Bigelow channeled most municipal contracts to Flinn's construction company. He awarded contracts to the lowest "responsible" bidder in spite of price—normally Booth & Flinn, Ltd.[82]

As large tracts of lands were subdivided, the city provided urban services and granted franchises to private companies for transit lines, telegraph and telephone systems, and electrical power. Streets, water pipes, and sewers remained the most important municipally supplied infrastructure. Because of spending excesses in the immediate postwar years, the new state constitution of 1874 imposed a municipal debt limit of 7 percent of the assessed value of taxable property, but various financial devices were utilized to make possible major construction. Municipal debt soared over the limit in these years primarily because of street and waterworks construction, most of which was performed by Booth & Flinn.[83]

The greatest amount of infrastructure was constructed in the East End, then emerging as a desirable residential area for the city's growing middle class as well as for the elite.[84] In April 1870, the city councils approved the so-called Penn Avenue act, under which approximately three-quarters of the total cost of Pittsburgh street improvements for the next seven years were incurred. Under the terms of the act, street commissioners elected by the property owners in improvement districts determined pavement types and entered into construction contracts. The city then issued bonds covering the contract obligations, and the commissioners assessed property owners for the costs of improvements in proportion to their property's street frontage. The act was originally intended only to apply to Penn Avenue, but it eventually provided for more than thirty streets, mostly in Oakland, East Liberty, and Shadyside. Total assessments under the act were approximately $5.6 million, more than half of which represented assessments on property owners who had not petitioned for improvements.[85]

The act played a significant part in opening up the East End, improving major thoroughfares such as Penn, Liberty, Highland, Forbes, and Fifth avenues as well as many smaller streets.[86] In 1875 the city engineer observed that it had greatly increased the value of suburban properties: "Rural homes assumed an attractiveness to the eyes of many who had never thought of going outside the old portion of the city. The paved roads provide easy and quick access to the 'Rural District,' where the enjoyment of the country could be combined with the conveniences of the city."[87] Many property owners objected to the assessments, however, and they challenged the Penn Avenue act in the courts. In 1879 the Pennsylvania Supreme Court declared the act unconstitutional, and in 1881 the city reached a compromise settlement that sharply reduced the assessments.[88] In addition to the East End, street openings and improvements also took place in other parts of the city. By 1887, 134 of 250 miles of city streets were paved; in 1910, the total had increased to 370 miles. The areas that were the slowest to receive paving were working-class neighborhoods, where streets remained unpaved well into the twentieth century.[89]

The city's rapid growth required increased supplies of water and the extension of water lines. By 1871, the city established a water commission, and in 1879 it opened a new waterworks that drew water from the Allegheny River into reservoirs on Highland Avenue and in Herron Hill. Enlargements in pumping capacity occurred in the following years to meet growing demand.[90] From 1889 to 1900, the city built a yearly average of 15.4 miles of pipe, as the water supply network increased in length from 268 miles in 1895 to 743 miles in 1915. The system was plagued by extensive waste and faulty pipes, resulting in frequent water shortages and campaigns by the Department of Public Works to induce citizens to cut water usage. In the 1890s the department began substituting technology for persuasion by installing water meters.[91]

More serious than the waste was the increasing pollution of the water supply. Pittsburgh drew its water from both the Allegheny and the Monongahela rivers into which more than three hundred fifty thousand inhabitants in seventy-five up-river municipalities discharged their untreated sewage.[92] The resulting pollution gave Pittsburgh the highest death rate from typhoid fever of the nation's large cities: 102.5 per 100,000 people from 1883 to 1907, and 130 per 100,000 people from 1899 to 1907. In contrast, in 1905, the average for northern cities was 35 per 100,000 persons. In the 1890s, investigations of the water supply were conducted using the new methods of bacterial science. These conclusively demonstrated the relationship between typhoid and the quality of the water, and in 1896 the councils approved an ordinance authorizing the mayor to create a Pittsburgh Filtration Commission. The commission's investigations reconfirmed the link between water and disease, and its report in 1899 recommended construction of a slow-sand filtration plant as the most economical means of dealing with the public health problem. In 1899 voters approved a bond issue for plant construction, but factional political battles over control of construction contracts necessitated a second vote in 1904 and delayed final completion of the filtration plant until the end of 1907. Once in operation, the filtration system had dramatic effects, and by 1912 Pittsburgh's death rate from typhoid fever equaled the average for the largest American cities.[93]

Construction of a sewer system followed the expansion of the water supply network. In 1870 the city had only five miles of sewers, mostly in the downtown; by 1875 mileage had increased to about twenty-five (thirteen miles of brick sewers and eleven miles of pipe sewers), mostly for storm water drainage. These sewers suffered from design faults and were often either undersized or oversized and subject to constant clogging. The city had no topographical maps until the 1870s, and sewers did not conform to topography; neither did they follow an overall engineering plan. Rather they were often built as a result of council members' attempts to meet the demands of their constituents. In 1881 a New York engineer, J.J.R. Croes, told city officials: "You have no sewers; you don't know where they are going, or where they are to be found."[94] Without sewers, the great majority of households in the city depended on cesspools and privy vaults for disposal of domestic waste.

Until the 1890s, sewers were almost completely lacking in the newly annexed areas of the East End and the South Side. An 1887 East End citizens' committee complained that "in warm weather many parts of the East End are absolutely unfit for habitation owing to the polluted atmosphere arising from open runs of filth of every description." Conditions in the South Side, where population density and industrial development were greater, were even more unsanitary. In 1890, for instance, the chief clerk of the Bureau of Health appealed to the city councils to install sewers in the South Side and thereby "number the lives which could be saved to the community, saying nothing else of the value of the thousands of days lost by the attendant sickness."[95]

Debate raged between different professional groups and politicians about the design of the sewerage system. Should it be a separate, small pipe system that carried only domestic and industrial wastes or a larger, combined system that could accommodate both waste water and storm water?[96] The city's public health and engineering professionals divided over this question. Physicians argued that the separate system was preferable because it would protect health by removing wastes from the household before they had begun to generate disease-causing sewer gas. Storm water was a secondary matter and could be handled by surface conduits. Engineers took a different position and maintained that sanitary wastes and storm water were equally important; therefore, a large pipe system that would accommodate both was more economical. The superior virtues of the combined system in terms of both health and storm water removal convinced city officials, and by the late 1880s Pittsburgh had begun to build a system of large combined sewers. Between 1889 and 1912, civil engineers from the new Bureau of Engineering of the Public Works Department constructed over 412 miles of sewers, almost all of the combined type.[97] The construction of this planned sewerage system signified a movement away from the "piecemeal, decentralized approach to city-building characteristic of the 19th century."[98]

City residents often attempted to keep their old privy vaults and cesspools and resisted connections to the new sewer lines because of costs but the Board of Health used the sanitary code to compel connection.[99] In 1888 the councils barred the construction of cesspools where sewer service was available, and in 1901 they outlawed water closets from draining into a privy vault and prohibited the connection of privy wells to a public sewer. The Bureau of Health ordered the cleaning or removal of thousands of privy vaults, although the effect of these orders was limited by a small staff of inspectors as well as collusion between inspectors and scavenger firms.[100]

The city financed sewer construction through several different fiscal instruments. Up to the middle of the 1890s, the city had paid for main sewers by assessing the charges to whole neighborhoods while charging the cost of lateral sewers to abutting property owners.[101] In 1895, however, the Supreme Court declared the city's assessment practices in regard to main sewers unconstitutional and forced it to assume the whole burden. To fund their liabilities, and to provide for future needs, the city resorted to bonds, some of which required the voters' approval (People's Bond Issues) and some of which were voted by the councils (Councilman's Bond Issues). Lateral sewers continued to be financed by assessing abutting property holders, although homeowners complained constantly about inequities in assessments.[102]

The quality of the built environment and of the infrastructure from neighborhood to neighborhood was very uneven. While the differences over infrastructure development prior to the Civil War often reflected a territorial split between old and new wards, the divisions in the late nineteenth century tended to follow class lines. That is, while middle-class areas often complained about delays in providing service, working-class neighborhoods actually suffered most severely from municipal underinvestment and poor services. Since abutters were required to pay for street improvements and water and sewer services, wealthy areas with a high proportion of homeowners benefited first, while working-class sections having largely tenants were slow to receive services. Many streets in working-class neighborhoods remained unpaved well into the twentieth century at a time when the growing middle-class districts were receiving new boulevards and smooth paving.[103]

Particularly costly were inequities in water and sewerage services. Working-class districts generally had poorer water supplies than did affluent neighborhoods.[104] The City Water Commission had ruled in 1872 that the size of the pipe laid on a particular street would be determined by the amount of potential revenue. This ruling resulted in either insufficient supply or no supply at all to poor neighborhoods. Many working-class areas relied on pumps for their water supply, and even when piped water was available, it was often accessed through a spigot in the back yard (frequently located near the privy vaults) rather than through indoor plumbing. The health of working-class families suffered from the insufficient water supplies. While the entire city had excess typhoid death rates before 1907, they were disproportionately high in working-class areas. Adding to the environmental health hazards in these districts was the lack of sewer services and reliance on cesspools and privy vaults well into the twentieth century.[105]

A similar pattern of achievement and inequities was reflected in the city's creation of a system of parks between 1867 and 1893, largely through the efforts of Edward M. Bigelow. Bigelow believed that a system of landscaped parks connected by spacious boulevards would make Pittsburgh more attractive. He considered parks, as did Frederick Law Olmsted, as "breathing spots" and important instruments for "the elevation of the people."[106] His two greatest achievements were Highland and Schenley parks, both located in the eastern part of the city in the land annexed in 1867. Used mostly by the upper and middle classes, these parks were difficult of access for working-class people who customarily did not use them except on major holidays. Bigelow attempted to persuade the councils to create small parks in working-class residential areas but was unsuccessful.[107]

The last third of the nineteenth century saw striking alterations in Pittsburgh and its built environment. In the late 1860s, although quite dynamic as an urban area, Pittsburgh was still primarily a pedestrian city of relatively small size and population and with limited urban services. By the beginning of the twentieth century the city had increased greatly in area and population, with extensive residential neighborhoods, a growing CBD, and sprawling industrial districts. Networks of pipes, tracks, and wires spread throughout the city, accommodating flows of water, sewage, people, and messages; bridges spanned the rivers and the many ravines that dotted the landscape; and a major park sytem had begun. The positive elements, however, were offset by the low quality and haphazard design of much of Pittsburgh's built environment. Poor planning and poor engineering, compounded by the city's topographical complexities, accounted for many of the inadequacies.

The Magee-Flinn political machine was responsible for much of the infrastructure. It centralized power within a decentralized governmental system and formed ties with entrepreneurs and businessmen willing to engage in payoffs in order to profit from city franchises and contracts. Corruption, therefore, was integral to the city-building process, but it was organized rather than unorganized corruption. City-building under the aegis of the machine was probably more systematic than before the machine took power, but it still raised questions of efficiency, equity, and ethics. The quality of life in Pittsburgh was often undesirable, especially in the working-class immigrant neighborhoods. Here polluted water, inadequate sewerage, unpaved streets, and unsanitary housing resulted in extremely high morbidity and mortality rates.[108] But not only the working class suffered from Pittsburgh's polluted environment and shoddy infrastructure. Corporate leaders and members of

Pittsburgh's growing professional class wanted change, and correcting the worst effects of industrialization and improving governmental planning and effectiveness became major tasks for both public and private sectors in the twentieth century.

Centralization, Decentralization, and the Automobile, 1900–1944

Between 1900 and 1930, Pittsburgh experienced further population and territorial growth, as it matured into an industrial and corporate center. In these years its population increased from 321,616 to 669,817 and its land area grew from approximately twenty-eight to fifty-four square miles. Great changes occurred in Pittsburgh's built environment during these decades, as the city took on the form of a modern metropolis. Pittsburgh was hard hit by the Depression, however, and after 1930 population and territory ceased to grow and conditions throughout the city disintegrated.

Roy Lubove's masterful *Twentieth-Century Pittsburgh*, which focuses on housing, environmental reform, and the planning process in a broadly interpretative framework, provides the major introduction to these years. Lubove describes how, at the beginning of the century, Pittsburgh's business and professional elite formed voluntary associations to push for the centralization of decision making, the further involvement of experts and professionals in the city-building process, and the elimination of the localistic and particularistic focus fostered by the machine. Reformers also sought to change the composition of the city councils from bodies composed of members of the working and lower-middle classes to representatives of the professions, large businesses, and the upper classes.[109] While both private and public sectors were involved in reform, the voluntary institutions "defined the issues and areas of intervention."[110] Volunteerism permitted the business elite to extend its influence by centralizing decisionmaking while emphasizing noncoercive change consistent with the values of limited government.

Countervailing these centralizing tendencies were the forces of decentralization and localism, reflected in the city's many neighborhoods, its ward governmental structure, and its political organizations. The effects of a radical new mobile transportation technology—the automobile—paradoxically increased the need for centralized planning of infrastructure and accelerated the dispersion of the residential population into decentralized areas of the county outside the city's boundaries (the county had 122 separate minor civil divisions). Much of the political history of twentieth-century Pittsburgh can be understood in terms of the tension between these forces of centralization and decentralization, and between the ideals of planning and the realities of political responsiveness to the demands of constituents and special interests.

Public works was a major element in the contest between the forces of centralization and those of decentralization. Reformers and urban professionals usually favored centralized planning conducted by experts and objected to infrastructure decisionmaking based upon political logrolling or favoritism. The machine used infrastructure construction as a means to secure patronage, to obtain money for itself and its followers, and to insure neighborhood political support. The reformers sought to end the political uses of infrastructure, and to secure a series of other governmental goals, by altering the structure of Pittsburgh government.

The charter reforms of 1901 and 1911 reflected the reform goals. The 1901 charter greatly increased the mayor's power, giving him administrative authority over the departments, including the Department of Public Works. The two councils lost much of their administrative power, although they were still elected by wards. The 1911 charter revision went further in the direction of centralization by abolishing the two-house, ward-elected council and replacing it with a single nine-member council elected at large. This step reduced the ability of the machine to control the council and supposedly changed its focus from neighborhood to citywide issues.[111] The 1901 charter revision, however, was not only the product of reformers attempting to assert the values of centralization over machine politics. It was also the result of political maneuvering between the Magee-Flinn Pittsburgh machine and the state machine of Senator Matthew S. Quay, and involved Edward M. Bigelow and the question of construction bids on the proposed water filtration plant. Although too detailed to discuss in this essay, the political battle serves as a reminder of the extent to which factional politics and reform goals could become intermingled.[112]

Voluntary associations sought a number of other governmental changes to improve the urban environment. Pittsburgh's deficiencies had been highlighted by the six volumes of the Pittsburgh Survey (1909–1914), a study characterized by its editor, Paul Kellogg, as an "appraisal, if you will, of how far human engineering had kept pace with mechanical in the American steel district."[113] In 1909 a reform mayor, George Guthrie, appointed the Pittsburgh Civic Commission, composed of leading businessmen and professionals, to set an agenda for environmental reform. The commission focused on city planning, housing, and transportation. Under its direction three leading urban professionals—a transportation specialist, Bion J. Arnold; a hydraulic engineer, John R. Freeman; and a city planner, Frederick Law Olmsted, Jr.—prepared a city planning document that provided an approach to dealing with questions such as transportation, water and sewerage, smoke and flood control, public buildings, and building code revisions. This study was followed by specialized reports dealing with aspects of the built environment and the problems of smoke and water pollution.[114]

These reports, despite their "expert" authorship, had limited results. Numerous debates and studies as well as legislation and the creation of commissions and departments followed their publication, but almost no substantial improvements occurred. In some cases a narrow definition of the problem, as in the case of smoke control, or a lack of action by the federal government, as in flood control, was partially responsible for the failure to act. In most cases, however, implementation failed because of the unwillingness of those who controlled the City Council and the county government to surrender their control of development or because of limitations of power and finances granted to the public agency. The areas of public transit and city planning illustrate these problems.

In 1909, Mayor William A. Magee asked Arnold to prepare a study of Pittsburgh public transit. His report, issued in 1910 under the title *Report on the Pittsburgh Transportation Problem*, noted deficiencies in Pittsburgh's system ranging from poor service to overcapitalization. His recommendations included the integration of the region's several transportation companies, subway construction, and electrification of the suburban railroad lines. Ultimately, Arnold argued, Pittsburgh's transportation problem would only be solved by complete "public control," if not public ownership.

In the years following Arnold's authoritative report, however, public and private groups could not agree on which transportation improvements to implement. During the 1910s and the 1920s, for instance, plans were discussed for both subways and elevated systems or for combinations of the two; for downtown loops versus through-routing; and for public ownership as opposed to private control. In 1919 voters approved, although by only a narrow margin, a $6 million bond issue for a subway that was never built. In 1917 the City Council created an Office of Transit Commissioner and then, in 1925, a Department of Public Transit with the purpose of constructing and managing a city-owned transit system. The council, however, could not agree on what type of system to build or which routes to use, and the department was eliminated in 1936. The Pittsburgh Railways Company, which was in receivership from 1918 to 1924, provided transit service, however unsatisfactory. Private ownership, although continually plagued with financial and service problems, persisted until 1964, when the Port Authority of Allegheny County consolidated thirty-three private companies under its direction.[115]

Frederick Law Olmsted's planning study, *Pittsburgh: Main Thoroughfares and the Down Town District*, prepared with the Committee on City Planning and issued by the Civic Commission in 1910, had more success than the Arnold report. Olmsted called for improved downtown traffic circulation, better CBD access, and street and roadway improvements throughout the city. He also proposed more aesthetic and effective use of Pittsburgh's riverbanks and steep slopes. Between 1911 and 1916 the city embarked on a program of street development that followed many of Olmsted's recommendations, including improved main thoroughfares from residential districts to the downtown, the elimination of grade crossings, reduction of grades, the widening of business thoroughfares, and the raising of areas subject to flooding. After the war, a $20 million bond issue provided funds for street widening, a limited access high-level roadway to the CBD from the east (the Boulevard of the Allies), and other improvements suggested by Olmsted. His imaginative ideas for riverbanks and steep slopes, however, as well as many of his other planning suggestions, were never implemented.[116]

While the council created city planning bodies in these years, they were ineffectual. The City Planning Commission, for instance, formed in 1911 following the recommendation of the Civic Commission, had only advisory powers and was little more than an "administrative eunuch."[117] A Municipal Arts Commission, approved in the same year, had veto power over the design of municipal structures and art purchased by the city, but limited itself to an educational role. A County Planning Department, formed in 1919, had even less power. Still, bureaucracies proliferated to deal with the complexities of a growing urban environment. Thus, the 1902 charter provided for creation of a Bureau of Electricity, a Bureau of Highways and Sewers, and a Bureau of Building Inspection; a Bureau of Surveys was created in 1907 (it became the Bureau of Engineering in 1914); a Bureau of Lighting in 1913; a Bureau of Tests in 1915; and a Division of Inspection in 1916, including inspectors of construction, explosives, elevators, boilers, fire escapes, and signs. These bureaus were often ineffectual and staffed with political appointees. Over the years they came and went, as politics clashed with attempts to create a more "rational" form of governmental organization and administration. Their weakness reflected not only the attempts of politicians to use them for partisan and factional purposes but also the reluctance of business and professional leaders to grant sufficient power to the public agencies to effectively regulate the private sector.[118]

Regional consolidation, another approach to dealing with problems of governmental regulations, was attempted during the first third of the twentieth century. In 1907, after a number of failed attempts, Pittsburgh succeeded in annexing its neighboring city of Allegheny, a prosperous industrial center of 150,000 population covering about eight square miles. The annexation, like the other Pittsburgh annexations, was actually forced consolidation (the "rape" of Allegheny), as the Allegheny vote was combined with that of its larger neighbor.[119] Other Allegheny County municipalities feared a similar fate, and in 1910 they formed the League of Boroughs and Townships of Allegheny County. In 1911 the league prevented passage in the state legislature of a bill for a Greater Pittsburgh that would have consolidated forty boroughs and townships, and in the 1920s it continued to prevent the city from securing wholesale annexation statutes. The strong localism of county communities also impeded cooperation in important areas such as sewage treatment and pollution control.[120] While Pittsburgh did annex almost twelve square miles of territory and all or part of eighteen contiguous townships and boroughs in the decade, the territory was scattered largely to the south and north of the city and was less significant than previous additions.[121]

Those interests, such as the Pittsburgh Chamber of Commerce, who favored a Greater Pittsburgh now embraced metropolitan government as an alternative to consolidation. Several bills providing for various forms of metropolitan or federated government were introduced in the legislature in the 1920s. Advocates of a Greater Pittsburgh favored a strongly centralized form of metropolitan organization, while those suburban political leaders willing to consider some form of cooperation supported a loose federation in order to fend off annexation. In 1929 a plan for a weak, federated consolidation of city and county reached the ballot. It included a provision permitting the formation of special districts to construct and maintain public improvements between governmental units. This plan, however, failed to secure the required two-thirds vote in a majority of the county's 122 governmental units even though it won 68 percent of the total vote. The more affluent white-collar suburbs joined with the city to favor the change, but the mill towns and outlying county districts opposed it. As Jon Teaford notes, "efficiency, economy, or metropolitan grandeur" might have been the values of the business and professional communities but they were not shared by working-class towns and semirural suburbs with a more localistic and parochial focus.[122]

The attempts to create a consolidated or metropolitan government occurred at a time when population was spreading beyond the city boundaries. Rapid growth of automobile ownership throughout the region after World War I was the prime cause. In 1910 automobile registration in Allegheny County lagged behind the national average, totaling 1,601 automobiles, or one to every 636 persons; by 1929 registration had increased to 203,866, or one car to every 6.7 persons, which was close to the national average.[123] Although it had ripple effects on commercial, industrial, and residential patterns throughout the region, the motor vehicle had its primary impacts on the CBD and on residential development in formerly underdeveloped areas of the city and on the urban periphery.

The great increase in motor vehicle use intensified traffic problems in Pittsburgh's small and constricted downtown. Between 1917 and 1927 the number of motor vehicles and streetcars entering and leaving the CBD in an average twelve-hour day increased from 21,644 to 100,343.[124] In contrast, the average number of horse-drawn vehicles in the CBD per day decreased from 8,370 to 1,823. The large increases in vehicular traffic reflected a sizable growth in uses of this district for employment, shopping, and entertainment. The lack of a bypass road for through traffic also increased street congestion, because cars and trucks destined for other parts of the city were forced to use downtown streets.

The automobile also greatly stimulated residential development in city and county. From 1910 to 1930, the wards located on the city's outskirts that had not previously developed because of access difficulties or because the land had been withheld from the market grew most rapidly. The Fourteenth Ward (the Squirrel Hill district), for instance, an affluent area which had the highest amount of automobile ownership and use in the city, had its fastest rate of development in the 1920s, especially after direct roadway connections with the downtown were made in 1922. Its population increased 85 percent compared to 13.8 percent for the city, and builders constructed over six thousand residential units (24 percent of the city total) to accommodate the population growth. In the suburbs, the automobile most affected townships and boroughs that had formerly been poorly served by public transit. South of the city, for instance, townships that were connected with the CBD via the Liberty Bridge and Tunnel grew from 35,505 to 55,802 in the decade. Automobile ownership was high in the newer suburban towns, and cars were heavily used to commute to work.[125]

Widespread use of the automobile in Pittsburgh and Allegheny County produced demands for roadway construction and posed challenges to rational planning. City and county governments formed new bureaucracies to address these questions, but like the older planning agencies they had only limited effects.[126] Frustrated by the absence of a strong planning focus, in 1918 a group of influential Pittsburgh businessmen organized the Citizens' Committee on City Plan of Pittsburgh (CCCP), a body through which they attempted to influence the "evolution of the physical environment."[127] Frederick Bigger, an architect and planner who believed in comprehensive planning and centralized decision making, directed the CCCP. Its main product was a series of six planning reports issued from 1920 to 1923 on Pittsburgh's needs in the areas of streets, playgrounds, transit, parks, railroads, and waterways. The CCCP was also the leading advocate of a Pittsburgh zoning ordinance, finally approved by the City Council in 1923.[128]

In 1922 the Pittsburgh City Council officially adopted the CCCP reports on streets and playgrounds as guides in "the expenditure of public funds" and the City Planning Commission endorsed them as "a measuring stick for further improvements." Adoption, however, did not mean implementation, and hostility to the idea of comprehensive planning as well as city and county political realities limited adoption of the planning recommendations. Frustrated by the city's failure to follow the comprehensive plan, Bigger complained in 1929 that "therein lies our community stupidity, for *piecemeal planning* leads us nowhere."[129]

The lack of comprehensive planning, however, does not mean that infrastructure was neglected. Quite to the contrary, for the 1920s witnessed the expenditure of large sums of public money on highway, bridge, tunnel, and sewer construction throughout the city and the county, as politicians responded to the demands of automobile owners, neighborhoods, and special interests. Funding for these projects was provided through large bond issues. In 1919, for instance, Pittsburgh voters approved a $20 million bond issue providing for highways, bridges, tunnels, and other infrastructure improvements.[130] The county commissioners also planned an ambitious road building and bridge program, reflecting both growth outside the city boundaries and the increasing importance of county government. In 1920 the newly elected county commissioners submitted a $35.5 million bond issue, primarily for roads and bridges, to the electorate. The voters approved the bond issue in 1924 after defeating it twice, apparently because of a lack of confidence in county government and its Public Works Department, although factional struggles in the Republican party also played a role. The bond issue included money for twenty-one bridges, over $8 million for roads, and over $1 million each for a land tunnel (the Armstrong Tunnel) and a new county office building.[131] In 1928, the voters approved a $43 million bond issue that included funds for public parks and an airport in addition to roads and bridges.

Constructing these projects often required cooperation and coordination between departments from different governmental jurisdictions. Highways and bridges involved five agencies: the City and County Departments of Public Works, the Civic Art Commission, the County Planning Commission, and the so-called Joint Planning Conference (created in 1922 with members from County Planning, City Planning, and CCCP). River bridge construction brought in the U.S. Corps of Engineers. Such fragmentation of authority frequently delayed construction. After 1911, for instance, the county had begun acquiring ownership of twenty-four major toll bridges with the intention of providing free passage. The costs of the project, however, were greatly increased in 1915 when the Corps of Engineers ordered the county to raise bridge levels to an average of forty-seven feet above full pool within the city limits. In some cases this necessitated new bridges, and the county reluctantly included funds for these projects in its various bond issues. Bridge design and aesthetics was another area that provoked interagency disputes, especially in regard to the Sixth, Seventh, and Ninth Street bridges over the Allegheny. Here, the Civic Art Commission blocked construction until the involved agencies agreed to build three identical suspension bridges and to include architects as well as engineers in the planning.[132]

Aside from bridges, the other major area of county construction involved roads and highways. Here again, planning was limited and government coordination difficult. In 1924, for instance, the County Works Department enlisted the aid of Frederick Bigger of the CCCP to help design a Major Highway Plan. The County Planning Commission approved the resulting plan but the commissioners frequently ignored it, and many of the new county roads completed between 1924 and 1931 were not in the plan.[133] Another example of a planning failure concerned attempts to mitigate downtown congestion caused by through vehicular traffic, estimated by the CCCP as 18 percent. The CCCP urged creation of an Inter-District Traffic Circuit, a proposal adopted by the City Planning Commission in 1925, but it was ignored by city and county officials, leaving local and through traffic clogging downtown streets.[134]

During the 1930s a sharp shift occurred in regard to the public provision of infrastructure in cities around the nation as well as

Pittsburgh. From approximately the 1850s through the 1920s, except for some state and county programs primarily related to roads and bridges, municipalities had themselves provided for the construction of public works. During the 1930s, however, the fiscal strains caused by the Depression nearly bankrupted many cities, and municipal spending for infrastructure construction and maintenance sharply declined. This gap was partially filled by the federal government, which became the chief supplier of public works such as roads, bridges, water systems, and sewers in order both to provide employment and to stimulate recovery.[135] The federal Public Works Administration and the Works Progress Administration were active in Pittsburgh during the Depression, but their activities and accomplishments have not yet been explored in detail. They played an important role, however, in helping the Democratic party end Republican dominance in city and county.[136]

Tensions between the interests representing centralization and those representing decentralization have existed for much of Pittsburgh's history, but they were particularly acute during the first three decades of the twentieth century. From approximately 1901 to 1918, the forces promoting centralization and governmental reform were triumphant. The political machine lost power, and reformers accomplished changes that centralized authority, promoted the values of rational administration, expertise, and efficiency, and deprived the neighborhoods of power. In other areas such as smoke control, public transit, and planning, however, gains were limited as business leaders opposed surrendering authority to the municipality. Achievements were also limited by the inability of politicians to reach consensus on programs and goals or their willingness to surrender control over development to experts and planners.

The coming of the automobile accentuated the tension between centralization and decentralization. The motor vehicle increased the need for rational and centralized planning, but also accelerated residential and business decentralization. The fragmentation of the county continued when proposals for limited metropolitan government were narrowly defeated. City planners drew up plans, and voluntary associations such as the CCCP urged their adoption, but politicians paid them little attention. Decisions regarding the built environment, especially those involving roads and bridges, were often based on the distribution of power and interests within the dominant Republican party. The decade was thus a "bleak era for those who advocated comprehensive planning and constructive public intervention in the physical environment."[137]

Focusing on planning failures, however, does not necessarily tell us how and why construction took place. The 1920s witnessed the largest amount of infrastructure construction in Pittsburgh and Allegheny County to that time. This included the longest automobile tunnel in the nation (the Liberty Tunnel), the paving and construction of many miles of city roads, a limited access roadway to the downtown, and ninety-nine bridges built by the county between 1923 and 1931. These projects were debated extensively by factions within the dominant Republican party as well as by various interest groups. Because historians have focused on negative planning experiences, however, they have largely neglected the study of the politics of infrastructure or the rationale for the choices that were made. Pittsburgh and Allegheny County were not unique in regard to the inability of the planners to control their environment or to resist the demands of politicians. They shared these problems with city planners throughout the nation. Planners had limited powers, and they could do little to alter the positions taken by powerful economic and political interests or to decisively shape the character of technological change.[138] It would not be until after World War II, with the forming of a unique public-private partnership to accomplish urban renewal in Pittsburgh, that the obstacles to planning and centralization would be largely overcome.

Post-Script: Pittsburgh Renaissances I and II, 1945–1988

In the decades after the end of World War II, Pittsburgh underwent a vast amount of urban reconstruction. Known as the Pittsburgh Renaissances, I and II, this reconstruction was accomplished through a unique public-private partership. Renaissance I extended approximately from 1945 to 1969, Renaissance II from the late 1970s to the late 1980s. These were prodevelopment movements that in some ways replicated the activities of the coalition of forces involved in infrastructure building during the second quarter of the nineteenth century.

The Renaissance grew out of an acute sense of crisis concerning Pittsburgh's future and the realization by the city's business leaders that, in a reversal of the historical pattern, only public power rather than voluntary institutions could stem the decline. The key individual from the private sector was the banker Richard King Mellon, the city's most influential business figure. Equally important in the renewal was Mayor David L. Lawrence, leader of the powerful Democratic organization, who joined with Mellon to spearhead the formation of a progrowth coalition. The Allegheny Conference on

Community Development (ACCD), formed in 1944, primarily represented the private sector in the coalition. The ACCD and its affiliated research and planning organizations, the Pittsburgh Regional Planning Association and the Pennsylvania Economy League, took the lead in goal-setting, planning, and implementation, especially in regard to the city's physical reconstruction. The public sector's role was primarily to facilitate the achievement of agreed-upon goals through the passage of legislation and its implementation, often through the power of eminent domain. Lawrence dominated the City Council, and it gave consistent approval to the agenda set by the ACCD.

The Renaissance programs were multifaceted, including both rebuilding old infrastructures and creating new ones. Environmental improvements, especially smoke and flood control, renewal of the downtown—including two new parks, Point Park and Mellon Square, and a civic arena—and the construction of freeways to the center of the city constituted the heart of the initial program. Many of the infrastructure developments had been anticipated in Robert Moses's *Arterial Plan for Pittsburgh* (1939), sponsored by the Pittsburgh Regional Planning Association, which also incorporated ideas set forth by earlier planners such as Frederick Bigger. A broad range of infrastructure projects extending from a regional sewage collection and treatment system to a new public transit system and airport were part of the agenda. Important initiatives from the voluntary sector were made in the areas of housing (Action Housing) and economic development (Regional Industrial Development Corporation). Major urban renewal efforts involved not only the Gateway Center complex in the Golden Triangle, but also large projects in the lower Hill (the Civic Arena), East Liberty, and the North Side (Allegheny Center). Like urban renewal projects elsewhere, they disrupted neighborhoods and uprooted sizable numbers of people.

The organizational key to implementing and providing for long-term management and financing of renewal was the creation of authorities that included both public and private representatives, especially the Urban Redevelopment Authority, the Parking Authority, Port Authority Transit (PAT), and the Allegheny County Sanitary Authority. These provided an extension of the bureaucratic approach to infrastructure begun in the late nineteenth century, but also provided a means to surmount the politics that often interfered with planning by professionals. While both public and private sources supplied funding for new construction, in the years after 1949 federal programs increasingly provided important contributions.[139]

The first Pittsburgh Renaissance represented the triumph of centralization over decentralization and the willingness of the private sector to accede to unprecedented levels of public intervention. But behind this willingness lay an agreement between the two sectors about the need for downtown and regional renewal as well as trust between corporate and political leaders. Historians, in particular Roy Lubove, have sketched the overall outlines of the first Pittsburgh Renaissance, and Michael P. Weber has published a major biography of David Lawrence.[140] In addition, we now have in-depth studies of projects and organizations such as the Allegheny County Sanitary Authority, Point State Park, and Action Housing.[141]

While agreement on goals and policy between the major public and private sector actors produced the Renaissance, a number of other urban actors—such as those displaced by urban renewal, neighborhoods who resented the emphasis on bricks and mortar downtown development, and citizens who objected to higher taxes—began to question the priorities of the public-private partnership in the 1960s. In 1969 this discontent came to a head with the election of an independent Democrat, Pete Flaherty, as mayor on a platform of "I'm nobody's boy." Flaherty claimed to be without ties to the Democratic organization, the business establishment, or to labor. He promised to balance the budget, to restore the neighborhoods, and to provide new and independent leadership. Flaherty's agenda was clearly different from those of his immediate predecessors. Development was not high on his list of priorities, and he consistently underinvested in infrastructure, limiting new construction and maintenance of the old. He successfully blocked one major addition to the transportation infrastructure proposed by the public-private partnership—the so-called Skybus, a fully automated, soft-tired vehicle that ran on an elevated concrete runway. (In 1976, however, Flaherty agreed to a compromise transit plan that resulted in construction of a light rail system with a downtown subway loop that was completed in the latter half of the 1980s). Rather than being viewed simply as an aberrant antidevelopment force, however, Flaherty represented a revival of that strain of decentralized politics that had always been a force in Pittsburgh. His administration is worthy of study from this perspective as well as for its uniqueness in post–World War II Pittsburgh politics.[142]

The most recent chapter in the Pittsburgh city-building and rebuilding process is Renaissance II, which dates through the administrations of Mayor Richard Caliguiri, 1977–1988. Caliguiri had originally run on a platform that promised to restore the public-private partnership that had been responsible for Renaissance I and to move the city into another era of progress. Pete Flaherty had actually taken steps in 1975 and 1976 toward a partial restoration of the partnership, indicating how difficult it was to maintain a politics of independence in a city with a powerful, development oriented corporate and downtown business establishment. While Renaissance II represented a new phase of the old partnership, it also had different dimensions. The public sector, for instance, had a much more important role in goal-setting than in Renaissance I. In addition, while there was a renewed emphasis on downtown development, Caliguiri also sponsored many neighborhood programs. Renaissance I had only a few dominant actors on both sides of the partnership, but leadership in II was much more diverse, with no equivalent to either Richard King Mellon or David Lawrence. Thus, in a sense, it can be argued that Renaissance II under Mayor Caliguiri attempted to combine and represent both the forces of centralization (the business establishment) and of decentralization (the neighborhoods).[143] From the perspective of infrastructure, the major accomplishments of the second Renaissance were the completion by PAT of the light rail system and the opening of busways to the south (1977) and to the east (1983), the formation of the Pittsburgh Water and Sewer Authority to begin reconstruction of a badly neglected water system, the rebuilding of a major downtown thoroughfare (Grant Street), and the signing of an agreement providing for construction of a new airport for the city and region.

This essay has reviewed the history of city-building in Pittsburgh and especially the creation of urban infrastructure from the late eighteenth century through the 1980s. For much of the period a rich but largely unpublished literature exists, making it possible to recreate the history's major outlines. Each of the four time periods discussed has a general unity involving the interaction of technology and spatial change, and politics and policy. A number of areas, however, especially in the more recent period, still need to be investigated. These include urban politics and voting patterns in regard to city-building, municipal and county budgetary and fiscal patterns, and studies of the several authorities formed in Renaissance I, especially the Urban Redevelopment Authority and Port Authority Transit. In addition, studies are needed of the differential effects of public policies on neighborhoods in both Renaissances, of major infrastructure projects such as the airport and parkways, and of the basic patterns of infrastructure extension and maintenance. On the private side we still lack comprehensive studies of the Renaissance's major private-sector organization, the ACCD, or of infrastructure networks such as those for energy and communications.

The existing literature reveals different sets of tensions through-out Pittsburgh's history that affected the city-building process. One set related to the conflict between those who already controlled a portion of the city's resources and those who wanted to capture some for themselves. This tension was reflected in infrastructure politics and policies throughout the history, especially in relation to divisions between the old and new areas of the city and between the downtown and the neighborhoods. Another set of tensions derived from the struggle between those groups who favored centralization of institutions and decision making and those who favored decentral-ization. This tension, for instance, was reflected in the contest be-tween the political machine and reform groups about who should control development. Class interests have also often played a role in the allocation of Pittsburgh infrastructure, while at other times po-litical factors involving inter- or intraparty competition have been decisive. Thus, this review of infrastructure and city-building in Pittsburgh demonstrates the extent to which a plurality of urban actors and forces representing different constituencies rather than any single group or class has determined the city's form and shape and, in turn, the environment in which many thousands of Pitts-burghers have lived.

NOTES

This essay has benefited greatly from the comments provided at different stages of its development by John Rowett, John Modell, Charles Jacobson, and Mark Rose, in that order. None should be held responsible for the gaucheries of the author.

1. See, for instance, the discussions of urban infrastructure in Royce Hanson, ed., *Perspectives on Urban Infrastructure* (Washington, D.C.: Na-tional Academy Press, 1984), and Jesse H. Ausubel and Robert Herman, eds., *Cities and Their Vital Systems: Infrastructure Past, Present, and Future* (Washington, D.C.: National Academy Press, 1988).

2. See, for instance, the valuable but limited study of James H. Thomp-son, "A Financial History of the City of Pittsburgh, 1816–1910" (Ph.D. diss., University of Pittsburgh, 1948).

3. For a bibliography of this literature, see Suellen M. Hoy and Michael C. Robinson, comps, and eds., *Public Works History in the United States: A Guide to the Literature* (Nashville, Tenn.: American Association for State and Local History, 1982).

254 JOEL A. TARR

4. Samuel P. Hays, "The Development of Pittsburgh as a Social Order," *Western Pennsylvania Historical Magazine* 57 (Oct 1974): 435.

5. Christine M. Rosen, *The Limits of Power: Great Fires and the Process of City Growth in America* (New York: Cambridge University Press, 1986).

6. For a summary of infrastructure and service developments in other cities, see Joel A. Tarr and Josef Konvitz, "Patterns in the Development of the Urban Infrastructure," in *American Urbanism: A Historiographical Review*, ed. Howard Gillette, Jr., and Zane L. Miller (New York: Greenwood Press, 1987), pp. 195–226.

7. Quoted in City of Pittsburgh, Department of Public Works, *The City of Pittsburgh and Its Public Works* (Pittsburgh, 1916), p. 35.

8. The best examination of the transition from the mercantile city to the industrial city is James C. Holmberg, "The Industrializing Community: Pittsburgh, 1850–1880" (Ph.D. diss., University of Pittsburgh, 1981). Holmberg's primary concern, however, is shifting community patterns of association and integration rather than the city-building process, although much about this process can be gleaned from his text.

9. Richard C. Wade, *The Urban Frontier: The Rise of Western Cities, 1790–1830* (Cambridge, Mass.: Harvard University Press, 1959), pp. 43–49; Catherine E. Reiser, *Pittsburgh's Commercial Development, 1800–1850* (Harrisburg: Pennsylvania Historical and Museum Commission, 1951).

10. John Swauger, "Pittsburgh's Residential Pattern in 1815," *Annals of the Association of American Geographers* 68 (1976): 265–77; Joel A. Tarr, *Transportation Innovation and Spatial Change in Pittsburgh, 1850–1934*, Essays in Public Works History, no. 6 (Chicago: Public Works Historical Society, 1978), pp. 2–4.

11. See Philip Bateman, "Early Reapportionment in Pittsburgh: 1845–1846" (seminar paper, Department of History, Carnegie Mellon University, 1983); Douglas Brown, "Local Competition and Internal Improvements" (seminar paper, Graduate School of Public and International Affairs, University of Pittsburgh, 1971), pp. 6–7.

12. Holmberg, "Industrializing Community," pp. 14–15.

13. John Teaford, *City and Suburb: The Political Fragmentation of Metropolitan America, 1850–1970* (Baltimore, Md.: The Johns Hopkins University Press, 1979), pp. 32–63.

14. Richard L. McCormick, "The Party Period and Public Policy: An Exploratory Hypothesis," *Journal of American History* 66 (Sept. 1979): 283–89. See also, Paul Kantor with Stephen David, *The Dependent City: The Changing Political Economy of Urban America* (Boston: Scott, Foresman/Little, Brown, 1988), pp. 37–61.

15. Louis Hartz, *Economic Policy and Democratic Thought: Pennsylvania, 1776–1860* (Chicago: Quadrangle Books, 1968), pp. 37–180.

16. For discussion of the "mixed enterprise policy," see ibid., pp. 82–96.

17. For property ownership by Monongahela Bridge stockholders, see Philip R. Werner, "The Transformation of Urban Land Use by Bridges: Pitts-

burgh, 1800–1860" (seminar paper, Department of History, Carnegie Mellon University, 1982), p. 9.

18. An attempt was made in 1846 to make the Allegheny River bridge free, but it failed. See Thompson, "Financial History," pp. 40–41. For the Roebling bridges, see D. B. Steinman, *The Builders of the Bridge: The Story of John Roebling and His Son* (New York: Harcourt, Brace, 1950), pp. 90–100, 205–15. A further example of state investment in private enterprise involved the Monongahela Navigation Company, chartered in 1828. Here the state subscribed to $30,000 worth of stock compared to a private subscription of $18,360. The state legislature often inserted requirements into banks' charters that they assist specified transportation companies. The Monongahela Navigation Company and three turnpike companies connecting to Pittsburgh were assisted in this manner. See Hartz, *Economic Policy*, pp. 46, 85.

19. Jon C. Teaford, *The Municipal Revolution in America: Origins of Modern Urban Government, 1650–1825* (Chicago: University of Chicago Press, 1975); Roger Lane, *Policing the City: Boston, 1822–1885* (Cambridge, Mass.: Harvard University Press, 1967), pp. 6–8.

20. Wade, *Urban Frontier*, pp. 79–82.

21. Thompson, "Financial History," pp. 7–8.

22. John Dankosky, "Pittsburgh City Government, 1816–1850" (seminar paper, Department of History, University of Pittsburgh, 1971), pp. 11–14.

23. Thompson, "Financial History," pp. 14–25.

24. Ibid., pp. 12–13.

25. For a pathbreaking book on urban finances, see Terrence J. McDonald, *The Parameters of Urban Fiscal Policy: Socioeconomic Change and Political Culture in San Francisco, 1860–1906* (Berkeley and Los Angeles: University of California Press, 1985); Terrence J. McDonald and Sally K. Ward, eds., *The Politics of Urban Fiscal Policy* (Beverly Hills, Cal.: Sage Publications, 1984).

26. Wade, *Urban Frontier*, pp. 83–87. The city government also handled paving and cleaning of the wharves and regulated boat traffic. See Reiser, *Pittsburgh's Commercial Development*, p. 133.

27. The first street paved was Market Street, in 1802. A visitor to the town in 1806 complained that aside from Market, all other streets were so "extremely miry, that it is impossible to walk them without wading over the ankle." See Wade, *Urban Frontier*, pp. 83–84.

28. Viewers were appointed by the Court of Quarter Sessions.

29. Michael Holt, *Forging a Majority: The Formation of the Republican Party in Pittsburgh, 1848–1860* (New Haven, Conn.: Yale University Press, 1969), pp. 65–66; Thompson, "Financial History," pp. 104–05.

30. Thompson, "Financial History," pp. 24, 38–40, 81–83.

31. The city was given the right to purchase the works after twenty years. Ibid., pp. 38, 47–49.

32. Reiser, *Pittsburgh's Commercial Development*, p. 129; Leland D. Baldwin, *Pittsburgh: The Story of a City* (Pittsburgh: University of Pittsburgh Press, 1938), p. 156; and Wade, *Urban Frontier*, p. 95.

33. Richard A. Sabol, "Public Works in Pittsburgh Prior to the Establishment of the Department of Public Works" (research paper, Department of History, Carnegie Mellon University, 1980), p. 5; Frank Kern, "History of Pittsburgh Water Works, 1821–1842" (research paper, Department of History, Carnegie Mellon University, 1982), pp. 1–4.

34. In 1823 seventy-eight prominent residents of the city had formed an "emergency committee" to guard against fire. See Wade, *Urban Frontier*, p. 292.

35. Kern, "History of Pittsburgh Water Works," p. 6. The initial cost was $40,000.

36. Wade, *Urban Frontier*, p. 296; Sabor, "Public Works in Pittsburgh," p. 5.

37. Thompson, "Financial History," pp. 44–45.

38. Ibid., pp. 44–45.

39. Paul Studenski and Herman E. Krooss, *Financial History of the United States* (New York: McGraw-Hill, 1952), p. 134.

40. Tarr and Konvitz, "Development of Urban Infrastructure," p. 199.

41. For a brief discussion of these conditions, see John Duffy, "Hogs, Dogs, and Dirt: Public Health in Early Pittsburgh," *Pennsylvania Magazine of History and Biography* 87 (July 1963): 294–305.

42. An 1802 statute directed supervisors to build channels of brick, but wood was also used extensively.

43. Wade, *Urban Frontier*, pp. 96–97, 285.

44. Ibid., p. 97–98. Many Pittsburgh "tenements" did not even have privy vaults.

45. The barge was towed downriver and emptied when full. See Terry F. Yosie, "Retrospective Analysis of Water Supply and Wastewater Policies in Pittsburgh, 1800–1959" (Doctor of Arts diss., Carnegie Mellon University, 1981), p. 15.

46. Charles E. Rosenberg, *The Cholera Years: The United States in 1832, 1849 and 1866* (Chicago: University of Chicago Press, 1962), pp. 1–100.

47. Jacqueline Karnell Corn, "Municipal Organization for Public Health in Pittsburgh, 1851–1895" (Doctor of Arts diss., Carnegie Mellon University, 1972), pp. 15–16.

48. Ibid., pp. 16–17; Yosie, "Retrospective Analysis," pp. 23–29; John Duffy, "The Impact of Asiatic Cholera on Pittsburgh, Wheeling, and Charlestown," *Western Pennsylvania Historical Magazine* 47 (July 1964): 205, 208–09.

49. Thompson, "Financial History," p. 38; Tarr and Konvitz, "Development of Urban Infrastructure," p. 200.

50. Holt, *Forging a Majority*, pp. 28–37; and Dankosky, "Pittsburgh City Government," pp. 36–37.

51. Bateman, "Early Reapportionment," pp. 6, 11.

52. Dankosky, "Pittsburgh City Government," pp. 18–21.

53. The ordinances permitted the railroads to lay their tracks on certain

streets, but they were then responsible for paving and repairing the streets, for limiting the speed of trains, and for paying city taxes. They were also required to burn coke rather than coal or wood and to use horses rather than locomotives on certain streets. See City of Pittsburgh, *A Digest of the Acts of Assembly and a Code of the Ordinances* (Pittsburgh: Errett, Anderson, 1869), pp. 358–86.

54. Holt, *Forging a Majority*, pp. 13–17, 220–62. Holmberg suggests that the issue of debt repudiation caused a split along sectional rather than political or class lines. See "Industrializing Community," pp. 231–34, 244.

55. In 1852 the city had eight plank toll roads.

56. Ads specifically aimed at businessmen appeared in the Pittsburgh press for "country residences" located near omnibus lines. Tarr, *Transportation Innovation*, pp. 4–5.

57. Ibid., pp. 4–6.

58. From the downtown three lines went east to the neighborhoods of Lawrenceville, Oakland, and Minersville; other lines went to the independent communities on the opposite banks of the rivers. See ibid., pp. 6–7.

59. Holmberg, "Industrializing Community," pp. 280–81. For 1850, see also Harold L. Twiss, "The Pittsburgh Business Elite, 1850–1890–1929" (seminar paper, Department of History, University of Pittsburgh,, 1964), p. 4.

60. I have adapted the terms, "Infrastructure for production" and "for consumption" from David Harvey's concepts of a "built environment for production" and "for consumption." See David Harvey, "The Urban Process Under Capitalism: A Framework for Analysis," in *Urbanization and Urban Planning in Capitalist Society*, ed. Michael Dear and Allen J. Scott (London: Methuen, 1981), pp. 96–114.

61. See Bernard J. Sauers, "A Political Process of Urban Growth: Consolidation of the South Side with the City of Pittsburgh, 1872," *Pennsylvania History* 1 (July 1974): 265–88; Holmberg, "Industrializing Community," pp. 236–39.

62. Thompson, "Financial History of Pittsburgh," p. 189.

63. Tarr, *Transportation Innovation*, pp. 7–14; Dennis E. Lawther, "The Impact of the Monongahela Incline Plane on the Social, Economic and Physical Development of Mt. Washington, 1870–1910" (seminar paper, Department of History, Carnegie Mellon University, 1977); William A. James, "The History of Urban Transportation in Pittsburgh and Allegheny County with Emphasis on the Major Technological Developments" (MA thesis, Department of History, University of Pittsburgh, 1947); and William Kenneth Schusler, "The Economic Position of Railroad Commuter Service in the Pittsburgh District: Its History, Present, and Future" (Ph.D. diss., University of Pittsburgh, 1959).

64. Tarr, *Transportation Innovation*, pp. 16–17.

65. Salvin Schmidt, "The Telephone Comes to Pittsburgh" (MA thesis, Department of History, University of Pittsburgh, 1948), pp. 5–116; Joel A. Tarr, "The City and the Telegraph: Urban Telecommunications in the Pre-

Telephone Era," *Journal of Urban History* 14 (Nov. 1987): 60–68. The fire and police departments also adopted the telegraph and the telephone at relatively early dates compared to other municipalities.

66. For introductions to the subject of energy development, see Will Bachand, "Municipal Lighting in Pittsburgh, 1880–1920" (research paper, Department of History, Carnegie Mellon University, 1973); Ronald Lasser, "The Electrification of the City of Pittsburgh, 1880–1930" (research paper, Department of History, Carnegie Mellon University, 1982), which is useful on corporate development; Gerry Clarke, "Natural Gas in Pittsburgh" (seminar paper, Department of History Carnegie Mellon University, n.d.), which is useful on industrial uses; and Pam Lettrich, "The Natural Gas Boom in Pittsburgh" (research paper, Department of History, Carnegie Mellon University, 1982). Local supplies of natural gas were largely exhausted by 1900.

67. For an excellent study of the effect of improved transportation on two Pittsburgh suburbs, see Sven Hammar, "Wilkinsburg and Edgewood: Commuter Suburbs" (seminar paper, Department of History, Carnegie Mellon University, 1972).

68. Tarr, *Transportation Innovation*, p. 18.

69. Ibid., pp. 13–19. Reflecting the rising demand for land in this area, tax assessments in wards 1–4 increased 104 percent from 1903 to 1907. See Howard V. Storch, Jr., "Changing Functions of the Center-City: Pittsburgh, 1850–1912" (seminar paper, Department of History, University of Pittsburgh, 1966), p. 9.

70. Tarr, *Transportation Innovation*, pp. 10–11.

71. In Shadyside, for example, a suburban neighborhood in the city's new Twentieth Ward, land subdivision took place in the 1860s and 1870s, with development occurring primarily in the late 1880s and the 1890s. Hundreds of individual property owners rather than large developers were primarily responsible for building until larger builders took over in the 1890s. See Robert J. Jucha, "The Anatomy of a Streetcar Suburb: A Development History of Shadyside, 1852–1916" *Western Pennsylvania Historial Magazine* 62 (Oct. 1979): 307–19. In contrast, in the growing industrial area of Hazelwood, which had housed a number of upper-class residents earlier in the century, new construction largely consisted of one- or two-story row houses of brick or frame. See Joel A. Tarr and Denise Di Pasquale, "The Mill Town in the Industrial City," *Urbanism Past and Present* 7 (Winter/Spring 1982): 1–14.

72. A practice supposedly followed by those interests whose projects were rejected by the city councils was to appeal to the state legislature to create a commission to administer their project, with the commission having the power to impose debt on the municipal treasury. The 1873 constitution barred this practice and prohibited legislative interference in urban street patterns. See Rosalind L. Branning, *Pennsylvania Constitutional Development* (Pittsburgh: University of Pittsburgh Press, 1960), pp. 97–98.

73. Reorganization and rationalization of Pittsburgh government began with the formation of full-time professional police and fire departments.

The police force was formed in 1868 when the city councils merged what was called the day and the night police into one unit under the mayor's direction. Shortly after, in 1870, the councils created a professional fire department to replace the volunteers who had often been a disruptive force in the city. For the police, see C. Garn Coombs, "Law Enforcement in an Urban Setting: The Struggle for a Police System in Pittsburgh, 1802–1868" (seminar paper, Department of History, Carnegie Mellon University, 1969). For fire, see Ronald M. Zarychta, "Municipal Reorganization: The Pittsburgh Fire Department as a Case Study," *Western Pennsylvania Historical Magazine* 58 (Oct. 1975): 471–86.

74. See, for instance, Martin Shefter, "The Emergence of the Political Machine: An Alternative View," in Willis D. Hawley et al., *Theoretical Perspectives on Urban Politics* (Englewood Cliffs, N.J.: Prentice-Hall, 1976), pp. 14–44; Joel A. Tarr, *A Study in Boss Politics: William Lorimer of Chicago* (Urbana: University of Illinois Press, 1971), pp. 22–47, 65–77.

75. For a discussion of the functions of the political machine, see Samuel P. Hays, "The Politics of Reform in Municipal Government in the Progressive Era," *Pacific Northwest Quarterly* 55 (Oct. 1964): 157–89. For an analysis of the business links of machine politicians see Joel A. Tarr, "The Urban Politician as Entrepreneur" in *Urban Bosses, Machines, and Progressive Reformers* ed. Bruce M. Stave (Lexington, MA.: D. C. Heath, 1972), pp. 62–74.

76. Bruce M. Stave, *The New Deal and the Last Hurrah: Pittsburgh Machine Politics* (Pittsburgh: University of Pittsburgh Press, 1970), p. 27.

77. John C. Adams, "Machine Politics in Action: Pittsburgh in the Early 1890s" (seminar paper, Department of History, Carnegie Mellon University, 1972), pp. 3–18.

78. In 1878, for instance, the superintendent of the waterworks complained that the councils' Water Extension Committee, which controlled water supply extensions, had not provided him with any information for over a year. See Sabol, "Public Works," p. 8. The standing committees were especially lax in their handling of public expenditures and were often accused of corruption.

79. Sabol, "Public Works," pp. 5–6; William G. Willis, *The Pittsburgh Manual: A Guide to the Government of the City of Pittsburgh* (Pittsburgh: University of Pittsburgh Press, 1950), p. 8; and Thompson, "Financial History," p. 121.

80. For a discussion of the growth of professionals in municipal government in this period see Jon C. Teaford, *The Unheralded Triumph: City Government in America, 1870–1900* (Baltimore, Md.: The Johns Hopkins University Press, 1984), pp. 142–73.

81. Barbara Judd, "Edward M. Bigelow: Creator of Pittsburgh's Arcadian Parks," *Western Pennsylvania Historical Magazine* 58 (Jan. 1975): 53–67. Bigelow's programs were so expensive that he won the title of "the Extravagant."

82. Phillip S. Klein and Ari Hoogenboom, *A History of Pennsylvania* (New York: McGraw-Hill, 1973), pp. 328–29.

83. Thompson, "Financial History," p. 112. In 1877, Pittsburgh defaulted on debt interest payments.

84. Holmberg, "Industrializing Community," p. 237.

85. Thompson, "Financial History," pp. 134–35.

86. Ibid., pp. 178–79.

87. Quoted in Jucha, "Anatomy of a Streetcar Suburb," pp. 305–06.

88. The Supreme Court had modified its position on the act's unconstitutionality in 1880 and held that assessments by the front foot were acceptable in cases where property owners petitioned for the improvement.

89. On one occasion in the 1890s, the city constructed major East End roads at municipal expense while abutters in other areas continued to pay. See Susan J. Kleinberg, "Technology's Stepdaughters: The Impact of Industrialization Upon Working-Class Women, Pittsburgh, 1865–1890" (Ph.D. diss., University of Pittsburgh, 1973), chap. 3, pp. 22–26, published as *The Shadow of the Mills: Working-Class Families in Pittsburgh, 1870–1907* (Pittsburgh: University of Pittsburgh Press, 1989), pp. 86–87.

90. Yosie, "Retrospective Analysis," pp. 84–85; Erwin E. Lanpher and C. F. Drake, *City of Pittsburgh: Its Water Works and Typhoid Fever Statistics* (Pittsburgh: City of Pittsburgh, 1930), p. 25. Parts of the South Side had their water needs supplied by the Monongahela Water Company, a private firm.

91. Yosie, "Retrospective Analysis," pp. 85–87.

92. Mark J. Tierno, "The Search for Pure Water in Pittsburgh: The Urban Response to Water Pollution, 1893–1914," *Western Pennsylvania Historical Magazine* 60 (Jan. 1977): 25.

93. Ibid., 5–36; Yosie, "Retrospective Analysis," pp. 119–32.

94. Quoted in ibid., p. 59.

95. Ibid., pp. 82–83.

96. For a full discussion of this issue, see Joel A. Tarr, "The Separate vs. Combined Sewer Problem: A Case Study in Urban Technology Design Choice," *Journal of Urban History* 5 (May 1979): 308–39.

97. Yosie, "Retrospective Analysis," pp. 88–103; Pittsburgh, *Pittsburgh and Its Public Works*, p. 41.

98. Jon A. Peterson, "The Impact of Sanitary Reform Upon American Urban Planning," *Journal of Social History* 13 (Fall 1979): 84–89.

99. During the 1880s the Board of Health attempted to force scavenger firms to utilize a new technology called the "odorless excavator" to clean cesspools. See Yosie, "Retrospective Analysis," pp. 107–08.

100. Ibid., p. 104–06.

101. The so-called Street acts, passed by the city councils in 1887–1889, provided that street and lateral sewer improvements would be made on the petition of one-third of the abutting property owners in a neighborhood. All abutters, however, would be assessed for improvements. In 1891 the State Supreme Court declared these acts unconstitutional. See Thompson, "Financial History," pp. 178–79.

102. Yosie, "Retrospective Analysis," pp. 112–113.

103. Kleinberg, *Shadow of the Mills*, pp. 86–87.

104. There were some exceptions to this generalization. On Mt. Washington, for instance, in 1900, housing density rather than class determined sewer service. See Lawther, "Impact of the Mononghalea Incline," p. 33.

105. Kleinberg, *Shadow of the Mills*, pp. 87–93; and Clayton R. Koppes and William P. Norris, "Ethnicity, Class, and Mortality in the Industrial City: A Case Study of Typhoid Fever in Pittsburgh, 1890–1910," *Journal of Urban History* 11 (May 1985): 269–75.

106. Koppes and Norris, "Ethnicity, Class, and Mortality," pp. 53–66.

107. Francis G. Couvares makes the interesting point that Bigelow's parks served to "define public space" for mass meetings as well as celebrations. See *The Remaking of Pittsburgh: Class and Culture in an Industrializing City, 1877–1919* (Albany: State University of New York Press, 1984), pp. 109–11.

108. These inequities formed one of the important themes of the Pittsburgh Survey of 1907–1908. See, especially, Paul U. Kellogg, ed., *The Pittsburgh District: Civic Frontage* (New York: Survey Associates, 1914). Couvares, *Remaking of Pittsburgh*; John Bodnar, Roger Simon, and Michael P. Weber, *Lives of Their Own: Blacks, Italians, and Poles in Pittsburgh, 1900–1960* (Urbana: University of Illinois Press, 1983), pp. 13–87; and Roy Lubove, *Twentieth-Century Pittsburgh: Government, Business and Environmental Change* (New York: John Wiley, 1969), pp. 1–9.

109. Hays, "Politics of Reform," p. 60.

110. Lubove, *Twentieth-Century Pittsburgh*, p. 29.

111. Ibid., pp. 20–22, 37–40; Janet R. Daley, "Zoning: Its Historical Context and Importance in the Development of Pittsburgh," *Western Pennsylvania Historical Magazine* 71 (April 1988): 104–07. A further reform goal was accomplished in 1913 when the legislature approved the enactment of a graded tax that imposed a heavier burden on land rather than on improvements.

112. The importance of these political factors is overlooked in several treatments of the period. See, for instance, Daly, "Zoning," p. 104. For a discussion of the factional political struggles see Klein and Hoogenboom, *History of Pennsylvania*, pp. 326–27, 377–78. The 1911 reform was also linked to conflict between city and state Republican factions.

113. Quoted in Lubove, *Twentieth-Century Pittsburgh*, p. 10.

114. Also important are developments in regard to sewage treatment. This controversy, however, was primarily a conflict among experts. See George Gregory, "A Study in Local Decision Making: Pittsburgh and Sewage Treatment," *Western Pennsylvania Historical Magazine* 57 (Jan. 1974): 25–42; Yosie, "Retrospective Analysis," pp. 225–65, 325–436; and Joel A. Tarr, Terry Yosie, and James McCurley III, "Disputes Over Water Quality Policy: Professional Cultures in Conflict, 1900–1917," *American Journal of Public Health* 70 (April 1980); 427–35.

115. For discussions of transit politics in this period, see Peter M. Farrington, "The Pittsburgh Subway" (research paper, Department of History, Car-

negie Mellon University, 1981); Marc Warner, "Where the Pittsburgh Subway Hid: An Analysis of Seventy-five Years of Indecision" (research paper, Department of History, Carnegie Mellon University, 1981); and Tarr, *Transportation Innovation*, pp. 23–24.

116. Tarr, *Transportation Innovation*, pp. 26–30.

117. Lubove, *Twentieth-Century Pittsburgh*, p. 53.

118. Ibid., p. 57. See also Daly, "Zoning," pp. 104–25. The new bureaus can be found listed in City of Pittsburgh, *Digest of the General Ordinances and Laws . . . to March 1, 1938* (Pittsburgh, 1938).

119. The Allegheny annexation still needs to be studied in detail. For an introduction, see Barbara C. Owens, "The Consolidation of Allegheny" (seminar paper, Department of History, Carnegie Mellon University, n.d.).

120. Yosie, "Retrospective Analysis," pp. 330–36.

121. J. Steele Gow, Jr., "Metropolitics in Pittsburgh" (Ph.D. diss., University of Pittsburgh, 1952), pp. 9–12; and Lubove, *Twentieth-Century Pittsburgh*, pp. 27–28, 97–98.

122. Teaford, *City and Suburb*, pp. 167–70.

123. Tarr, *Transportation Innovation*, pp. 24–25. Over 32,000 commercial vehicles and 1,155 buses also served the county.

124. The breakdown was from 9,911 to 68,133 automobiles; from 6,432 to 22,623 motor trucks; and, from 5,301 to 9,587 streetcars.

125. Tarr, *Transportation Innovation*, pp. 24–38.

126. In 1926 the city created a Bureau of Traffic Planning that had authority over traffic regulation, but it possessed no authority to alter basic street and roadway patterns.

127. Lubove, *Twentieth-Century Pittsburgh*, p. 87.

128. See Daly, "Zoning," pp. 118–25; and Anne Lloyd, "Pittsburgh's 1923 Zoning Ordinance," *Western Pennsylvania Historical Magazine* 57 (July 1974): 289–306.

129. Lubove, *Twentieth-Century Pittsburgh*, pp. 90–92. Bigger's frustration and his professional elitism was also reflected in his comment that "public ignorance" blocked many "meritorious projects." See Robert Phipps, "The Building of the Boulevard: A Roadway to the Suburbs" (research paper, Department of History, Carnegie Mellon University, 1972), p. 5.

130. Peter M. Farrington, "The Allegheny County Highway and Bridge Program, 1924 to 1932" (MS thesis, Department of Civil Engineering, Carnegie Mellon University, 1981), pp. 31–34; Lubove, *Twentieth-Century Pittsburgh*, pp. 90–92, 96.

131. Lubove, *Twentieth-Century Pittsburgh*, pp. 37–79.

132. Farrington, "Highway and Bridge Program," pp. 82–111. The building of the Point Bridge raised similar complications.

133. Ibid., pp. 58–77; Lubove, *Twentieth-Century Pittsburgh*, pp. 92–93. Parts of the Major Highway Plan were constructed in the 1930s as the Belt System.

134. Tarr, *Transportation Innovation*, pp. 30–31; Farrington, "Highway and Bridge Program," pp. 68–77.

135. Tarr and Konvitz, "Patterns in the Development of the Urban Infrastructure," pp. 212–13.

136. During the first two years of its existence, for example, the Works Progress Administration spent nearly $70 million in Allegheny County, constructing hundreds of miles of highways, sewer and water lines, and building and renovating playgrounds. See Michael P. Weber, *Don't Call Me Boss: David L. Lawrence, Pittsburgh's Renaissance Mayor* (Pittsburgh: University of Pittsburgh Press, 1988), p. 68; Stave, *New Deal.*

137. Lubove, *Twentieth-Century Pittsburgh*, p. 96.

138. Mark S. Foster, *From Streetcar to Superhighway: American City Planners and Urban Transportation, 1900–1940* (Philadelphia: Temple University Press, 1981), pp. 91–115.

139. Tarr and Konvitz, "Patterns in the Development of the Urban Infrastructure," pp. 216–17.

140. Lubove, *Twentieth-Century Pittsburgh*, pp. 106–76; Weber, *Don't Call Me Boss*, pp. 197–292.

141. Donald Stevens, "The Role of Nonprofit Corporations in Urban Development: A Case Study of Action Housing of Pittsburgh" (Ph.D. diss., Carnegie Mellon University, 1987); Yosie, "Retrospective Analysis," chap. 7 (on the Allegheny County Sanitary Authority); Robert C. Alberts, *The Shaping of the Point: Pittsburgh's Renaissance Park* (Pittsburgh: University of Pittsburgh Press, 1980). The major environmental programs have also been studied. See Roland M. Smith, "The Politics of Pittsburgh Flood Control, 1908–1960," *Pennsylvania History* 42 (Jan. 1975): and 44 (Jan. 1977): 5–24, 3–24; Joel A. Tarr and Bill C. Lamperes, "Changing Fuel Use Behavior and Energy Transitions: The Pittsburgh Smoke Control Movement, 1940–1950," *Journal of Social History* 14 (Summer 1981): 561–88.

142. For an introduction, see Shelby Stewman and Joel A. Tarr, "Four Decades of Public-Private Partnerships in Pittsburgh," in *Public-Private Partnership in American Cities*, ed. R. Scott Fosler and Renee A. Berger (Lexington, Mass.: Lexington Books, 1984), pp. 78–79, 89–94.

143. For a discussion of Renaissance II until 1982 and a comparison with Renaissance I, see ibid., pp. 94–112.

The Point, where the Allegheny and Monongahela rivers join to form the Ohio, has been through the years Pittsburgh's most dramatic physical feature.

Pittsburgh in the eighteenth century was little more than a frontier outpost. This engraving, published in 1826, appeared in the atlas that accompanied the account of General Victor Collot's travels in America in 1796. *Darlington Memorial Library, University of Pittsburgh*

An engraving by Charles Graham, published in *Harper's Weekly* for February 27, 1892, shows Pittsburgh as seen from Duquesne Heights. *Archives of Industrial Society, University of Pittsburgh*

A view of Pittsburgh from Mount Washington during the urban renewal program known as Renaissance I, c. 1953. *Archives of Industrial Society, University of Pittsburgh*

The Pittsburgh "Triangle," including Point State Park and the skyscrapers of Renaissance I and II, 1988. *Arthur G. Smith*

The Workplace. Steel, coal, food processing, glass—four of Pittsburgh's major industries—provided jobs for thousands of workers, newcomers as well as older residents, in the industrial era.

A steelworker taking a sample of molten metal from a ladle at the U.S. Steel Homestead Works in the 1950s. *Archives of Industrial Society, University of Pittsburgh*

A coal miner in Library, Pa., 1946. Blacks worked in the region's mines as well as its mills. *Teenie Harris*

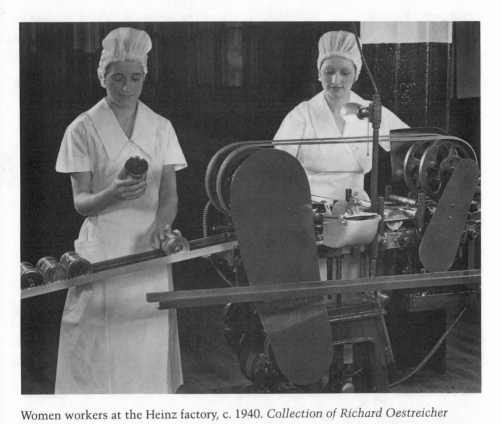

Women workers at the Heinz factory, c. 1940. *Collection of Richard Oestreicher*

Pittsburgh glassworkers, c. 1900. *Philadelphia Commercial Museum Collection, Pennsylvania State Archives*

Families. A stark contrast between families of affluence and families of poverty; between those who benefited much from the city's economic growth and those who benefited less.

The children of Charles Hart Spencer line up for a family portrait at their home in Shadyside, c. 1900. *Archives of Industrial Society, University of Pittsburgh*

An unidentified mother and children in the lower Hill district at the turn of the century. *Archives of Industrial Society, University of Pittsburgh*

Black Women. The lives of black women in Pittsburgh varied enormously: some found in education the road to opportunity; others faced a bleaker future amid the constraints of slum life.

"Coeds" at the University of Pittsburgh, c. 1947. L to R: Edith Johnson, Mary Louise Wray Stewart, Esther Dalton Hawkins, Mary Jane Mitchell Page. *Teenie Harris*

An outdoor privy in the mining town of Library, Pa., c. 1947. Substandard housing was not confined to the Hill district. *Teenie Harris*

Upper Class and Working Class. For the affluent, home was a place of relaxation and leisure which enhanced the quality of life; for the great majority of families with blue-collar occupations, home was often a place to carry out economic production—a sweatshop.

The family of Charles Hart Spencer seated at the dinner table in their Shadyside home. *Archives of Industrial Society, University of Pittsburgh*

The proprietor of a Hill district sweatshop and two of her employees who are stripping tobacco leaves. This photograph was made for *Women and the Trades* (1909), the first volume of the Pittsburgh Survey. *Archives of Industrial Society, University of Pittsburgh*

Young Boys. The lives and opportunities of children varied sharply by their socioeconomic class. For the affluent, tennis was a typical recreation, and for some of the less fortunate, summer camp was a possibility.

A boy from an upper-middle-class family (son of Charles Hart Spencer) dressed for tennis, c. 1907. *Archives of Industrial Society, University of Pittsburgh*

A summer camp for boys during World War II, sponsored by the *Courier,* was one of that black-owned paper's many efforts to promote better race relations. *Teenie Harris*

Housing was the most obvious indication of inequality; it varied widely, community by community, with extremes and mixtures in between.

"Cairncarque," the Shadyside residence of Robert Pitcairn, as it appeared in *Palmer's Views of Pittsburgh and Environs* in 1903. Pitcairn was an official of the Pennsylvania Railroad. *Archives of Industrial Society, University of Pittsburgh*

A group of middle-class homes in the East End, 1920. *Archives of Industrial Society, University of Pittsburgh*

The Pittsburgh Bureau of Health photographed these tenements in the Hill district in 1918. *Archives of Industrial Society, University of Pittsburgh*

Two-story brick and frame houses lined south Eighteenth Street on the South Side in 1911. *Archives of Industrial Society, University of Pittsburgh*

Communities. The street was often the center of community life, even when the intrusions of transportation forced people onto the sidewalks. Buildings, signs, and storefronts reflected the variety of custom and condition.

Racial diversity in a Hill district grocery store, c. 1945. *Teenie Harris*

The intersection of Logan and Colwell streets in the lower Hill district, 1929. *Archives of Industrial Society, University of Pittsburgh*

A view down Phelan Way in Polish Hill, looking toward Immaculate Heart of Mary Church, 1908. *Archives of Industrial Society, University of Pittsburgh*

▪ *Steel City Aristocrats*

JOHN N. INGHAM

HISTORICAL TREATMENT of urban elites, in Pittsburgh and elsewhere, has not fared very well. Compared to the richness, the theoretical sophistication, and the very real fascination which labor historians have developed in studying the working class, studies of elites have been disappointing. This cannot be because the study of elites is fundamentally less interesting or less important than the study of labor. It is due instead to the differing approaches used by the two groups of historians. Generally guided by Marxist analysis, labor history has focused upon the concepts of class formation and class consciousness among the workers, examining areas of class conflict in order to re-create a dynamic and exciting context within which to understand them.[1]

Historical treatment of urban elites, on the other hand, has been marginally successful, at best. Informed by Max Weber, rather than Marx, these studies have also examined the issue of class, but generally in a far more sterile theoretical context. Whereas the Marxist historians posit the importance of class consciousness, elite historians have instead viewed class as a largely statistical category. Weber believed that class was simply a group of people with similar "life chances," and that status distinctions were more significant. Since most elite studies have focused on issues derived from these Weberian ideas, it has minimized their impact. With class viewed as the rather prosaic accumulation of property, much time has been spent examining the issue of mobility into the upper class. That is, how many people rise and fall from one statistically defined income or wealth level to another over a period of time? The conflict, such as it was, in these studies generally revolved around issues of status: who gets into which club, who is in the Social Register and which families intermarry with one another. Little attention was paid either to the issue of class consciousness, although the upper class has been arguably the most class conscious of all groups, or to the issue of class as opposed to status conflict. That is, little attempt has been made in existing studies to expand the range of analysis into areas of conflict the elite might have with other classes or the community as a whole over fundamental economic, social, or political issues. As a result, elite studies have generally been sterile, lifeless exercises in statistical manipulation and examinations of minute status differences, with little sense of how these might relate to situations of power and influence in the broader community.[2]

Elite studies, then, have not been the most productive area of urban history. The study of Pittsburgh's elites has had a rather strange history of its own. The breathtaking rise of Pittsburgh to industrial prominence on the shoulders of the iron and steel industry captured the imagination of the American masses in the early twentieth century. Fascinated with the audacious image of hard-driving industrialists like Andrew Carnegie, Henry Clay Frick and Andrew Mellon, journalists wrote a number of exciting popular histories of these men and their times.[3] This fascination was not shared by historians and other academics, however. Although Andrew Carnegie has been studied by a number of historians, the rest of Pittsburgh's elite have been largely ignored. This is curious, since it is generally conceded that Pittsburgh was virtually the prototype for the industrial city of the late nineteenth and early twentieth centuries. What better place to test the impact of the emergence of heavy industry on social, political, and economic arrangements than in America's Birmingham?

This chapter will examine several studies of Pittsburgh, most of which remain unpublished. I will attempt to answer a number of important questions: First, what is distinctive about Pittsburgh? Second, what does the study of Pittsburgh's elite tell us about the nature of power relationships in the city, particularly as they relate to profound changes in the basic economic structure? Third, to what extent is the history of Pittsburgh's elite a story of continuity and to what degree of change? Fourth, to what degree can we define an elite or upper-class "culture" in the city that is similar to, but distinct from, ethnic or religious cultures? Finally, in what ways can the study of Pittsburgh's elite be improved so that it more clearly demonstrates important issues of the city's history?

A major question, as noted above, is Pittsburgh's distinctiveness—what makes it stand out in comparison to other cities, and how might elite studies shed light on this feature? On the superficial level this seems quite simple. Pittsburgh, of course, is the Steel City. True enough, but it did not begin producing steel in large amounts until the late 1870s, nearly a century after the city was founded, and largely moved out of steel production in the decades after World War II. So steel production characterized the city for only about one-third of its two hundred years. Pittsburgh's renown in early years came not from the nature of its manufacture but from the heritage of its citizens—it was the heart of the Presbyterian Scots-Irish culture which in so many ways characterized the American frontier. It was this group that gave the city its earlier moral fibre of strict no-nonsense Calvinism; they set the cultural tone in the early years.

Pittsburgh's elite studies, unfortunately, have not developed this Scots-Irish cultural stamp very clearly. Part of the reason is that most of them, as with most other works dealing with Pittsburgh, are fascinated with its industrial progress and pay little attention to its early roots. One study which is an exception is an analysis of the "founding families" of Allegheny County by Joseph Rishel.[4] Rishel took twenty of the families delineated by Frank W. Powelson in an earlier study.[5] Tracing these individuals across at least four generations, Rishel analyzed a total of 1,006 individuals, who were divided into "core" and "non-core" groups. The core group were the direct lineal descendents of the founders, whereas the non-core were those who married into that group. He found that "Northern Irish" made up 71 percent of the new wealth, but just 25 percent of the old upper class.[6] This observation raises important questions which have yet to be explored: How did the Scots-Irish achieve prominence in the early years? Was it strength of numbers? Was it an aptitude for a particular kind of business or industry? What difference did the large Scots-Irish Presbyterian population make to the overall character of Pittsburgh? Was the fact of Pittsburgh's frontier origins significant? Although Rishel looks at occupation, career lines, religion, and several other factors, we do not yet have a "portrait" of this important founding group nor any sense of what made Pittsburgh distinctive at this time.

We also know little of Pittsburgh's elite in the early nineteenth century. By 1820 Pittsburgh had become a small, thriving mercantile city of over seven thousand individuals. Richard C. Wade, in his book *Urban Frontier*, asserts that merchants were the bellwether of Pittsburgh's economy and society by this time, and that most of them were Scots-Irish Presbyterians.[7] To this point, however, we do not have a useful study of what seems to be a young, ethnically distinct aristocracy consolidating its power as the city began to mature. An important question in this regard is the nature of political power in early Pittsburgh. Sam Bass Warner found in Philadelphia at the same period that it was ruled by a powerful set of merchants, interlocked by marriage and business interests.[8] Were the Pittsburgh families that Rishel studied the "ruling elite" of the young city? Or was there a split between those who pursued their narrow economic interests and those who were enamored of political and military affairs?

Studies of Pittsburgh's elite thus lack the proper historical and cultural foundation to provide an understanding of what happened during the city's early years. This all becomes painfully clear when we examine the nature of Pittsburgh's elite at mid-nineteenth century. By this time the city had become a major manufacturing center and the emerging hub of iron production in America. It was also the birthplace of the Republican party and an important cog in the evolution of Civil War politics. The city's importance at this time should have resulted in a number of investigations of elite members who played such a crucial role in these transformations. Unfortunately, this has not been the case. No study has analyzed Pittsburgh's elite during this time period in any systematic manner, and the little we do know comes from two studies done for different purposes.

By 1850 Pittsburgh and Allegheny City combined had a population of nearly seventy thousand. Although mercantile activities predominated in Pittsburgh in 1820, as early as the 1830s manufacturing began overshadowing commerce. By 1860 about 20 percent of the work force was engaged in industry. Glass and cotton production were important, but ironmaking was by far the city's major business. This profound transformation of the city's economic base could provide a valuable testing ground for theories dealing with the concepts of continuity and change in elite leadership. Harold C. Livesay and Glenn Porter, in their study of the connections between merchants and manufacturers in the early years of industrialization in America, addressed the question of how manufacturers obtained capital to expand their industries. They believed that merchants were able to gain a significant voice in new industrial ventures because of the capital they had available for investment purposes. One prominent example was the American Iron Works in Pittsburgh (Jones & Laughlin). In this firm, the practical iron manufacturers who founded the concern soon turned to Benjamin F. Jones and Samuel Kier, commission merchants, for capital. These two men later drew upon the greater capital resources of yet another merchant, James Laughlin. In this manner, the continuity of elite structure was preserved in the midst of a profound economic transformation.[9]

This conclusion is substantiated by Michael Holt's observations on the political situation in the city during the same time period.

Holt extracted from the 1850 census 149 men who owned real property worth more than $25,000. He found that half of them lived near
one another in stately old mansions on or near Penn Avenue in the
Fourth Ward. A majority (57 percent) were Presbyterian, and most of
these were members of the First or Third Presbyterian churches. A
large minority (41 percent) were Episcopalian and attended Trinity
Church. Just 2.5 percent were Catholic. Even at this relatively early
stage, 70 percent of the elite were native born; of the foreign born, 83
percent came from Northern Ireland, and 10 percent from Germany.
The Scots-Irish foreign born had lived in the community for a long
time by 1850 and had achieved social acceptance as well as economic success.[10]

Like Livesay and Porter, Holt found that the merchants and
manufacturers among this elite had mutual rather than antagonistic
interests. He provides several examples of merchants who had invested in manufacturing endeavors.[11] These economic ties were abetted by strong social ties. Many went to the same churches and lived
in the same neighborhoods. They also joined forces to support similar charities and hospitals, went to fancy dress balls together, attended the same teas and parties and intermarried with one another
extensively. By mid–nineteenth century, the Pittsburgh elite had
created a strong, tightly knit local aristocracy which evidently managed to integrate the older mercantile elite with a newer manufacturing group.[12]

The only elite study dealing with this period is by Harold Twiss.
He developed a list of forty-five directors in downtown Pittsburgh
banks in 1850. Forty-seven percent of them were engaged in mercantile pursuits, well above the 25 percent Holt found, and 33 percent
were in manufacturing or mining, just slightly above Holt's 27 percent. Another major difference between the two studies was the
larger number of lawyers in Holt's work (16 percent) compared to
Twiss's (2 percent). In most respects, however, they differed little.
Each had about 30 percent of the group born abroad, and each had
about the same percentage of Presbyterians and Episcopalians. One
important and interesting dimension of Twiss's study was his analysis of the first job held by these men. Nearly two-thirds got their
start in a family business, indicating that few of them were "self-
made." Of those who entered family businesses, 38 percent were in
mercantile concerns, 25 percent were in manufacturing, and the rest
in law, banking, or other professions. This evidence gives strong
support to the sense of old family aristocracy developed by Holt in
his study. It was, in a word, a privileged group, one whose eminence
and positions on bank boards derived at least as much from their old

family status as from their own business achievements. This old
family status probably also accounts for the somewhat larger number of merchants among Twiss's group. Their mercantile fortunes,
which they used to help fund industrial expansion, also made them
attractive as bank directors.[13]

Most of Pittsburgh's elite studies have focused on the period
from 1880 to 1920. It was during this period, of course, that Pittsburgh emerged as the greatest steel producing center in the world.
As the city's industrial productivity grew enormously during these
years, its population grew apace. With two hundred thirty-five thousand inhabitants in 1880, forty years later there were nearly six
hundred thousand. What had been a small industrial city on the eve
of the Civil War, was a large, thriving, industrial metropolis during
the late nineteenth and early twentieth centuries. Again, we see that
the city was undergoing profound transformation in two significant
areas. First, the industrial base of the city changed profoundly. Steel
was different from iron. The city's prewar iron industry was largely
unintegrated. Most Pittsburgh ironmasters saw little benefit in
smelting their own pig iron, so their mills were only finishing
plants, producing merchant bar iron, nails, and tools. These mills in
most instances were quite small, producing fifteen to twenty-five
thousand tons annually. Steel, on the other hand, demanded integration of facilities and economies of scale. Massive steel enterprises,
headed by the gigantic Carnegie Steel Company, came to the fore in
Pittsburgh during the late nineteenth century. The steel industry by
the 1890s was utterly different from the iron industry of thirty years
before. Then, in the decades between 1895 and the outbreak of
World War I, the steel industry was hit by the great merger wave, as
hundreds of small firms were absorbed into huge new consolidations, such as U.S. Steel Corporation. Surely these developments
must have had a dramatic impact upon the structure of Pittsburgh's
elite.

Conventional wisdom has always held that this was the case.
Herbert Casson, who wrote an account of the rise of the steel industry
in 1904, said that the city's motto should be From Rags to Riches,
since there were so many "shirt-sleeve millionaires" there.[14] This
impression was adopted by later historians. Samuel Eliot Morison
and Henry Steele Commager, in their influential textbook published
in the 1950s, asserted: "The most typical figure of the industrial age
was undoubtedly Andrew Carnegie."[15] Yet systematic studies of Pittsburgh's elite do not substantiate that view.

In my own studies of Pittsburgh's iron and steel elite between
1874 and 1901 I found that, despite the great transformation of the

industry in the late nineteenth century, there was little change in the management of the iron and steel firms. Eighty-six percent of the 339 iron and steel manufacturers I investigated were native born, and two-thirds of their fathers were also born in America. Further, their fathers were already part of the city's economic elite, since 70 percent of them were businessmen and another 14 percent were professionals. Of the fathers who were businessmen, 80 percent had been engaged in manufacturing, the great majority of those in the antebellum iron industry. There is little indication, then, that the image evoked by Casson, and reiterated by Morison and Commager, was factual. The immigrant, poor-boy-to-riches pattern made prominent by Carnegie and Thomas Mellon was not typical of Pittsburgh steel manufacturers. This also further substantiates the image of continuity that Livesay and Porter and Holt found in the prewar period. Just as older merchant families were able to use their financial resources to enter the fledgling iron industry, older iron manufacturers were able to retain a firm hold on the postwar steel industry.[16]

Nor had Pittsburgh's iron and steel elite changed much in other respects from 1850. The greatest number were from Northern Ireland, with a large minority of English and Welsh heritage. One important change, however, was that a significant minority of 21 percent were of Germanic origin. Although there were a small number of Germans in the elite group in 1850, they had grown impressively during the intervening years. The religious configuration remained about the same as before. Nearly 50 percent of the iron and steel manufacturers were Presbyterian, with 29 percent Episcopalian, 4 percent Catholic and 1 percent Jewish. There is every indication that this larger elite group continued to function as a cohesive upper-class unit. To an impressive degree, as will be detailed below, they continued to attend the same churches, the same clubs, live in similar neighborhoods, and, most importantly, to intermarry with one another. Vast economic and population expansion and profound technological transformation did not seem to disrupt the small city aristocracy of a half-century before.[17]

There were, however, other parallel elite groups in Pittsburgh by this time. As noted above, German immigrants became important members of the Pittsburgh elite at an early date. Two of Rishel's "family founders" were German, and Holt found a small minority of German immigrants (4 percent) among his 149 wealthy Pittsburghers in 1850. By the late nineteenth century, a far larger German elite community was residing in the city. Layne Peiffer sheds some interesting light on the German Christian elite in the city. Within his sample of twenty-five individuals, he found that men of means of

Germanic extraction began arriving in the city in the 1840s, and by
the 1870s had created a fairly large, if not particularly cohesive,
group.[18] Many had left their homeland because of the political revolu-
tions of 1830 and 1848. They were thus a "better class" of Germans
who were emigrating to protect their property and status. These men,
however, were not wealthy; they instead possessed modest capital
and were armed with skills which they could put to productive use in
a young and growing Pittsburgh. Of his German family founders, two-
thirds had been born in Germany, while the rest were born in Allegheny County of immigrant parents. The great majority made their
mark in mercantile or manufacturing pursuits, but few of them were
involved in the burgeoning iron and steel industry. Most of his manufacturers were brewers, with only Joseph Vilsack, who was primarily
a brewer, engaged in steel production.[19]

Peiffer also finds that all twenty-five of his men were "self-
made," although most had "surprisingly" good educations for the
nineteenth century. This fact, combined with the results of my own
study, indicates that these men, if not from elite origins, did have
some very helpful advantages, thus calling into question the accuracy of his "self-made" label. A more than rudimentary education,
particularly in nineteenth-century Europe, was difficult to achieve,
and usually denoted advantages of money or status. Nonetheless, it
is important to recognize that Peiffer is examining a relatively *nou-
veau riche* group. Even if a few men, like the Negleys and Anshutzs,
were of old money, most of the German-American fortunes were
newly minted.

To a fairly significant degree, then, these men and their families
were marginal in Pittsburgh. Most were neither fully accepted
among the older British elite, nor were they able to build their own
self-standing parallel aristocracy, principally because although all
were Christian, they were composed of two antithetical groups—
Roman Catholic and Protestant—and the latter group was even further divided. As a consequence, Peiffer found that the Germans
never coalesced into a cohesive upper-class group. Nevertheless,
there were new groups rising to challenge the older aristocracy in
Pittsburgh, and that latter group had to find ways to respond to the
challenge.

Michelle Pailthorp's study of German Jews introduces a group
which provided an even greater contrast to old-stock Pittsburghers.
Pailthorp took the membership list of the elite Jewish Concordia
Club as her core group, generating a cast of individuals who arrived
in Pittsburgh over a period of thirty to forty years in the late nineteenth and early twentieth centuries. Since they were all of recent

German origin, they superficially resembled Peiffer's group, but the Jewish religion of her group set them apart. There were other respects in which her group was different also. Although Peiffer found a significant number of merchants among his Germans, most of these were engaged in wholesale trade. The great area of opportunity for German Jews in Pittsburgh, as well as elsewhere in nineteenth-century America, however, was in retailing. Well over half of Pailthorp's Jews made their fortunes in some aspect of the retail trade. Far fewer were involved in manufacturing, and those who were so engaged tended to produce various kinds of light machinery. None of her Jews were in Pittsburgh's mighty iron and steel industry, and few had professional occupations, although about 5 percent were attorneys.[20] Perhaps even more surprising, given patterns of German Jewish success in New York City and certain European cities at this time, was that few were involved in banking. It should be recalled that it was private investment banking, not commercial banking, in which Jews had generally had their success, and the Mellon family seemed to dominate the investment banking field in Pittsburgh. In any event, German Jews in Pittsburgh in the late nineteenth century were outsiders. They did not live in the same neighborhoods, belong to the same clubs, or intermarry with the older Pittsburgh core elite. To better understand how that older elite managed to maintain its hegemony over the city's social system, it is necessary to examine some of the mechanisms it employed for this purpose.

Among the most interesting and oft-studied institutions of upper-class maintenance is that of neighborhood and residence. We have observed that in the 1850s the great majority of Holt's wealthy Pittsburghers lived in what is now downtown Pittsburgh, on or near Penn Avenue. Rishel also found that the vast majority of his founding families lived in an area bounded by Fifth Avenue, Water Street, Market Street, and Wood Street.[21] The problem with this downtown neighborhood, however, was that it was not an elite area in the classically restrictive sense. That is, cheek by jowl with the elite residences were a number of commercial and industrial areas, along with the county jail and working-class housing. For that reason, wealthy families began looking for a somewhat more isolated and homogeneous residential area. Many of them found what they were looking for across the river in Allegheny City, then legally separate from Pittsburgh.

Thomas Kelso notes that Allegheny began developing elite residential areas in the early 1880s. A key decision occurred in 1819, when the city set aside 100 acres of land to be reserved for common pasturage. First known as the Commons, a section of it later was called West Park. On the streets surrounding the park an elite area emerged over the next several years. This elite district was able to resist the incursion of industrial, commercial, and working-class residential elements, although they soon ringed the area, making it an island of privilege in a sea of urban malaise. In fact, just east of Federal Street, the eastern boundary of the area, was the so-called Dutchtown or Swiss Hole made up of working-class Germans who were themselves chased from the area in the late nineteenth century by invasions of Eastern and Southern Europeans. During its heyday, however, Allegheny was a highly compact, densely urban, upper-class area, not unlike similar enclaves in Philadelphia, Boston, New York, and other seaport cities.[22]

Kelso chose to study a "representative sample" of sixty-one (actually sixty-four) families who fulfilled the following characteristics: they were listed in the Blue Book of 1887; they lived on one of eleven adjoining streets in Allegheny; and they were listed in the 1880 census as having three or more servants. His figures indicate that although a few of the families arrived in the area during the 1850s, the first large migration came during the 1860s, when nearly one-third of the families arrived. Then, during the 1870s, thirty-one families, nearly half the total, moved in. Since the pre–Civil War elite lived mostly in the Penn Avenue area, the question arises as to whether these old families were the ones moving into Allegheny after the Civil War. This is not easy to answer. If we look again at Rishel's families, we find that very few moved to Allegheny. As late as 1880 just about 10 percent were living there, and that was the peak.[23] As I shall observe below, when the Penn Avenue families moved, they tended to go to the East End. If the pre–Civil War elite were not the ones moving into Allegheny in the 1870s, then perhaps they represented new industrial wealth—men who were making their fortunes in the massive new iron and steel enterprises emerging in the city. This proposition will be tested in more detail below.

In terms of basic social characteristics, the Allegheny elite did not seem to differ in any significant respect from our picture of the pre–Civil War elite or from the iron and steel manufacturers in my own study. Nearly all of Kelso's family heads were born in the United States, as were 75 percent of their fathers. If they represented new money, they certainly were not recent immigrants. Further, nearly all were Protestant, mostly Presbyterian and Episcopalian, so they did not diverge in religious respects either.[24] In 1874, according to my findings, the single largest group (29 percent) of iron and steel magnates lived in Allegheny.[25] These men, of course, were the heart and core of an industrial elite who rose to wealth and prominence on the back of the revolution in iron and steel. In that sense they differed from Rishel's elite, which had older, landed wealth and more ancient pedigree in the area. Thus, although the wealthy elite in Allegheny did not differ in cultural ways from the older Pittsburgh elite, they did achieve wealth in fundamentally different economic endeavors. Therein, perhaps, lies the secret to Pittsburgh's stability in the nineteenth century. Although the economy was undergoing a revolution, the new aspirants for power and prestige in the city were largely replicas of the older elite. Sharing nearly identical ethnic origins, time of arrival in America, religion, and middle-class roots, they were absorbed into the elite structure of Pittsburgh with little fanfare or fuss. Allegheny remained an exceptionally popular urban residence for the city's new industrial aristocracy for some forty years. Then, about the time of World War I, it began to decline rapidly. Even before that happened, a new, more suburban area, became the center of gravity for the city's social elite.

This lush new area was known simply as the East End. Suburban, as opposed to urban neighborhoods like Penn Avenue and Allegheny, it was already attracting 35 percent of steel families in 1887. By the turn of the century over half the iron and steel elite were living there, and from 1915 to 1933 nearly two-thirds resided there.[26] It began a slow decline after 1940, concurrent with the growth and development of Sewickley. A factor of great significance about the East End, as opposed to Allegheny, is that it was equally successful in attracting members of Pittsburgh's pre–Civil War elite families. Although few of Rishel's founding families ever moved to Allegheny, they began migrating to the East End in large numbers in the 1870s. After 1880, the great majority of those upon whom Rishel had information lived in the East End.[27] We must, then, examine the rise of this elite suburban neighborhood in more detail.

The area which was to become the East End was annexed by the city of Pittsburgh in 1868. It remained, however, a rather bucolic rural retreat for a number of years thereafter. The crucial event was the building of a streetcar line on Fifth Avenue in the 1870s, followed in the 1880s by cable cars for the steep inclines. From the middle of the 1880s onward, a large number of palatial homes were built. Fifth Avenue, from Neville Street through the Shadyside area up into Point Breeze became a vast and opulent "millionaires' row."

Renee Reitman studied the wealthy in Shadyside, one of the main elite areas in the large and somewhat amorphous East End. She selected her families using the following criteria: first was listing in the Pittsburgh Blue Book (generally recognized as an elite index);

second was listing in the Social Register (a clearly upper-class index); and third was male membership in the Duquense Club, considered an important citywide "millionaires' club" and one of the most prestigious in the city. Using these criteria, Reitman generated a list of 110 families and found information on 75 percent.[28]

Re-analyzing these 82 families provides some fascinating insights. The early residents, that is those living in Shadyside prior to 1875, represented "old families" and had occupations which lay largely outside the burgeoning industries in the city. The heads of these old families were professionals, mostly lawyers, who came from landed or mercantile wealth, and had attended prestigious, eastern, ivy league schools. Only one, Henry Laughlin, Jr., was in the steel industry, and his family, as we have seen, made its money in mercantile ventures prior to the Civil War. Reitman divided those who arrived in Shadyside after 1878 into three categories: "self-made men"; men from inherited wealth or with a family firm; and professionals. With some minor adjustments to her data, we find that 42 percent of these later arrivals were probably self-made, and 38 percent came from inherited wealth. The remaining 20 percent were professionals, most of whom came from comfortable social origins.[29]

It is clear that whether new money or inherited wealth, these new arrivals in Shadyside were in the cockpit of Pittsburgh's industrial transformation. At least 40 percent were involved in the iron and steel industry, and nearly all the rest, except for the professionals, were in some other industrial endeavor. There was just one newspaper publisher and one banker. Further, those involved in mercantile pursuits were either iron or coke merchants, or were involved in shipping these materials. Unlike Rishel's families on Penn Avenue, this was an industrial elite. In this respect they resembled their counterparts in Allegheny. But since Reitman found that so many were self-made, they may represent a group of newer entrants to the social elite in the city: a group of new men who were carbon copies of the older elite in many respects, and who made their money in the new post–Civil War industries.

My own research offers some substantiation for this conclusion. Using the Social Register as a convenient indicator of upper-class status, I found that although a far larger percentage of the iron and steel manufacturers in my sample (210 versus 82) lived in the East End rather than Allegheny in 1908, the social status of the latter group was higher. Ninety-five percent of the Allegheny families were listed in the Social Register, compared to 83 percent from the East End.[30] These figures provide reasonable evidence that the East

End elite was just what Reitman indicated—largely men of later arrival and subsequently lower social status when compared to the Allegheny elite. Those in Allegheny, in turn, were probably of lower social status than Rishel's families on Penn Avenue, but there is no way to verify that conclusion.

In other social and cultural respects the elite in Shadyside differed little from the older Pittsburgh social elite. Of the eighty-two heads of families in Reitman's study in 1900, all but three were born in the United States. All except one were Protestant, and all were of Northern European ethnic stock. Of the Protestants, 56 percent were Presbyterian and 16 percent were Episcopalian. The remainder belonged to other Protestant denominations.[31]

Looking at where the Shadyside elite came from also provides some important clues about elite formation in the area. Of the forty-five families who moved there between 1878 and 1899, about 25 percent moved directly from Allegheny, while another 15 percent had earlier lived in Allegheny, but moved to another part of the East End before ultimately landing in Shadyside. The majority, however, moved to Shadyside from other parts of the East End, or came from the Penn Avenue area downtown.[32] A number of these latter families, as well as several of those from Allegheny, were of very great social prestige. What was being created in Shadyside and the East End in the late nineteenth century was a melting pot of older and newer money. All the wealthy residents were similar in ethnic and religious orientation, and all their families had been in America for some time with at least middle-class status, but this new suburban environment was one instrument used to create a new metropolitan aristocracy.

Some indication of that trend is seen in the families who moved to Shadyside between 1900 and 1910. A small proportion of this group represented "new money," probably just over one quarter. The majority had inherited wealth (60 percent) or were professionals (14 percent). Other than that, however, there was little difference in these new Shadyside residents. Of the same ethnic and religious background, most were entrepreneurs in iron, steel, coal, and railroads.[33] They were an industrial elite.

Francis G. Couvares provides in-depth analysis of how these old and new elites were integrated together in Shadyside. Couvares is one of few recent historians to examine the whole phenomenon of life-style, and the results of his analysis are pregnant with meaning. He points out that when the Pittsburgh social elite moved to the East End they not only changed residence but also profoundly changed their way of life. They went from a highly circumscribed

"Victorian" urban life-style to a freer, more cosmopolitan, and more open pattern of living. Couvares comments upon the "dour provinciality" of Pittsburgh's iron elite, to whom luxury was a sin and recreation did nothing but distract them from their proper business and religious pursuits. The ironmasters who dominated the upper reaches of Pittsburgh society in the years before the turn of the century had little interest in forging a "distinctive bourgeois cultural order or to link themselves spiritually with their class." Until the 1890s, therefore, Pittsburghers had little to choose from in terms of genteel amusement: "With no museums, no orchestra or musical societies, no night life or club society, no literary or artistic circles, opportunities for establishing bonds of sociability and solidarity among the 'natural kings of Pittsburg,' were severely limited."[34]

With the opening of vast suburban tracts in the East End, however, all of this changed. "In the pastures of the east end, middle-class Pittsburghers found escape not only from the vicious, but also from the confinement of their own narrow virtue."[35] The houses they built there were larger and more palatial, and in these new homes the Pittsburgh elite began giving lavish parties. At the same time, amateur theatrical companies were started, and art galleries and music societies also emerged. These cultural attractions were abetted by fine retail establishments, markets, theaters, Shadyside Academy, and the Pennsylvania College for Women.

The significance of these changes, according to Couvares, is that by the turn of the century a suburban point of view had emerged to match the new suburban life-style. This point of view stressed the exclusivity of the group from other classes and other "urban" environments. More importantly, it also stressed a strong sense of class identity. In conjunction with newly emerging social clubs, what Couvares calls an "official elite culture" began emerging in the late nineteenth century.[36] These clubs for the first time drew members from the entire metropolitan area, so that families from the East End, Sewickley, Allegheny, and other areas now began to identify with one another as class cohorts. It was this enhanced class feeling which made it possible for the newer elite Reitman studied in Shadyside to blend almost imperceptibly with the older upper class. This phenomenon of class consolidation and integration was carried on even further in the newer suburbs which emerged later in the twentieth century.

Two of the most important of these new upper-class suburban enclaves were Sewickley and Sewickley Heights. The elite area of Sewickley has been studied by Fred Wallhausser. A long-isolated

Allegheny County village about twelve miles from downtown Pittsburgh, Sewickley had its own indigenous upper class who had resided in the area for many years prior to the 1850s. In 1851 the railroad was built into the area, bringing an influx of new families. Wallhausser's study deals with one group that came between 1830 and 1850, which he calls the "old upper class," and a group that came after 1850, called the "new upper class." Since travel between Sewickley and Pittsburgh prior to the railroad took two and a half hours in the summer and was virtually impossible in winter, Wallhausser felt this made it possible for the local "old upper class" to develop prior to 1850.[37]

It is difficult to know just how many individuals were in the old upper class in Wallhausser's study, nor is it clear just how he determined who indeed were upper class. None of these individuals, for example, are among Rishel's founding families, although several, like the Nevins, are of undenied prestige and lineage. In any event, Wallhausser identifies the salient properties of this upper class, and they are traits by now familiar to us in Pittsburgh: Scots-Irish and Presbyterian. This older upper class fought against having their rural, isolated way of life disrupted by the coming of the railroad. One of their leaders, David Shields, tried to stop the building of the railroad, but was unsuccessful. With the completion of the line Sewickley was just a half hour from Pittsburgh, winter or summer, and large numbers of wealthy families began moving into the area. Most of these new men, unlike the old upper class, used Sewickley as a "bedroom suburb" as they continued to run their businesses in Pittsburgh. About fifty-five such families moved into Sewickley during the years from 1850 to 1895, most of whom appear to have moved from Penn Avenue in Pittsburgh. Just 22 percent came from Allegheny. As these new families poured into Sewickley, they evidently overwhelmed the older upper class. The confrontation became most acute in the Sewickley Presbyterian Church. After 1851 every elder in the church was a member of the new upper class. As a consequence, in 1864 the older upper class broke away to form the United Presbyterian Church of Sewickley. Shortly thereafter, another group of old upper-class individuals, mostly those living in the Edgeworth section of Sewickley, also left the older church to form Leetsdale Presbyterian Church. Yet another group of upper-class individuals affiliated with the Episcopal church. Although the two new churches appeared to garner the allegiance of a majority of the old upper class, a number (such as the Nevinses) remained members of the old church, while some of the new upper class joined both the Leetsdale and Sewickley United Churches.[38] Most probably, Sewickley, like the East End, was in the process of developing a series of broadly attractive suburban institutions and luring wealthy families from the entire metropolitan area.

Further evidence for this conclusion appears when we examine Sewickley Heights, a wooded area above Sewickley which grew very slowly throughout the entire nineteenth century. With the removal of the Allegheny Country Club from Allegheny to Sewickley Heights in 1899, and its consequent metamorphosis into the city's most important country club, a number of old Allegheny families, along with some from Pittsburgh itself, were attracted to the area. Evidence presented by Stephen Schuchman makes clear that the earliest arrivals in Sewickley Heights, those who came prior to 1900, were almost entirely from Allegheny. Three came from Sewickley and one from Shadyside. Although about a quarter of these men were in iron and steel, and another quarter engaged in other kinds of mining and manufacturing, the great majority were in banking, investments, mercantile pursuits and other somewhat more genteel ventures. This gives some indication that they were of older money and perhaps higher status than the families moving to Shadyside at the same time. Of the group which came to the area between 1900 and 1915, 40 percent came from Sewickley, 33 percent from Allegheny, and another 25 percent from Shadyside and the East End. A large portion of this group was engaged in the iron and steel industry (40 percent), while another 17 percent were in other manufacturing ventures. A large minority (43 percent) were engaged in banking, investments, the law, and other professional occupations.[39]

Thus, we can detect some subtle differences between Sewickley Heights and the East End. Both were suburban environments, and both served to integrate and consolidate Pittsburgh's social elite. But Sewickley Heights appeared to be of higher social prestige than the East End. Its elite families were older, were in more genteel occupations, and seemed to have a larger number of Episcopalians, as opposed to the more particularistic and "ethnic" Presbyterianism of nineteenth-century Pittsburgh. The enhanced status of Sewickley and Sewickley Heights can also be inferred from the following figures: in 1921, 96 percent of the forty-seven iron and steel families living there were listed in the Social Register. By 1933 this had increased to 97 percent. Then, in 1943 it rose to 100 percent, where it remained thereafter, the most uniformly upper-class residential environment in Pittsburgh.[40] Farther from Pittsburgh than the East End, the older elite was able to shape this nearly unpopulated area to its own specifications. And, although it had cosmopolitan institutions like the Allegheny Country Club to tie it to class cohorts in the East End and elsewhere, the Sewickley elite could always remain somewhat separate and isolated, just a bit less a part of the cosmo-politan life-style emerging in the East End.

These developments, and others, were part of a complex process whereby the Pittsburgh social elite became less provincial and at least somewhat more cosmopolitan. As noted above, one phenome-non of this transformation was the general movement away from Presbyterianism, which was closely associated with the Scots-Irish, and toward Episcopalianism, which became a virtually generic upper-class religion in America, regardless of region or the original ethnoreligious status of the family. This was part of a broader move-ment by these same families to send their children to eastern prep schools and colleges. Among the Pittsburgh iron and steel elite there was a decided shift away from local academies, such as Sewickley and Shadyside, toward schools like the Episcopalian St. Paul's in Concord, New Hampshire. At the same time, they began sending their sons to Yale and Princeton, with smaller numbers going to Harvard. This shift in college attendance became particularly pro-nounced after 1920.

Other studies have noted the shift toward Episcopalianism. Among Rishel's founding families, just over 13 percent were Episco-palian in 1820, but by 1900 the percentage had doubled to nearly 27 percent. They also increasingly patronized eastern schools. In 1860, 46 percent were educated at local schools, but by 1900 this had dropped to 33 percent. Many young scions of these founding families were by this time attending Princeton, with smaller numbers at Yale and Harvard. Reitman noted that in Shadyside there was an increas-ing trend after 1900 for families to send their sons to eastern prep schools rather than Shadyside academy. The trend at the college level was even more pronounced, with 93 percent sending their sons to ivy league schools in the early twentieth century.[41]

These religious, educational, and residential trends were allied with similar transformations in other upper-class social institu-tions, the most important of which were the social clubs. A number of important men's social clubs appeared in Pittsburgh in the late nineteenth century. Comparing membership in these clubs to Social Register listings, it is clear that they are rank ordered in terms of social prestige. Of the downtown men's luncheon clubs, the Pitts-burgh Club had the highest status, closely followed by the Duquesne and Union clubs. Of the suburban golf and country clubs which were designed to appeal to the entire family, the Allegheny Country Club and the Pittsburgh Golf Club had the highest prestige, with the Pittsburgh Country Club and the Oakmont Country Club ranked significantly lower.[42]

Couvares notes that the Pittsburgh and Duquesne clubs, both of which had been founded in the 1870s to appeal to the needs of small numbers of iron and steel manufacturers, "blossomed in the late 1880's and early 1890's into crucial organizations of official elite culture."[43] They drew their membership from the entire metropolitan area. Evidence from the other studies provide support for this. Kelso found that most of the men from the Allegheny elite belonged to these clubs, and Reitman found that the male heads of all but two of her families in Shadyside were members of the Duquesne Club, with somewhat smaller numbers in the Pittsburgh and Union clubs. Many also belonged to both the Pittsburgh Golf Club in the East End and the Allegheny Country Club in Sewickley. Wallhausser found that the males in 40 percent of his families belonged to the Duquesne, Pittsburgh, and Union clubs. A significant number also joined the Allegheny Country Club. The clubs also served as a means (along with prep schools and ivy league colleges) to link the male elite of Pittsburgh with their counterparts in other cities on the East Coast. The country clubs served similar functions for the entire family.[44]

Yet there were limits to the integrative functions of these clubs, and these limitations accentuate the important role they played in providing an exclusive sense of belonging, of special status, to their members. If one examines Peiffer's German Christians, the evidence appears to indicate that only about 20 percent belonged to the "mainline" social clubs. The Germans, as a somewhat marginal group in nineteenth-century Pittsburgh, were slowly being integrated into club and neighborhood. Marriage would follow at a still slower rate. The situation for the German Jews was even more dramatic. Since the men's clubs always defined themselves as "Christian" organizations in their charters, Jews were automatically excluded from membership. The same held true for the suburban "family" clubs. As a consequence, members of the German Jewish elite were forced to develop their own parallel institutions. The most important of these was the Concordia Club, built in Allegheny in 1874, a "city" men's club modeled after the Duquesne and Pittsburgh clubs. In the early twentieth century a "country club," the Westmoreland Club, was also organized for wealthy Pittsburgh Jews. Membership in the two clubs was nearly totally dominated by the older German-Jewish elite, with smaller numbers of Eastern Europeans gaining admission during the twentieth century.[45]

With the exception of groups like the Jews, and most likely various Catholic groups, it seems clear that Couvares is largely correct about the broader process taking place. A cohesive upper class was forming, it was becoming more cosmopolitan and less provincial, it was developing ties to families in other cities. As far as it goes, this phenomenon was similar to that taking place in other cities in America during this time. The sociologist E. Digby Baltzell first recognized the formation of a "national" upper class in America, seeing the turning point coming in the 1890s. In his book on the Philadelphia upper class, he viewed the creation of a centralized national upper class as following quite naturally from a centralized economy. The major elements that transformed a series of local, provincial aristocracies into a national, metropolitan upper class, were New England boarding schools and ivy league colleges which allowed young men and women from wealthy families all over the United States to interact and socialize with their class cohorts.[46] This theory is fine as far as it goes, and Couvares has adopted it for his analysis of the Pittsburgh elite. The rock upon which all of this founders for the Pittsburgh upper class, however, is marriage. It is in the area of its integration into a national upper class, and the role marriage to elites in other cities played in this that Pittsburgh appears to be truly distinctive.

In order to understand the role of marriage in upper-class formation and maintenance, we must consider the phenomenon in all its complexity. The most intimate area of social contact, it was also the one over which upper-class families had the greatest control. They could therefore monitor this institution far more closely than they could their neighborhoods, their churches, or even their social clubs. Both Rishel and I have used the principle of endogamy, that is, the tendency to marry within a specific group, as the key to the way marriage functioned for upper-class groups.

Rishel found that about 25 percent of the marriages of his founding families were endogamous in the early nineteenth century, but the figure dropped to about 10 percent by the turn of the century. Similarly, marriages to other Pittsburgh families also declined over time, so that by 1880 just 21 percent were to identifiable local families.[47] It is tempting to see this as part of the trend toward a national upper class as delineated by Baltzell and Couvares, but Rishel's figures do not show an increase in marriages to out-of-town families. Instead, it was the number of "unknown" marriage partners which increased from under 40 percent to over 60 percent. It is possible these were to members of out-of-town upper-class families, but there is no way to tell from the evidence. In my own research, I found that those families with the greatest social status and prestige, the "core families," had the lowest rate of marriage to elite families outside of Pittsburgh. Yet they were the most cosmopolitan in all other aspects. The largest percentage of them were listed in the

Social Register, which to Baltzell was a national upper-class ranking, and they also had higher percentages of membership in the most prestigious clubs in the city. They were more likely to become members of the Episcopal church and to send their sons to eastern prep schools and colleges. The other groups among the Pittsburgh iron and steel elite showed an inverse relationship of social class to the propensity to marry outside the city. That is, the lower the social prestige of the group, the more likely they were to marry into an elite family from another city. It was the most prestigious families who were the last to feel the impact of marital exogamy, not the first as implied by Baltzell and Couvares.[48]

Comparing the Pittsburgh situation with other cities, it seems not so much that Baltzell was wrong about what was generally happening to the upper class in America, but that Pittsburgh was different. There was something about Pittsburgh which made its wealthiest and most prominent citizens, those who were becoming cosmopolitan in nearly every other respect, retain a strong local orientation in their marriage patterns, and most probably in other close social contacts. Why was this the case? The answer is not obvious, but a clue was hidden in Baltzell's analysis—the idea that the creation of a national upper class paralleled the emergence of a centralized economy. Although it seems axiomatic that Pittsburgh, the very center of the vast steel industry in America, became an integral part of this centralized economy, especially with the creation of U.S. Steel, facts indicate otherwise.

The conventional image is that the great mergers cut a broad swath through the formerly independent Pittsburgh iron and steel companies. The facts appear to contradict this assumption. Of the independent iron and steel companies which had been founded in and around Pittsburgh prior to 1895, 73 percent were still independent when the dust settled after 1901. In fact, most of those which had disappeared had gone bankrupt or been absorbed by consolidations controlled by the Pittsburgh elite. Only seven firms were actually absorbed by U.S. Steel. Further, another sixteen independent iron and steel firms were organized between 1898 and 1901. Thus, in 1901 there were a total of forty independent steel firms in the Pittsburgh area, and almost all were controlled by members of the older social elite. Eschewing the large batch production techniques of Bessemer and open hearth furnaces, and the bigness and national markets that went with them, they instead stressed product differentiation and small-scale, often local markets. Even as war clouds gathered in 1940 eleven highly successful independent steel firms controlled by old upper-class families continued to function with great success.

For reasons that are still not wholly clear, the Pittsburgh social elite decided to remain somewhat separate from the broader national upper class. Rather than selling their firms to national corporations (which probably would have increased their fortunes), they continued to run profitable and successful smaller and more localized enterprises. Similarly, they continued to marry into familiar Pittsburgh families, even while sending their children to eastern schools, joining resort colonies and clubs with the upper classes of other cities. Their cosmopolitanism was perhaps more of a veneer than Couvares suspected, and this relative provinciality, this rootedness and still rather prudish Calvinism, may have continued to dominate the cultural style of Pittsburgh long after New York, Philadelphia, and Boston had succumbed to an undifferentiated cosmopolitanism.

Another critically important element of any elite study, one which needs to be more thoroughly developed for Pittsburgh, is the question of power relationships. That is, what is the relationship of the elite to the city's economy and the populace as a whole, and what are the relationships of various subgroups of the elite to one another? Do we find in Pittsburgh a single, integrated upper class with a common culture? And, are they able to function as a "ruling class" in the city? These two questions are not unrelated. The greater the cohesiveness of the upper class, the greater its sense of class consciousness, the more likely it will be able to function as a ruling class, rather than as one of a number of competing elite groups in some sort of "brokerage" situation.

Couvares suggests that members of Pittsburgh's elite developed a strong "official elite culture," but were unable to extend their dominion over the city's working classes. Although they were able to mold themselves into a cohesive upper-class group with a unified class culture, they were not able to become a ruling class.[49] What light do the various elite studies of Pittsburgh shed on this issue?

A study by George Bedeian has dealt directly with these questions. A major objective of his study is to find out just how eligible candidates were sorted out and socially integrated into the upper class. This in turn relates to the question of the various channels of recruitment into the upper class. Finally, he attempts to analyze the different formal and informal "influence" and "power" structures within the Pittsburgh upper class. He bases his analysis on a comparison of three indices of social status—the Social Directory of 1904, the Social Register, and the Blue Book. The Social Directory, which Bedeian believes is the most exclusive in terms of social status, listed 762 families and individuals in the single year it was published. The Social Register, which was published annually after

1894, listed 1,200 families and individuals in 1908, and the Pittsburgh Blue Book, published yearly from 1887, included over 4,000 families in 1904. Bedeian assumes the latter volume was the least exclusive. Once the overlaps were eliminated, Bedeian had a total of 4,675 individuals and families to analyze.[50]

Bedeian found patterns of college education among his three groups similar to those observed earlier. Thirty-five percent of the men and families listed in the Social Directory and Social Register went to ivy league colleges in the late nineteenth century. Just 9 percent of those listed in the Blue Book attended these schools. His data is a little early to catch the most dramatic shift toward Episcopalianism, since only about 25 percent of those listed in the two more prestigious directories were of that religion (about 50 percent were Presbyterian). But that total assumes more significance when compared to the 8 percent Episcopalians in the Blue Book. The Blue Book contained greater numbers of men who belonged to lower status Methodist, Lutheran, and Baptist denominations, and was also where the prominent German Catholics were listed. Ninety-five percent of all Catholics in Bedeian's study were in the Blue Book, while only 5 percent were in the other two directories. The situation for Jews was somewhat different. Although none was listed in the prestigious Social Directory, they made up 3 percent of the listings in both the Social Register and Blue Book.[51]

Bedeian also looks at the date of arrival of the various family groups in Pittsburgh. These data reveal that listees in the Social Directory represented significantly older Pittsburgh families, with one-third arriving in the city prior to 1830. It is clear that a large number of them were among the founding families of Pittsburgh. Social Register families had been in the city nearly as long, with one-quarter coming before 1830. Blue Book listees were comparative newcomers, with just about 14 percent arriving before 1830. Even more dramatically, fully half the Blue Book families came to Pittsburgh after 1870, compared to 20 percent of the Social Directory families and 35 percent of the Social Register families. These figures, however, mask further important distinctions. The recent arrivals to Pittsburgh among the Social Register listees came mostly from the East Coast, and many were well-educated and prestigious members of prominent families in these areas. The recent Blue Book arrivals, on the other hand, tended to come from small towns and rural areas in Western Pennsylvania, or were German and Irish Catholics who came from abroad or smaller cities in the area, hoping to make their mark in Pittsburgh.[52]

Bedeian links arrival time with wealth. He finds that those families listed in the Social Directory, socially the most exclusive index, were the wealthiest, followed by those in the Social Register, and finally the Blue Book listees. He also found that within each directory group there was an overwhelming concentration of wealth among a few families. The degree to which wealth was skewed, even among the upper class, was staggering. He found that 7 percent of the families held 50 percent of the upper-class wealth. Of those with the greatest wealth and status—that is, those listed in the Social Directory—nearly 40 percent were engaged in manufacturing, while 23 percent held professional positions, 13 percent were merchants, and 12 percent were bankers. The pattern of those in the Social Register was slightly different. Of more recent arrival in Pittsburgh, with less wealth than those in the Social Directory, the greatest number (34 percent) were engaged in the professions, with 28 percent in manufacturing, 11 percent in banking, and just 9 percent merchants. Those in the Blue Book showed an even greater divergence—by far the greatest number (38 percent) were professionals, with 22 percent merchants, and just 15 percent engaged in manufacturing.[53]

Bedeian thus provides a somewhat clearer picture of just what factors—time of arrival in Pittsburgh, ethnicity, religion, occupation, and so forth—were involved in determining the social status of various groups in the city. He rightly asserts that those families at the upper levels of society "comprised an important, but interlocking group, welded together through marriage and business, who formulated the crucial corporate and financial decisions which affected the modern and technological growth of twentieth-century Pittsburgh."[54] Missing, or only suggested in his and other studies to date, is how this upper-class social network influenced the social and economic development of Pittsburgh. The question of power relationships is crucial to elite studies, yet has been almost totally neglected.

What makes Couvares's study of Pittsburgh so interesting and provocative is that he attempts to go beyond the rather prosaic analysis of elite categories. By asking the fundamental question (derived from Marxist analysis) of whether the Pittsburgh social elite was a ruling class in the late nineteenth and early twentieth centuries, he raises the analysis to a new and important level. In the simplest terms, Marx asserted that the ruling class in advanced societies held its dominant position by virtue of its possession of the major instruments of economic production, along with political dominance (which was consolidated by its control of the military), and its control over the production of ideas. Clearly, the issue of control of military power is not germane to the Pittsburgh situation, but the degree to which the social elite was able to control the corporate, political, and cultural life of the city is of fundamental concern.

In the realm of economic power, I noted above that the great consolidation movement at the turn of the century had little effect on the city's steel industry. Steel manufacturers evidently made a conscious decision to reject attempts to include their plants in larger consolidations, possibly out of fear that their ability to influence the economic, and ultimately political, social, and cultural life of the city, would be diluted if they lost their economic base. They therefore oriented their firms to serve smaller, more specialized markets with differentiated products, allowing their families to retain control of their firms, and ultimately of Pittsburgh's industry, for at least another generation.

This is suggestive, but we need to understand power relationships in a broader and more sophisticated framework. We do need to know if the social elite retained control of their own independent firms, but we also need to know the relationship of these firms to the rest of the industry. That is, to what degree does the balance of power in the city's steel industry shift to firms and families located outside Pittsburgh? We also need to know the nature of labor relations in their plants. That is, to be a ruling class, the first step for the Pittsburgh elite was to control its own workers. Then we have to develop an understanding of the manner in which the elite tried to extend its hegemony over the general community. To develop this, future studies will have to examine social welfare practices, cultural programs, and, most importantly, the political framework of the city. Let us look at each of these areas in more detail in order to determine what already has been done, and what else needs to be done.

Two of the best and most important elite studies of the last decade have used a regional focus in order to determine the relative power and influence of elite groups from various areas. In their studies of regions in eastern Pennsylvania, Burton W. Folsom and Edward J. Davies II have shown how the structure of a city's elite can determine its ability to grow, develop, and then retain power as a regional center of trade and manufacturing.[55] The differing fortunes of cities in these regions better illustrate which entrepreneurial and elite factors are important. No studies of this type have been done for Western Pennsylvania, yet they would seem to be crucial for understanding how Pittsburgh emerged as the dominant entity in the region. Such studies would help us to better understand the power relationships based upon firm, kinship, ethnoreligious group, and social clubs.

Labor relations are another important area. Much work has been done by labor historians on working-class culture and the role that strikes and other job actions played in creating this culture, but they have largely ignored labor relations. Historians of the elite have examined control of firm and industry, but have generally paid little attention to the phenomenon of labor relations. We need to know more about how class position affected the nature of labor relations in Pittsburgh, particularly in the all-important steel industry. Labor historians have developed a very clear picture of the powerful, autonomous "craftsmen's empires" which existed in Pittsburgh iron and glass plants in the mid–nineteenth century. But we have no sense of how this affected upper-class Pittsburgh manufacturers' ability to run their plants, nor, in broader terms, how it might have affected their ability to serve as a ruling class.

Some significant patterns do emerge. High status, old family ironmasters generally seemed to accept unions as part of their way of doing business and were even able to incorporate them into a largely paternalistic pattern of mill management and labor relations. Ironmasters of more insecure social status, and especially those who had recently risen from the working class themselves, were most likely to directly oppose the unions, refusing to grant union recognition, locking out the workers, and bringing in scabs. Then came the great "union busting" period in the steel industry and a new phase of paternalistic labor relations characterized by "welfare capitalism." Couvares views this later phase of labor-management relations as part of the elite's attempt to become a ruling class at the turn of the century.[56] The problem is that we don't know much about it. The most extensive information we have on a local Pittsburgh industry is for the H. J. Heinz plant, but its experience does not appear typical. U.S. Steel had one of the first and most important welfare plans, but, as we have seen, the corporation's direct impact on Pittsburgh was less than formerly believed. We do need to know more about how the various independent Pittsburgh steelmasters responded to the nonunion era in the city. Did they use welfare plans? Did they have older, paternalistic practices which survived into the twentieth century? Or were they just rigid, unbending, small-scale entrepreneurs?

The important question concerning the hegemony of the Pittsburgh elite over their workers also extends to their ability to influence the general population. One way in which they might do this is through social welfare policies. This is an area which has not yet been explored by historians of the city's elite. There are two aspects which must be examined: first, what were the social welfare policies of the municipality, and what role did the Pittsburgh elite play in formulating these policies? In order to answer this question, politics and the role of the elite in the political system must be examined. The second, related question is: What private, philanthropic policies did the elite develop to deal with this issue?

Couvares has addressed this issue in an interesting and provocative way. He states that the Pittsburgh elite in the nineteenth century essentially isolated itself from the working classes and provided little or nothing in the way of welfare programs, either public or private. Then, with the turn of the century, they used the reform ethic of Progressivism to inaugurate a series of welfare reforms which were designed to give the elite hegemony over the city's lower orders. Couvares believes they were largely unsuccessful in this attempt.[57] In order to understand the issue more fully, however, it will be necessary for future historians of the elite to develop the political structure of the city more completely. In addition, we must look at the role played by upper-class women and their organizations. Up to now, elite historians have examined only male roles in business and society. But particularly in the area of social welfare, the actions of female members of the social elite are critical. We do know that upper-class women were important in the Civic Club in Pittsburgh. This became a key social welfare organization in the city, but we need to know more about the organization—what it did, and why—and also to tie the upper-class women involved more concretely into the broader upper-class networks discussed above.

Finally, the whole area of culture, and of elite influence on this culture in the days before the development of mass communications, is vitally important. Again, Couvares has developed the most interesting analysis of this phenomenon. He records the attempt of Pittsburgh's upper classes to structure the environment of recreation, leisure, and culture at the turn of the century, actions which would enhance the role of the elite as a ruling class in the city. As he notes, the working class had its own leisure activities in the nineteenth century, but the elite attempted to establish theaters, art galleries, museums, libraries, and other cultural institutions in order to achieve hegemony over the cultural life of the city.[58] In this attempt it was largely unsuccessful, but a more in-depth analysis of the goals and tactics of the social elite is necessary for our complete understanding of the issues involved.

Future historians of the elite in Pittsburgh must be far more sensitive to the issue of power and to the class relations which accompany that power. They must therefore ask new questions of the evidence and look for different sorts of evidence to more completely understand the milieu of the Pittsburgh elite. We must look at economic decisions and relate them to class status and background. We must also transcend the economic environment to understand the attempts (if indeed there were such attempts) of this social and economic elite to be a ruling class. What was its relationship to the political structure? Did its members run for political office? Did they attempt to influence politics in more informal ways? Were they successful in this? What role did women play in the creation of a hegemonic ruling class? What was their specific role as actors in the upper-class system itself? Were they the real "gatekeepers" of upper-class society through their influence on marital patterns and other social institutions? What was their role as ambassadors of the social elite to the broader community, especially their own social reform and welfare organizations and church groups? What role did leisure, amusements, sports, and popular culture generally play? Couvares has given us some suggestive insights along this line, but they need to be enhanced. And we must particularly attempt to tie the newer cultural developments of the late nineteenth century into the older Scots-Irish Presbyterian culture. What groups stimulated the emergence of activities which were so antithetical to the straitlaced Calvinism of mid-nineteenth-century Pittsburgh? Was it the young members of these aristocratic families? Did they get these new ideas about sport and recreation from their schooling in the East?

Finally, we must remain aware of Pittsburgh's distinctiveness. Evidence from elite studies indicates that Pittsburgh's upper-class families were more reluctant than those in other cities to join fully into a homogenized, cosmopolitan, national upper class. They were even apparently willing to suffer a reduction in their fortunes and a diminution of their economic influence in the national steel industry in order to retain their economic hegemony. Did they also attempt to extend or retain their hegemony in social, political, and cultural areas? Evidence from Couvares indicates that they attempted to do this, but failed. Undoubtedly, we need more evidence in order to answer this question. What is clear is that we cannot take the experience of other cities as proof of what was happening in Pittsburgh. The Steel City, with its traditions of Scots-Irish Presbyterianism and its profitable, locally dominant steel industry, was able to pursue a separate and distinct pattern for its elite for a number of years longer than was true in many other cities. It may be that the Pittsburgh elite "stooped to conquer." Further innovative work on Pittsburgh's elite should do much to help us understand the distinctive life-style which came to characterize the city in the twentieth century.

NOTES

1. Mike Davis, "Why the U.S. Working Class Is Different," *New Left Review* 123 (Sept.–Oct. 1980): 3–44; and David Montgomery, "To Study the People: The American Working Class," *Labor History* 17 (Fall 1980): 458–512, provide good summaries of this work, along with relevant critiques.

2. Some of the best of these elite studies include Frederick Cople Jaher, *The Urban Establishment* (Urbana: University of Illinois Press, 1982); Peter R. Decker, *Fortunes and Failures* (Cambridge, Mass.: Harvard University Press, 1978); Clyde Griffen and Sally Griffen, *The Ordering of Opportunity in Mid-Ninteenth Century Poughkeepsie* (Cambridge, Mass.: Harvard University Press, 1978). I have also contributed to this genre with *Iron Barons: A Social Analysis of an American Urban Elite* (Westport, Conn.: Greenwood Press, 1978).

3. Some of the most significant of these works include: Herbert Casson, *The Romance of Steel* (New York: A. S. Barnes, 1907); Stewart Holbrook, *Iron Brew* (New York: MacMillan, 1939); John K. Winkler, *Incredible Carnegie* (Garden City: Garden City Publishing, 1931); Burton Hendrick, *The Life of Andrew Carnegie* (New York: Doubleday, Doran, 1932); George Harvey, *Henry Clay Frick: The Man* (Privately printed, 1936); Harvey O'Conner, *Mellon's Millions: The Biography of a Fortune* (New York: McLeod, 1953). Marcia Davenport, *The Valley of Decision* (New York: Scribner's, 1942), dealt with the rise to prominence of a Scots-Irish iron family. A movie starring Gregory Peck and Greer Garson was released in 1945.

4. Joseph Francis Rishel, "The Founding Families of Allegheny County: An Examination of Nineteenth-Century Elite Continuity" (Ph.D. diss., University of Pittsburgh, 1975).

5. Frank W. Powelson, "Founding Families of Allegheny County," Pennsylvania Division of the Carnegie Library of Pittsburgh (Pittsburgh, 1963).

6. Rishel, "Founding Families," p. 45.

7. Richard C. Wade, *The Urban Frontier: Pioneer Life in Early Pittsburgh, Cincinnati, Lexington, Louisville, and St. Louis* (Cambridge, Mass.: Harvard University Press, 1959).

8. Sam Bass Warner, *The Private City: Philadelphia in Three Periods of Its Growth* (Philadelphia: University of Pennsylvania Press, 1968), pp. 82–83.

9. Harold C. Livesay and Glenn Porter, *Merchants and Manufacturers: Studies in the Changing Structure of Nineteenth-Century Marketing* (Baltimore: Johns Hopkins University Press, 1971), pp. 65–69.

10. Michael Holt, *Forging a Majority: The Formation of the Republican Party in Pittsburgh, 1848–1960* (New Haven: Yale University Press, 1969), pp. 31–35.

11. Ibid., pp. 33.

12. Ibid., pp. 35.

13. Harold L. Twiss, "The Pittsburgh Business Elite, 1850–1890–1929" (seminar paper, Department of History, University of Pittsburgh, 1964), pp. 6, 9, 16.

14. Casson, *Romance of Steel*, p. 267. Holbrook, *Iron Brew*, p. 275, makes a similar appraisal.

15. Samuel Eliot Morison and Henry Steel Commager, *The Growth of the American Republic*, vol. 2 (New York: Oxford University Press, 1950), p. 135.

16. John N. Ingham, "Iron and Steel Families of Pittsburgh, 1875–1960" (seminar paper, Department of History, University of Pittsburgh, 1964); "Elite and Upper Class in the Iron and Steel Industry, 1874 to 1965" (Ph.D. diss., University of Pittsburgh, 1973); *Iron Barons*; "Rags to Riches Revisited: The Role of City Size and Related Factors in the Recruitment of Business Leaders," *Journal of American History* (Dec. 1976): 615–37; "The American Urban Upper Class: Locals or Cosmopolitans?" *Journal of Urban History* 2 (Nov. 1975): 76–87; "Robber Barons and the Old Elites: A Case Study in Social Stratification," *Mid-America* 42 (July 1970): 190–204.

17. Ingham, *Iron Barons*, pp. 20–23.

18. Layne Peiffer, "The German Upper Class in Pittsburgh, 1850–1920" (seminar paper, Department of History, University of Pittsburgh, 1964). To generate his sample, Peiffer culled the names from the city's Blue Books or other sources, but in the process missed a number of important individuals. The Anshutz, Fahenstock, and Negley families, all of whom had been in Pittsburgh since the turn of the century, were not on his list, nor were the Shoenbergers, who came in the earlier nineteenth century.

19. Ibid., pp. 11–13.

20. Michelle Pailthorp, "The German-Jewish Elite of Pittsburgh: Its Beginnings and Background" (seminar paper, Department of History, University of Pittsburgh, 1967), pp. 14, 22, 24.

21. Holt, *Forging a Majority*, p. 31; Rishel, "Founding Families," pp. 145–54.

22. Thomas Kelso, "Allegheny Elites, 1850–1907" (seminar paper, Department of History, University of Pittsburgh, 1964), pp. 2–6.

23. Ibid., pp. 7–12; Rishel, "Founding Families," pp. 150–54.

24. Kelso, "Allegheny Elites," p. 12.

25. Ingham, *Iron Barons*, p. 109.

26. Ibid.

27. Rishel, "Founding Families," pp. 152–54.

28. Renee Reitman, "The Elite Community in Shadyside, 1880–1920" (seminar paper, Department of History, University of Pittsburgh, 1964), pp. 2–3.

29. Ibid., pp. 3–14.

30. Ingham, *Iron Barons*, pp. 109–13.

31. Reitman, "Elite Community," pp. 9–10.

32. Ibid., p. 5.

33. Ibid., pp. 10–14.

34. Francis G. Couvares, *The Remaking of Pittsburgh: Class and Culture in an Industrializing City, 1877–1919* (Albany: State University of New York Press, 1984), pp. 34–35, 36–37.

35. Ibid., p. 96.

36. Ibid., pp. 104–05.

37. Fred Wallhausser, "The Upper-Class Society of Sewickley Valley, 1830–1910" (seminar paper, Department of History, University of Pittsburgh, 1964), pp. 4–11.

38. Ibid., pp. 12–30.

39. Stephen J. Schuchman, "The Elite at Sewickley Heights, 1900–1940" (seminar paper, Department of History, University of Pittsburgh, 1964), pp. 5–6, 23–26.

40. Ingham, *Iron Barons*, p. 113. During these same years between 79 and 84 percent of the iron and steel families in the East End were listed in the Social Register.

41. Rishel, "Founding Families," p. 189; Reitman, "Elite Community," pp. 17–19.

42. Ingham, *Iron Barons*, pp. 119–27.

43. Couvares, *Remaking of Pittsburgh*, p. 99.

44. Kelso, "Allegheny Elites," pp. 12–13; Reitman, "Elite Community," pp. 19–21; Wallhausser, "Upper Class Society," pp. 24–25.

45. Peiffer, "German Upper Class," p. 26; Pailthorp, "German-Jewish Elite," pp. 1–2, 4, 20, 22–23, 40–44.

46. E. Digby Baltzell, *Philadelphia Gentlemen: The Making of a National Upper Class* (New York: Free Press, 1958), chap. 2.

47. Rishel, "Founding Families," p. 168.

48. Ingham, *Iron Barons*, pp. 127–53.

49. Couvares, *Remaking of Pittsburgh*, pp. 105–19.

50. George Bedeian, "Social Stratification Within a Metropolitan Upper Class: Early Twentieth-Century Pittsburgh as a Case Study" (seminar paper, Department of History, University of Pittsburgh, 1974), pp. 12–20.

51. Ibid., pp. 36–65.

52. Ibid., pp. 71–75.

53. Ibid., pp. 79–112.

54. Ibid., p. 110.

55. Burton W. Folsom, Jr., *Urban Capitalists: Entrepreneurs and City Growth in Pennsylvania's Lackawanna and Lehigh Regions, 1800–1920* (Baltimore: Johns Hopkins University Press, 1981); and Edward J. Davies II, *The Anthracite Aristocracy: Leadership and Social Change in the Hard Coal Regions of Northeastern Pennsylvania, 1800–1930* (DeKalb: Northern Illinois University Press, 1985).

56. Couvares, *Remaking of Pittsburgh*, pp. 117–18.

57. Ibid., chaps. 6 and 7.

58. Ibid., pp. 105–19.

• *Pittsburgh and the Uses of Social Welfare History*

ROY LUBOVE

THE MOST SIGNIFICANT insight into the history of the history of social welfare in Pittsburgh might be that there is little of it. It would be worth speculating about the reasons. Clarke Chambers has suggested that the generally marginal status of social welfare history derives, in large measure, from the failure to utilize the "new" social history. Welfare history, instead, has continued to favor traditional perspectives—institutional, organizational, reformist—which usually conceived of welfare as a benefit from "the top down." The new social history offered clues to an alternative approach: the experience of blacks, women, workers, immigrants, and the poor as historical actors and not mere passive victims or beneficiaries of largesse from their betters. A subsidiary explanation for the marginal status of welfare history, according to Chambers, is the difficulty most historians might have in identifying with the subjects of social welfare—the dependent and delinquent classes (as they put it in the nineteenth century). He concludes that if welfare historians would incorporate the perspectives and methodology of the new social history, it would "restore welfare history as an important and central part of the historical experience of the American people."[1]

Chambers's analysis of the issue is both perceptive and constructive. I would, however, add another explanation for the marginality of welfare history. For most historians, if they think about it at all, welfare seems to be a marginal historical force. It is not, ostensibly, a prime mover in contrast to ethnicity, race, gender, immigration, class, labor, urbanization, agriculture, economics, politics, literature, or art. Social welfare, by contrast, is a kind of salvage operation, propping up the wreckage produced by more important change agents. Most historians, understandably, prefer to identify with the hammer more than the anvil. They want to deal with the forces that make history, not with the debris.

Pittsburgh: Economic Development

These considerations have particular relevance to Pittsburgh. It is not surprising that historians have been most attracted to its flamboyant and dramatic economic development, its immigrant and working-class life. What insights can social welfare development in Pittsburgh provide comparable to those in the economic or labor sphere? As much as any nineteenth-century city, Pittsburgh sheds light on the character of the industrial revolution in America, especially the relationship between technology, locational advantage, and community development (regional as well as city).

Locational factors which had benefited Pittsburgh before the Civil War diminished with the coming of the railroad, the decline of the river trade, and competition from cities closer to western markets. Pittsburgh responded with maximum exploitation of its competitive superiority in access to raw materials (notably the mineral fuels needed for heavy industry). The city's preeminence as an iron-and-steel center after the 1880s was associated with the coking coal of the nearby Connellsville coal fields; these produced the finest metallurgical coke in the United States. In essence, the Pittsburgh industrial empire of the nineteenth century was built on access to one particular fuel. This advantage was eventually undermined by the development of by-product coke ovens, more economical when situated nearer the furnaces, and which led to the use of competing coals. Once again, access to rapidly growing markets become more important than a previous locational advantage.

The regional mining industry produced a scattering of isolated, barren villages in the various counties of the Pittsburgh industrial area and a comparatively immobile labor force tied to the mines. The same industry also produced a desecrated landscape as a result of topographical alterations and acid discharge. Supplementing the mining villages was a concentration of population in the mill towns strung along the river valleys. These communities were organized around one or more large plants, whose transportation and water needs led to a usurpation of the flatlands, relegating homes to the hills and slopes. The population of the region, dispersed by topography, was further fragmented by ethnic allegiances, notably among the Southern and Eastern European immigrants who flooded the region after the 1880s. They had responded to the rapid expansion of the heavy industries and, not least, the technological improvements which reduced the need for highly skilled labor. Along with its locational advantage in the late nineteenth century, the large-scale influx of immigrant labor made possible the Pittsburgh region's meteoric rise as the iron-and-steel center of the world.

Economic Development: Social and Civic Consequences

Pittsburgh is not only a prime exhibit for the historian interested in the economic and industrial development of cities in the paleotechnic era; it is equally significant in illustrating the social and civic consequences of nineteenth-century urbanization and industrialization. In his account of Philadelphia, Sam Bass Warner claimed that the "failure of urban America to create a humane environment . . . is the story of an enduring tradition of privatism in a changing world." He defined privatism as "concentration upon the individual and the individual's search for wealth." Warner contends that our decisive urban experience occurred in the period 1830–1860, when the city and privatism first collided; the relationship between the private and public spheres was established, and the "communitarian limits of a city of private money makers were reached and passed."[2]

Superficially, Pittsburgh at the turn of the century would seem to confirm Warner's interpretation. It surely lacked social cohesion, and "privatism" or materialism was no doubt instrumental. The lack of cohesion, of civic integrity, was reflected in the "archaic social institutions such as the aldermanic court, the ward school district, the family garbage disposal, and the unregenerate charitable institution, still surviving after the conditions to which they were adapted have disappeared." Ultimately, the social and civic dereliction was expressed, in the opinion of the editors of the Pittsburgh Survey, in the stark contrast "between the prosperity on the one hand of the most prosperous of all the communities of our western civilization . . . and, on the other hand, the neglect of life, of health, of physical vigor, even of the industrial efficiency of the individual." No community, in Europe or the United States, had ever "applied what it had so meagerly to the rational purposes of human life."[3]

Although plausible, I would contend that a line of analysis extending from privatism or materialism to social pathology and civic fragmentation is incomplete and misleading. Privatism in Pittsburgh, or in American community life generally, was itself a functional response to the pluralism and heterogeneity of American society. The few examples in our urban tradition of ideological and social cohesiveness, of integrated environmental and social development, were associated with homogeneous, compact, small-scale communities (New England farm villages of the seventeenth century, religious and secular communitarians of the nineteenth century). It is one thing to be critical of privatism, economic individualism, materialism, and their social or environmental consequences. It is something else to explain just how civic cohesiveness could be attained in a context of religious, class, ethnic, and racial heterogeneity. Large-scale, heterogeneous civic entities do not nurture much sense of identity with the whole, not to mention the spirit of familism or mutual aid. It is not surprising that the ideal of material progress filled the ideological and social vacuum; it was a unique basis for consensus in a pluralistic, balkanized civic culture. An entrepreneurial, voluntaristic ideology shaped urban development in the United States, resulting in a disjunction between economic decision making and social organization, a commodity conception of land and housing, and a presumption that social organization and social welfare were necessarily incidental by-products of commerce.

So the enduring challenge of American urban culture was established in the nineteenth century: how to achieve collective goals, or satisfy collective needs in a large-scale, pluralistic urban system, but with minimal coercion or government by fiat. Equally fundamental was the elementary problem of defining collective needs in a pluralistic society. What happened, I would contend, is that our national urban culture followed the path of least resistance; it exploited the one collective ideal on which there was consensus and license for government to act—the ideology of material progress and aggrandizement. The difficulties and dangers of government coercion in relation to social goals were bypassed by adopting the impersonal market as the arbiter of economic, environmental, and social development. A kind of competitive pluralism became the basis for order and structure in a society which lacked a European-style ruling class, and which viewed government as a threat to liberty rather than a source of justice and social cohesion.

Market Discipline and Spatial Organization

If, historically, we bypassed the dilemmas of achieving consensus and centralized leadership in a polyglot society by reliance upon market disciplines, it was also the case that the market disciplines were expressed, partly, in spatial organization. To put it in the form of a question: what were the sources of stability in an urban situation which was unstable because of explosive growth and a friction-generating social pluralism? What conditioned and stabilized social relationships? Nowhere are these questions more pertinent than to Pittsburgh in its nineteenth-century incarnation.

The organizing system was an impersonal, decentralized ecological process which reduced friction between disparate groups by spatial separation. Economic, class, and social differentiation were expressed in a spatial dimension: like groups factored out and clustered on the subcity level of the neighborhood and even the single block. This may not have been a particularly egalitarian solution to problems of conflict, communication, and consensus in a pluralistic society, but it was effective in reducing friction and promoting stability. One might say, in short, that the nineteenth-century solution to the problem of group relations was to organize space in a way which reduced contact and thus friction between potentially antagonistic groups.

Stabilization was also promoted, in ecological terms again, by differential group mobility and occupational clustering. Religious and ethnic groups tended to concentrate in selected enterprises like the Irish in politics, the building trades and contracting, or the Slavs in mining and heavy industry. Again, ecological imperatives promoted differentiation and separation. The potential for conflict increased whenever economically comparable but ethnically different groups competed for the same jobs, housing, or neighborhoods.

The Pittsburgh Survey and Social Pathology

She does not use the language of spatial ecology, but Margaret Byington, in her masterpiece of community analysis for the Pittsburgh Survey, *Homestead: The Households of a Mill Town*, nonetheless perceived the significance of ecological elements in the community's social system. She found that 53 percent of the 6,772 steel mill employees were Slavs (28.3 percent were English-speaking or native born). The separation between Slavs and the rest of the community was dramatic and total, even greater than between whites and blacks: "Neither in lodge, nor in church, nor, with a few exceptions, in school, do the two mingle. Even their living places are separated." Transients were numerous and their presence reduced further any "effective civic force of the community." Other than the home, social interaction centered on the ethnically homogeneous fraternal societies, churches, and other voluntary associations. The most powerful social institution among the Slavs was the church (Greek Orthodox and Roman Catholic), whose influence was enhanced through control of the parochial education system. The church, in Byington's view, reinforced the isolation of the Slavic population and, therefore, the civic fragmentation and impotence of Homestead: "as many priests speak no English and are little more in touch with American ideas than their people, the church life tends to preserve rather than to remove national distinctions."[4]

If the multifaceted Pittsburgh Survey can be said to have a single theme, it would be, in sharp contrast to the city's economic sector, the fragmentation of its civic, political, and social systems. Democracy, the community, "must overhaul the social machinery through which it operates if it would bring its community conditions up to standards comparable to those maintained by its banks, its insurance companies and its industrial corporations." Human and social engineering lagged pitifully behind the mechanical. Although the Pittsburgh Street Railway Company had carefully plotted the whole area—populations, streets, wards, boroughs—as a basis for extending its lines, the Pittsburgh Bureau of Health had not even issued an annual report from 1899 to 1907.[5]

The Pittsburgh Survey is one massive documentation of the marginality of social welfare institutions in shaping the development of the city. Even by contemporary standards of "scientific philanthropy" as embodied in charity organization, Pittsburgh was behind the times. I would not presume to go to an opposite extreme and claim that in social welfare history one has a neglected key to American urban history. Even if more closely aligned to the "new" social history, I doubt that social welfare will dominate the historical agenda. Still, it has more value than its almost total neglect (in the case of Pittsburgh) would suggest. Social welfare can, potentially, serve as a basis for exploration of many dominant themes and issues.

Social Welfare: The Colonial Era and Early Nineteenth Century: Class Factors

One would be hard pressed to illustrate that argument in the case of colonial or early national Pittsburgh. It is virtually a blank as far as welfare history is concerned. For some insight into how social welfare can illuminate broader, more controlling issues in the period, one must turn instead to a study of Philadelphia by John K. Alexander. The author uses poverty and social welfare to explore the effects of the Revolution upon social and political institutions. Attitudes toward the poor, he claims, do not support the proposition that the "late colonial and revolutionary periods were marked by a high degree of social unity . . . and simple humanitarianism."[6] Responses to poverty, before and after the Revolution, were characterized neither by humanitarian nor subsistence objectives, but by aspirations for deference and social control on the part of the elite. Class stratification was the norm, and responses to poverty were governed by that reality.

Alexander develops his argument in chapters dealing with living and working conditions of the poor in the latter eighteenth century; the Constitution of 1776 and its effect on the criminal code, particularly imprisonment for debt; liberalization of the franchise; education reform; public and private poor relief. To the degree that the Revolution had any liberalizing effects, such as extension of the franchise, it only reinforced the determination of the upper classes to use relief programs as a means to restore and maintain deference patterns and class domination.

Alexander attempts to clarify the relationship between social class and social welfare through an extended analysis of the public welfare sector. Prior to 1776, poor relief was controlled by the overseers of the poor. Legislation that year authorized the establishment of a House of Employment, financed and managed by private "Contributors to the relief and employment of the poor." Class differences between overseers (mechanic-artisan) and contributors (Quaker mercantile elite) at the outset were expressed in conflicting relief policy; managers of the House of Employment consistently condemned outdoor relief in contrast to reform and reformation through institutionalization. The object was to transform character, to change the attitudes and habits which had engendered poverty. Although the overseers long resisted this policy, they eventually succumbed and endorsed abolition of outdoor relief by the early 1790s.[7]

Private relief agencies in the post-Revolutionary era also favored a relief strategy designed to "control or mold the needy in the image of the worthy poor"; the best way to do this would be to curb relief to the unworthy poor. This might be accomplished by requiring applicants for relief, medical assistance, or employment to obtain a recommendation from a respectable citizen. This requirement promoted the illusion, if not the reality, of lower-class deference.

In a paper focusing on public and private relief in Pittsburgh during three business depressions—1837–1843, 1857–1859, and 1873–1878—Cecelia F. Bucki essentially favors the same social control framework: "The relief systems discussed . . . reveal much about interclass hostility and attempts at control . . . of one class over another." Ideologies and policies concerning relief suggested that "members of the benevolent class found it necessary to differentiate between the worthy and unworthy poor and conclude that poverty was a moral, not an economic, issue."[8] Further research in this early period in Pittsburgh might deal with the influence, if any, of Scots-Irish Presbyterianism upon welfare policy.

Voluntary Institutions and the State

These accounts of social welfare in Philadelphia and Pittsburgh not only involve class issues but a second central theme in American history: the role of voluntary institutions, and their relationship to the public sector. The controlling concept is relationship; through social welfare history one deals with the interaction between public and private, the state and society. Some historians and sociologists have recently criticized the neglect of the state on the part of social historians who have favored the study of the more private and routine spheres of life. "The state," claims William Leuchtenburg, "is enmeshed in the very warp and woof of our national culture." Even at the pinnacle of the laissez-faire political economy, the state was "deeply implicated in fostering economic growth."[9] Social welfare history can contribute considerably to an understanding of state and society in American life. By the nature of their subject, social welfare historians have never lost sight of the state's presence. Even the most superficial immersion in welfare history since the rise of the modern state forces one to deal with the public sector and the voluntary association.

Social Welfare and the New Social History

Along with class, the state, and the role of the voluntary association in American life (as well as the relationship between state and voluntary association), welfare history can illuminate significant themes in contemporary social history. Although Bucki uses a social control explanation for the development of relief systems, she also suggests, tangentially, that the "structure of the day-to-day survival patterns of working class communities needs to be researched."[10] Chambers develops this theme more fully: "It is the continued existence and practice of extended-kinship responsibilities, especially among black, immigrant, and working-class families, that welfare historians have most neglected. To begin their analyses only at the point at which the state or agencies of the private sector began to exert their powers is to overlook the historical reality of kinship networks and thus not fully to appreciate the larger context of welfare."[11]

Given its ethnic and working-class mix, Pittsburgh could serve as an outstanding historical laboratory for this neglected dimension of welfare. The Pittsburgh Survey emphasized the community's social pathologies, the fragmentation of its civic institutions, but it also provided much evidence of a working-class subculture of fraternal and mutual aid organizations. In Homestead and other mill towns, benefit organizations were pervasive. Byington found announcements for fifty fraternal order meetings scheduled within a week. She secured information for twenty-three lodges out of as many as one hundred. Their membership totaled 3,663 (predominantly male), and almost all offered benefits and insurance as well as social activities. In contrast to the town, the fraternal lodges endeavored to arouse some sense of "fraternity and common interest."[12]

Although the authors of the Survey and other social workers or reformers in the period acknowledged the reality of the working-class social and fraternal subculture, this awareness was not translated into any kind of supportive strategy. To the contrary, these institutions were usually viewed as parochial obstacles to Americanization and civic cohesion. They had to be transcended in favor of a more elite and paternalistic approach to social melioration: greatly expanded government regulatory, service, and welfare functions, on one hand, and a private welfare sector committed to scientific and professional methods, on the other.

Elites and Gender in Social Welfare

This social welfare and reform ethnic was fully compatible with the cosmopolitan and centralizing aspirations of the elites described by Samuel P. Hays in his essay, "The Politics of Reform in Municipal Government in the Progressive Era." The business and professional leaders who dominated social agencies and reform movements "were all involved in the rationalization and systematization of modern life; they wished a form of government which would be more consistent with their objectives." Specifically, they sought structural reforms which would reduce the domination of "local and particularistic interests."[13] The parallel effort in the voluntary sphere was embodied in the quest for rationalization, professionalization, and centralization of social services—an end to anarchy and parochialism in favor of charity organization and expert administration.

In Pittsburgh and elsewhere, by the period of the Survey, women were conspicuous in the activities of welfare and reform organizations (not least in the work of the Survey itself). Along with businessmen, professionals (including a corps of public health-oriented physicians), and the socially prominent, women sought to shape the civic culture to their liking. An intriguing question concerning social betterment in Pittsburgh and elsewhere would deal with the degree to which the structure or policies of social agencies were gender related. Social welfare offered women, among other things, a means by which to acquire political influence and a channel of self-support and upward career mobility when opportunities for careers were limited. Social welfare, in short, functioned as a source of liberation from constricted domestic roles.

Women were prominent in the work of the Civic Club of Allegheny County. Organized in 1895, it became a leading reform agency along with the Chamber of Commerce and settlements such as Kingsley House and Irene Kaufmann. The Civic Club's first vice-president and chairman of its strategic Legislative Committeee was Lucy Dorsey Iams. Her participation sheds light not only on social welfare and the role of women in American society, but on other themes I have mentioned, such as the relationship between state and society (voluntary associations). Married to a lawyer active in Democratic party affairs, she served as his secretary, read extensively in the law, and became an accomplished court stenographer. As a member of the legislative committees of diverse organizations—the State Federation of Pennsylvania Women; the Consumers' League of Western Pennsylvania; the Pennsylvania and Allegheny County Child Labor Associations; the Associated Charities of Pittsburgh—she became virtually the unofficial coordinator of reform legislation for Western Pennsylvania. Mrs. Iams's activities suggest an important basis for linkage between the public and private welfare sectors. Individuals like herself circulated between them, or promoted legislation which expanded the responsibilities of the public sector. Her leading role in the enactment of housing legislation illustrates the point. Chairman of the Civic Club's Tenement House Committee, she was instrumental in the drafting and enactment of the state tenement house law of 1903 and in winning an appropriation for city inspectors to enforce the measure.[14]

Mrs. Iams explained that the key to effective reform was skill in political maneuver. The 1903 housing law, for example, was not the product of widespread public interest or pressure. Advised by one politician, she nurtured the measure through the legislature with the help of another. Although social welfare, social reform, or social justice is usually advocated in the name of the public interest, or the "people," it is often best understood as the creation of an elite: professional, bureaucratic, upper-class, or political. In fact, the study of social welfare is as much an exercise in the analysis of elites in American society as it is the study of poverty and marginality. Pittsburgh, again, can serve as a potentially outstanding research laboratory in developing this perspective. Social welfare and social control are not the issue. It goes beyond that into the realm of social reconstruction, if not utopianism. Social welfare and philanthropy were a substitute for party politics, increasingly a lower-class domain, and a basis for definition of the self and the social order.

Play, Recreation, and Social Reconstruction

An ostensibly marginal issue like playgrounds and recreation can illustrate how social welfare served these purposes. The Civic Club's Education Department opened a summer playground in 1896. This had expanded to a dozen by 1899, when the Board of Education started to contribute operating funds. The Pittsburgh Recreation Association was established in 1906, and its second superintendent, W. F. Ashe, became director of the city's Bureau of Recreation, created in 1915. The "play spirit," according to Beulah Kennard of the Civic Club, was the path to "civic unity" and a "bond of followship" uniting rich, poor, and worker.[15] The broader significance of the play movement, of social welfare as a basis for self-definition and social reconstruction, is captured in a study of the relationship between organized play and urban reform by Dominick Cavallo.[16]

Progressive Era concepts of environmental causation and "scientific" reform coexisted with ideals of character-building and a kind of moral imperialism (best exemplified in the prohibition movement). The playground movement sheds light on this dualism, as well as the degree to which progressive reform centered on child welfare. Not least important, the playground movement contributes to an understanding of the transference effect of social reform—the tendency to shift functions from family to state and professional caretakers. Cavallo describes his study as a "history of efforts made by urban social reformers during the years 1880 to 1920 to transfer control of children's play from the children and their families to the state." Play, in the eyes of the social workers, educators, and psychologists who promoted the movement and established the Playground Association of America in 1906, was "too serious a business to be left to children and parents."[17]

Cavallo devotes considerable attention to theories of child psychology and development which, in the late nineteenth and early twentieth centuries, provided intellectual support for the playground movement. The most influential was the recapitulation theory of G. Stanley Hall (ontogeny recapitulates phylogeny). It was supplemented by James Mark Baldwin's theory of imitation, John

Dewey's theory of knowledge as a function of encounter between child and real life objects or situations, and the psychology of habit training identified with Edward L. Thorndike.

Although these interpretations of child development differed significantly, they all represented a post-Darwinian perspective on child nurture, substituting evolutionary principles for the religious, static, and dichotomized views which had prevailed. Man was no longer the instrument of a divine purpose, and mind and body no longer occupied separate spheres (with mind occupying a loftier niche in the cosmic hierarchy). Instead, mind, body, emotion, and instinct were conceived as an interpenetrating continuum. It followed that physical or muscular conditioning in an appropriate setting—the supervised playground promoted by playground leaders such as Joseph Lee or Luther Gulick—could shape character and morality. In other words, "muscular conditioning was the key to mental and moral efficiency."[18] The contemporary urban child, like his primitive ancestors, was a physical creature and his moral and cognitive growth could be influenced through an organized play which channeled physical instincts and responses into socially desired behavior.

Play advocates devoted much attention to the adolescent, hoping to transform his peer-group tribalism into behavior appropriate to an industrial, bureaucratized society. Their ideal, specifically, was to create a viable balance between the individualism of the past and the cooperative imperatives of an interdependent modern community. They aspired to the creation of a kind of ideal-typical personality who combined the "male's rational and morally autonomous nature" with the "female's emotional and morally empathic nature." Thus gender was equated with bi-polar moral attributes, and play organizers aspired to a synthesis which tempered male aggressiveness or inner-direction with female capacities for empathy and socialization. Cavallo uses Jane Addams to illustrate the nature of the synthesis: both Addams and the ideal team player "wedded characteristics that in the nineteenth century had been sundered into idealized and stereotypical masculine or feminine attributes." This new, synthesized personality type would, hopefully, combine the individuality, adaptability, and social sensitivity required by an industrial society.[19]

It seems outlandish, in retrospect, that anyone could believe that a few hundred or a few thousand supervised municipal playgrounds could function as the primary instrument of personality development or socialization, and thus the solvent for the conflicts and contradictions of an urban civilization. The subject provides an object lesson in historical analysis: significance is not necessarily synonymous with effectiveness or major impact. Treated imaginatively, even an ostensibly marginal historical phenomenon like the playground movement can offer significant insights into social reform and ideals of personality and social organization.

The Social Settlement: Pittsburgh's Kingsley House

Recreation and play were central to the character-building aspirations of the social settlement in the late nineteenth and early twentieth centuries. According to Lillian Wald of New York's Henry Street Settlement, the "young offender's presence in the courts may be traced to a play impulse for which there was no safe outlet."[20] But recreational opportunity was only one element in the broader settlement scheme of neighborhood reconstruction. There has never been a social institution quite comparable to the settlement in its prime from about 1890 to 1918. It functioned as a kind of general caretaker of society, with special emphasis upon the life of the neighborhood in which it was situated.

Whether in New York, Chicago, Pittsburgh, or Jersey City, the settlement formula was simple. Members of the privileged classes, usually college-educated and often women, would live in poor neighborhoods. While "dipping into the current of life" and becoming enriched themselves, they would, in turn, stimulate the growth of American mainstream culture and ideals among the immigrant and working-class poor.[21]

In order to achieve its objectives of building character and promoting assimilation, and to nurture a neighborhood cohesiveness which would transcend religious, class, and ethnic differences, the settlement had to deal with the working and living conditions which threatened to sabotage its efforts. This outward thrust of the settlement, the commitment to environmental melioration, is illustrated by the housing and sanitary reforms inspired by Pittsburgh's Kingsley House in the early twentieth century.

Kingsley House also clarifies the religious impulse which inspired the reform movements of the Progressive Era. The spirit of Evangelical Protestantism had permeated the charitable institutions and national reform crusades of the antebellum era. In the late nineteenth century, a new spirit of liberal Christianity embodied in the Social Gospel greatly influenced ostensibly secular instruments of reform like the settlement. Thus Richard Ely, the economist, declared that the city was "destined to become a well-ordered household, a work of art, and a religious institution in the truest sense of the word 'religious'." And Josiah Strong proclaimed that "when the social spirit has been Christianized we shall have, not a fraternity of convenience, but a genuine brotherhood of love sprung from a common fatherhood."[22] The social settlement was, in good measure, a secularized expression of the Christian idealism embodied in the Social Gospel. The history of social welfare, again, is woven into a central theme of American life—religious thought and institutions.

Established by a minister, and supported by affluent patrons, Kingsley House embodied two key aspects of the national settlement movement: religious idealism and dependence upon the social elite for funding and support. As an expression of contemporary social work, the social settlement also played an important role in expanding channels of career mobility for women (and, in time, for the children of immigrants and minorities). Social work emerged as an alternative to the traditional female careers of teaching and nursing. The first three head workers at Kingsley House were women, and in the settlements of Pittsburgh and other communities women were conspicuous as volunteers as well as professional staff.[23]

Established in 1893 through the initiative of Reverend George Hodges, Kingsley House was originally located at Penn Avenue and Seventeenth Street. It relocated in 1901 to a residence (the gift of Henry Clay Frick) at the corner of Bedford Avenue and Fulton Street in the Hill district. Another affluent patron, Charles L. Taylor, donated a sixty-five-acre farm in Butler County which became the Lillian Home, the site of summer outings.[24] The original location on Penn Avenue exemplified the kind of neighborhood that settlement pioneers selected to launch their experiments in Christian brotherhood. Close to iron and steel mills, glass and cork works, the area housed a population of factory workers and laborers. Irish-Americans predominated, but Germans, Russians, and Austrians had been settling along Penn Avenue. There were few homeowners. The social settlement was not akin to the Salvation Army, church mission, or welfare agency which sought out a skid row or lower-class clientele. It reached out to the immigrant working class (and especially their children). Kingsley House's first head worker explicitly repudiated any notion that college settlement work "is mainly for the very poor or for the degraded."[25]

The spirit of the Social Gospel dictated the creation of a social order based on Christian brotherhood and justice. A socialized Christianity made no distinction between the sacred and secular realms; to the contrary, it implied the translation of the Sermon on the Mount into everyday life, the introduction of the Golden Rule into economic and political affairs. Thus the Kingsley House of the 1890s yearned to "break down class distinctions, to bring together in real friendship those whom the accident of habitation has entirely separated, to realize that great ideal of social equality for which this century stands." Kingsley House might become a social center, "not only for the neighborhood, but for the city, where workman and capitalist, cultured and uncultured, rich and poor, radical and conservative, can meet freely and interchange ideas."[26]

The religious idealism which motivated the early settlement worker was dramatically expressed in the person of William H. Matthews, resident head worker from 1902 to 1910. Born in England, Matthews came to America at the age of nine. His family settled in New England, and the child's schooling ended at age twelve, when his father became ill. Matthews spent the next seven years working in the textile mills. Befriended by a Congregational minister who aroused a thirst for education, Matthews entered a private seminary in Massachusetts, and then Williams College. After graduation, he attended Union Theological Seminary in New York and took over a boys' club at Union Settlement. He gained additional experience as leader of a boys' club at Bethany Mission in Manhattan. Appointed assistant director of Union Settlement in 1901, he was persuaded to head Kingsley House the next year (it was not only women for whom the social settlement served as a source of upward mobility).

The English counterpart of the American Social Gospel was Christian Socialism, and Matthews was greatly influenced by reform-minded English clerics like Charles Kingsley, Frederick Denison Maurice, and Samuel L. Barnett. "To all of them," Matthews claimed, "the belief in the Fatherhood of God and the Brotherhood of Man was most real." Equally important, they were practical idealists who not only dreamed of a better world "but also worked to make things better." Matthews described the settlement as an instrument of social Christianity, "an effort to bring Christianity back to its earlier humanitarian aspects, an attempt to bring the Kingdom of Heaven down here on this earth."[27]

More than any single individual, Matthews shaped Kingsley House in his image. He conceived of the settlement not only as an embodiment of Christian idealism, and a character-building agency through its clubs and classes, but as an instrument of social reform in Pittsburgh. The logic was simple: how could the immigrants, workers, and children that Kingsley House sought to influence be receptive if they were constantly dragged down by a degraded environment?

Matthews had barely arrived in Pittsburgh before he plunged into an aggressive crusade for better housing. When the Board of Health neglected his complaints, he bought a camera and began to document the "filthy, unsewered alleys and courts" in the neighborhood and other parts of the city.[28] He spoke at local clubs, churches, anywhere he could get a hearing. He brought Jacob Riis in from New York to help out. Matthews condemned the corrupt, graft-infested city government for its indifference to the unsanitary conditions and vice in the Hill district. And in 1907–1908 Kingsley House became a base for the investigators of the Pittsburgh Survey, thus linking the settlement to the greatest experiment in social research ever undertaken in any American city. John A. Fitch resided at Kingsley House while preparing his volume *The Steel Workers* for the Survey.

Matthews filled the pages of the settlement newsletter, *Kingsley House Record*, with articles, reports, and pictures concerning housing and sanitary problems and the settlement's efforts to correct them. He emphasized the vital connection between environmental improvement and the daily work of the settlement—classes in English, typing, telegraphy, manual training, needlework, housekeeping, and others. The test of the settlement's success was "development of character," but what kind of character could emerge from "unsanitary, germ-laden tenements . . . producing a physical and moral degeneracy full of danger to any community"?[29] What kind of citizens would be nurtured by filthy alleys and courts where the garbage cart never visited, or houses infested with smallpox, tuberculosis, and diphtheria? Could children living in these conditions ever become good Americans, endowed with "self-respect, honor and decency"? In short, the home "has ever been the key to good citizenship. . . . without decent homes we cannot have decent citizens."[30]

Kingsley House was, in microcosm, the American settlement movement in the late nineteenth and early twentieth centuries. It incorporated the spirit of the Social Gospel, the educational, club, and recreational activities expected to nurture character, and the environmental reform objectives which would support the character-building and Americanization objectives.[31] Kingsley House also exemplified the uniqueness and contradictions of the broader social settlement movement in the early decades. Unlike any social agency, before or since, it functioned as a kind of general caretaker of humanity. It professed a commitment to brotherhood and equality, but was managed by members of the social elite and an educated professional staff who would set an example of character and culture. It was concerned with the faulty character of the worker, the poor, and the immigrant, but it became identified with the correction of environmental pathologies, social legislation, and the expansion of government welfare and service functions. Perhaps the early settlement like

Kingsley House can be understood best as a unique experiment in constructive paternalism based upon a combination of religious idealism, socialization aspirations, and social legislation efforts rooted in a simplistic environmental determinism.

Andrew Carnegie and the Gospel of Wealth

What the playground and settlement movements had in common with Andrew Carnegie's Gospel of Wealth was a faith in the uses of philanthropy to reconstruct the social order. Equally important, it was through philanthropy that society's elite of businessmen, and the professional, educated classes, could influence the social order independently of politics with its aura of corruption and lower-class dominance.

If science and large-scale business enterprise exemplified the cosmopolitan, rationalistic, centralizing tendencies of the era, these same values controlled the ideology and program of mainstream social work as embodied in charity organization. Like charity organization, in turn, the Gospel of Wealth can be understood as an effort to define the basis of constructive philanthropy, combining science (evolutionary biology) and business practice (efficiency, centralization, and hierarchy). Both charity organization (which came late to Pittsburgh in 1908) and the Gospel of Wealth were significant expressions of late-nineteenth-century conservative political economy, representing impulses deeper than social control in a narrow, simplistic sense.

The exponents of constructive philanthropy, like Carnegie and the leaders of charity organization, were attempting to define the basis for harmony, authority, and progress in contemporary society. Like the liberal, they were in quest of community.[32] But the attainment of their community depended upon the resolution of a fundamental contradiction in American life, the contradiction between political equality (one person, one vote—for those eligible to vote) and economic inequality (the basis of incentives and rewards). The American democratic state subscribed to an inherently unstable, and potentially explosive, contradiction between the practice and ideology of political equalitarianism and an economic system rooted in unequal outcomes and wide disparities of wealth. No greater challenge confronted America than the creation of the institutions and ideology which could resolve the dilemma, or reduce its potential for damage.

In essence, an important root of American conservative social theory can be located in nineteenth-century social welfare. Combining scientific and business values, it aspired through "constructive" philanthropy to reconcile the contradiction between political equality, and economic hierarchy and inequality. Not at all antiquarian or irrelevant to the subsequent evolution of American society, the political economy of nineteenth-century social welfare retains an enduring relevance in its effort to confront a central problem of governance in a democratic, but nonsocialist society.

One can thus interpret Carnegie's Gospel of Wealth as an expression of constructive philanthropy which attempted, through the application of scientific law and business principles, to resolve the contradiction and define a satisfactory basis for community. In keeping with Darwinian biology, he stressed the importance of the law of competition in social evolution: "We cannot evade it; no substitutes for it have been found; and while the law may be sometimes hard for the individual, it is best for the race, because it insures the survival of the fittest in every department." Evolution, through the laws of competition and survival of the fittest, dictated the inexorability and desirability of inequality in biology and social arrangements. "We accept and welcome," said Carnegie, "as conditions to which we must accommodate ourselves, great inequality of environment; the concentration of business, industry and commerce in the hands of a few." It was essential for the progress of the race that the homes of the few "should be homes for all that is highest and best in literature and the arts, and for all the refinements of civilization, rather that none should be so."[33] Much better a great discrepancy than universal squalor.

Along with the law of competition, the circumstances which would lead to a high order of civilization included obedience to the law of accumulation of wealth, respect for private property, and individualism. Exemplified in these laws, the evolutionary process operated benignly for the progress of society as a whole, whatever the consequence for specific individuals. In short, Carnegie's justification for wide disparities in condition, for the dichotomy between political equality and economic inequality, is rooted in Darwin's evolutionary biology extended into the social order. But social instability might arise out of these circumstances; not all would take the evolutionary logic and outcome as self-evident truth. It was necessary to go a step farther and consider the uses of concentrated, surplus wealth.

The proper administration of wealth promised to promote ties of brotherhood, binding rich and poor "in harmonious relationship." The law of civilization had concentrated it in the hands of the few; they were now obliged to define the appropriate laws of philanthropy, or principles of distribution compatible with evolutionary biology and good business practice. Carnegie dismissed consuming the wealth or bequeathing it to descendants as unworthy of a sacred trust, a stewardship. The only legitimate use of great fortunes was the organization of benefactions which would improve and dignify the lives of the masses. Such constructive philanthropy exemplified the "true antidote" for economic inequality and the basis for class harmony. Indeed, constructive philanthropy, so defined, was the basis for the ideal state "in which the surplus wealth of the few will become, in the best sense, the property of the many, because administered for the common good."[34] Constructive philanthropy, the stewardship of wealth, and not the machinery of government, was the force to elevate the race. Like business, philanthropy had to become large-scale, concentrated, rational.

Thoughtless, indiscriminate, frivolous charity had to be avoided because it thwarted the evolutionary process. Better, Carnegie insisted, that the "millions of the rich were thrown into the sea than so spent as to encourage the slothful, the drunken, the unworthy." Constructive philanthropy helped those who helped themselves; it assisted the motivated, it did not pauperize by means of unconditional handouts. One might conceive of the stewardship of wealth as a step-ladder ideal of social welfare: the way to reconcile philanthropy and evolution, political democracy and economic inequality, was to provide the ladders upon which the deserving could rise. In condemning ill-conceived philanthropy, Carnegie went so far as to argue that one person who lived comfortably by begging was "more dangerous to society, and a greater obstacle to the progress of humanity, than a score of wordy Socialists."[35]

Carnegie praised such benefactions as the Cooper Union in New York City and the Tilden bequest for the New York Public Library. Compatible with these precedents, the constructive philanthropy most acclaimed by Carnegie was the free library. Closely allied and attached where possible, should be an art gallery and museum, and a hall for lectures and instruction (as at Cooper Union). He also advocated the founding of universities and support of hospitals, medical schools, laboratories—all institutions which not only cured but prevented ignorance and disease. Other expressions of constructive philanthropy included public parks, meeting halls, public baths, and the inspirational value of attractive churches.

Political evolution in the twentieth century, the rise of the welfare state in particular, suggests the repudiation of Carnegie's stewardship or Gospel of Wealth ideals. As a succinct critique, one could not improve upon Margaret Byington. Carnegie had given Homestead a library, Charles Schwab donated a manual training school, and Henry Clay Frick contributed a small park. What did the workers think: "They appreciate what the library and manual training offer to them and their children, but they resent a philanthropy which provides opportunities for intellectual and social advancement while it withholds conditions which make it possible to take advantage of them." Rather than a library, the workers "would rather have had higher wages."[36]

The Elite and Cultural Capital

This critique raises a significant issue concerning the role of the elite and the creation of a community's (or society's) cultural capital. Certainly a community with the libraries, museums, parks, hospitals, educational institutions is better off than a community without them. Much of Europe's great reservoir of cultural institutions is the creation of a despotic state, church, and aristocracy. To no small degree, the much maligned captains of industry in this country served a comparable role. Without the concentrated wealth, would Pittsburgh have had a major art museum (and its latter-day Scaife addition), or would New York have the Metropolitan and its other great museums? Carnegie believed that Peter Cooper made a greater contribution to humanity in the form of Cooper Union than if he had dissipated the money in higher wages for his workers. In a sense the worker was an involuntary benefactor of future generations; his sacrifice of current consumption levels was transformed by Carnegie and his peers into the cultural sustenance of the future.

It would be difficult today to find many advocates of the stewardship of wealth as a basis for social welfare and communality, but the questions endure. First, is the elite and its concentrated wealth indispensable for the well-being of society, particularly in relation to its cultural capital? Latter-day expressions of this issue in Pittsburgh need to be studied. The Buhl, Mellon, and Heinz foundations have played an important role in the life of the community (like their counterparts elsewhere). The Buhl Foundation, for example, sponsored Chatham Village in the 1930s, and made an enduring contribution to our concept of residental subdivision and design.

Social Welfare, the Welfare State, and Conservative Social Theory

A second enduring issue—and a significant dimension of the conservative social theory embedded in nineteenth-century social welfare—concerns the conflict between political equality and economic inequality. The welfare state, if anything, exacerbates the problem. As the scope of state action increases, the temptation grows to use the political machinery as an instrument of economic redistribution and the creation of "rights." The potential for demagoguery, or government by fiat, is substantial. A tendency emerges to convert every conceivable moral imperative into coercive legal and economic mandates. Proclaimed "rights" escalate and are translated by a responsive political and judicial machinery into claims upon the society's economic resources and social institutions. We may not accept nineteenth-century solutions to the conflict, but it will not go away.

Social Welfare and Welfare Capitalism

Another late-nineteenth- and early-twentieth-century "solution" to problems of social harmony was welfare capitalism. Beginning in the 1880s, businessmen attempted to challenge unionism and win the loyalty of workers by providing a wide range of unilateral benefits: education, recreation, medical care, housing, old age pensions, profit sharing, stock ownership. Welfare capitalism suggests another significant use of social welfare history—the analysis of industrial relations. Using welfare capitalism as a point of departure, the social welfare historian can illuminate the complexities of capital-labor relations and related themes, such as the nature of company cultures and the social obligations of business.

A dramatic and comprehensive example of welfare capitalism, linking it to experiments in innovative community planning, was Pullman, Illinois. George Pullman, of Palace Car fame, founded his model industrial town along the west shore of Lake Calumet in 1880. A professional architect (Solon S. Beman) and a landscape architect (Nathan F. Barrett), large-scale management and planning, and concern for visual amenity all combined to produce an exceptionally attractive community. The town demonstrated that unified design did not necessarily conflict with variety and surprise. Despite the conventional gridiron street plan, the planners avoided monotony by skillful placement of a public square, building, or scenic view. The housing, row and multifamily, was plain but neat and spacious. Market house, school, arcade, and hotel provided for most routine domestic needs. Even the sewerage system was distinctive; collected in a large tank, it was pumped onto a 170-acre Pullman farm.

State labor commissioners, convening in Pullman in 1884, could not have been more sanguine. Welfare capitalism, as embodied there, might be the key to industrial harmony. Like most others, the economist, Richard Ely, writing for *Harper's* the next year, lauded the physical environment. But much as he admired the design, Ely differed from contemporaries in his ominous vision of social upheaval. A placid population was not necessarily content, let alone grateful for a benevolent despotism which discouraged dissent and self-expression. In taking for granted the satisfaction and gratitude of his subjects, Pullman ignored the need for communication—for mutual dialogue and understanding. Jane Addams, in the aftermath of the explosive strike of 1894, would incisively portray the great industrialist as a modern-day King Lear who felt betrayed by his "children."

Stanley Buder, in his study of Pullman, examined the business and social philosophy which inspired the great experiment in community planning.[37] Pullman had assumed that application of the management techniques of corporate capitalism to community development would result in a propitious blend of profit, environmental amenity, and social harmony. A key assumption was that men necessarily lived up to their environment. An ordered and attractive community would produce industrious and loyal workers. Buder's explanation for the failure of Pullman as a prototype of capitalist social engineering centers on the flaws inherent in a paternalistic, semifeudal relationship between company and men. Equally important, as Buder recognizes in chapters tracing the town's transformation into a lower-class ethnic enclave of Chicago, was the ecological challenge to authority in American life. It was difficult to keep any community uncontaminated, not least one situated near a great metropolis. Pullman could banish liquor from his community, but he could not squash the multitudinous saloons of nearby Kensington and Roseland.

But Pullman failed as an industrial utopia for another reason, one which had nothing to do with the founder's personal inadequacies or the inability to maintain the community as an isolated objet d'art. Apostles of welfare capitalism attempted to square the circle and exclude power conflict from industrial relations. From its formative years in the 1880s to its maturity in the 1920s, the critical assumption behind welfare capitalism was that the work force could be satisfied and loyal with no sacrifice of managerial authority. This was the logic, not only of the model industrial town, but the many other expressions of welfare capitalism. The Pullman strike challenged, but did not destroy the conviction. Given its potential as a strategy for industrial harmony and managerial authority, it was something its exponents wanted to believe in.

For apostles of unionism, and the self-determination that came with it, welfare capitalism was an abomination. John A. Fitch, the nation's leading critic of the twelve-hour day in the steel industry, and author of the volume on the steelworkers in the Pittsburgh Survey, argued that any plan which served "to make employees acquiescent is undemocratic even under good employers, because it stifles an independent spirit." What was important to understand about the labor policies of the steel industry was not the "pensions, sociological departments and Sunday Schools." It was the spirit "of arrogance and contempt for the rights both of their employees and of the public." Business benevolence was in "sinister opposition to democracy. It is a policy which . . . means feudalism, and the denial to workmen of rights that must be regarded in America as fundamental."[38] The United States Steel Corporation may have established plant safety committees in 1908, a centralized medical organization in 1909, an accident compensation program in 1910, along with disability and old age pensions, but its real labor policy was expressed in the Secret Service Department which monitored any expression of union sentiment.[39]

Critics of welfare capitalism were hardly mollified by Henry Ford's declaration of profit sharing and the $5 day in 1914. It was another expression of paternalism and autocracy: unilateral, conditional, and noncontractual. The company hired a flock of investigators who scurried about Detroit, visiting the homes of eligible workers. If they discerned careless habits, lack of thrift, or other departures from good character, the unfortunate worker forfeited his share of the "profits" and his wage remained at a lower level. Considering his explanation for the $5 day, one might have confused Henry Ford with a social worker: "Our investigators," he proclaimed, "find that there has been a remarkable epidemic of house cleaning. When a man gets a higher wage he will not only be a better workman, but he will be a better man and will carry the influence home to his family."[40]

What linked Henry Ford, George Pullman, the United States Steel Corporation, and most other exponents of welfare capitalism was the search for a system of industrial relations which would curb union sentiment. But what endows this ostensibly narrow and negative objective with broader significance was the implicit expansion of the scope of business enterprise. Welfare capitalism in the decades before the Great Depression established precedents for many benefit programs developed later through collective bargaining. Though arbitrary, paternalistic, and antiunion, welfare capitalism socialized American business in the sense that obligations to both worker and community were acknowledged—obligations which were neither legal mandates nor inherent in the operation of the business. As an alternative to government intrusion and welfare, as well as collective bargaining, welfare capitalism was a significant phase in the historical evolution of the business system.

Likewise, the substance of welfare capitalism in a specific firm or industry can serve as the basis for studying what one historian has called "the world views of businesspeople and the cultures of their firms."[41] The Cadbury family of Birmingham, England, chocolate manufacturers and founders of the widely acclaimed industrial garden suburb of Bournville, was governed by the Quaker ethic in its employee relations. This created a "distinctive managerial culture and strategy" which successfully combined welfare and efficiency objectives. The creation of Bournville would thus express "the ethics of social contract, not the ethics of social control."[42] Located a few miles outside Birmingham, financed and planned by George Cadbury, the community was donated in 1900 to the Bournville Village Trust (establishing local self-government). Like Port Sunlight, a contemporary industrial garden suburb founded by Sir William Lever near Liverpool, Bournville had great impact upon planners and housing reformers in the United States.[43] In these planned industrial communities of winding, tree-lined roads, handsome cottages, and spacious parks, they saw a dramatic alternative to the environmental and social pathologies of American working-class communities. A visitor to Port Sunlight had not realized that "there was anywhere in the world a village in which there was nowhere to be found one ugly, inartistic, unsanitary, or other demoralizing feature."[44] And in Bournville, the Cadburys created a miniature welfare state in addition to the environmental amenities.[45]

An American version of a capitalist welfare state could be found in the shoe factories of the Endicott Johnson Corporation in the Binghamton, New York, area. The far-ranging welfare system, built over the first three decades of the twentieth century, included profit sharing, medical care, recreation and athletic facilities, low-cost homes for purchase, and access to management. Company policy was unabashedly antiunion, but the welfare programs offered management an alternative to repression in countering union influence. The company emphasized (as in latter-day Japan) the spirit of mutuality and family. Portrayal of patriarch George F. Johnson as a father figure "was aimed at making industrial protest and rebellion the equivalent of patricide."[46] Union organizers, in fact, had little success there in the 1920s and 1930s. It would be illuminating, perhaps, to approach business culture and industrial relations from a perspective based on comparative welfare systems. George Pullman, for example, combined community-building with feudal-like paternalism and suffered "patricide"; U.S. Steel combined more limited welfare schemes with severe repression and successfully avoided unionism until the 1930s; Endicott Johnson apparently succeeded with a blend of comprehensive welfare and minimal repression.

The Pittsburgh industrial region offers no examples of welfare capitalism as inclusive as Endicott Johnson, let alone Pullman or Bournville. But it does provide opportunities to explore the subject in a variety of industries and companies. At one extreme was the coal and coke village. A company store anchored one end of the village or "patch"; a schoolhouse and church would likely appear at the other end. Homes, perhaps fifty to one hundred two-story, double, frame boxes spanned the hillside. Below the village was an engine house and coal tipple. Banked along the valley floor were the lines of coke ovens, spewing columns of flame and heavy, brown smoke. Often, the towns were close enough to the ovens to suffer from a miasma of thick, heavy smoke. Streets hardly merited the name. Gutters consisted of open, shallow ditches or deeper gullies, and served as repositories for rubbish and garbage. Residents generally stepped directly from house onto "street."

Since few mines were located close enough to cities to provide employees with alternatives to the company town, it was an example of enforced welfarism (company as well as worker). Besides housing, the coal operators invariably established a company store. Competition was not permitted, and the store usually was profitable. In some cases, employees were expected to spend a proportion of earnings there each week; sanctions for noncompliance might include unpleasant work assignments or inferior housing. Another mining town staple, forced upon the operators, was some provision for medical care. This was probably the most significant contribution of the coal industry to the development of welfare capitalism. Medical benefits ranged from full costs of treating injured workmen to little more than payment for initial treatment. Some mining companies established benefit societies (usually compulsory); in return for a monthly premium of 35 to 50 cents, an injured worker received a disability benefit of $5 to $6 a week, or a death benefit of $100.

If welfare capitalism in the coal and coke villages was minimal and disdained by worker and company, the model industrial town of Vandergrift, Pennsylvania, was a dramatic contrast. Hardly so grand as Pullman or Bournville, Vandergrift nonetheless greatly impressed contemporary observers. Located astride the Kiskiminetas River, forty miles east of Pittsburgh, the community was the creation of George G. McMurtry, president of the Apollo Iron and Steel Company. Forced to expand operations around 1895, McMurtry determined to test the proposition that workers "given an opportunity to live in a clean, healthy, beautiful town, which gradually they could own and govern, would become a permanent group of citizens working together like other citizen bodies."[47] To help insure the success of Vandergrift as an experiment in citizenship, the housing deeds stipulated that no liquor was to be sold for ninety-nine years.

Welfare capitalism in the Pittsburgh region usually favored the skilled worker. This was not only the case in the dominant steel industry, but in the model industrial community: "Vandergrift was to be an elite community for skilled workers, while the more expendable unskilled workers were left to fend for themselves." But as the unskilled working population increased, the company was forced to create an adjoining community, Vandergrift Heights. Lots here were cheap, but improvements were minimal.[48]

Other experiments in the Pittsburgh region in housing and community development included the towns of Midland and Aliquippa. Midland was the creation of the Pittsburgh Crucible Steel Company which constructed, initially, 120 homes on a site overlooking the Ohio River about twenty-two miles from Pittsburgh. An interesting feature of the subdivision plan was the ecological differentiation it imposed: separate sections were allocated to Southern and Eastern Europeans, blacks, Native American workers, skilled workers, and officials.[49]

Aliquippa, Pennsylvania, launched around 1907 by the Jones & Laughlin Steel Company, was developed on a site of fifteen hundred acres on the left bank of the Ohio River, nineteen miles below Pittsburgh. Three-and-a-half miles long, one-half mile wide, the community ran parallel to the river and was bisected into two strips by the Pittsburgh & Lake Erie Railroad. The steel plant occupied the section between river and railroad, the section beyond was occupied by employee homes.[50]

Aliquippa claimed a population of eight thousand by 1913. Tom Girdler, who was to become general superintendent of the Aliquippa Works and general manager of Jones & Laughlin, first arrived in 1914. Describing Aliquippa, unapologetically, as a "benevolent dictatorship," he praised its integration of physical and social planning: "Employing engineers, architects, and landscape men, the company had used the utmost foresight in laying out the town, always controlled by the desire to make it the best possible place for steelworkers to work and to rear their families." The company houses included all modern utilities, lawns, trees, gardens, and easy access to woodland. In a burst of romantic enthusiasm, Girdler claimed that some day the "supervised growth of Aliquippa, Pennsylvania, will take on glamour, too." Society would appreciate the "magnificence, the genuine splendor of that project, designed to make a decent, dignified community conforming to the best American standards."[51]

Stuart Brandes, surveying the evolution of welfare capitalism, contends that even if "an unpleasant step in the process of American industrialization, it was probably a necessary one." A society could not simultaneously afford heavy capital investment and "good housing, good schools, good medicine, and generous pensions for all its workers." And the welfare state which succeeded the experiments in welfare capitalism was not so different in certain respects—business paternalism and bureaucracy were superseded by government paternalism and bureaucracy likewise "aimed at imposing solutions to social problems from above."[52] Certainly, the diverse and noncentralized experiments in welfare capitalism over half a century could never outdo the welfare state in the scope and impact of coercive paternalism.

Referring to the steel industry, especially, John A. Fitch asserted that the "only dangerous agitators are those who attempt to build an industry on a foundation of wages too low to admit of decent standards of family life, of hours too long to admit of proper rest or relaxation, and of silence and acquiescence as the price of a job."[53] Framed in these terms, no amount of welfare capitalism could compensate for the absence of self-determination; it was essentially an element in a system of industrial feudalism and autocracy. But from a different, latter-day perspective, welfare capitalism was part of an ongoing quest to define the character of industrial relations, the social obligations of business enterprise, and alternatives to the welfare state.[54]

Conclusion: Social Welfare History and the Abandonment of Social Control

In this chapter I have examined the proposition that social welfare history has much greater value than its marginal status among historians would suggest. Used imaginatively, it can serve as a point of departure for the exploration of central themes in American history: class; voluntary associations and their relationship to the state and public sector; self-help and survival strategies in the immigrant, working-class, and black communities; the role and status of women; religious thoughts and ideals; concepts of social reconstruction and philanthropy; aspects of conservative as well as liberal social theory; and the social responsibilities of business.

This perspective is fully compatible with the emphasis by Clarke Chambers upon the incorporation of the new social history into welfare history. But social welfare history also needs to be liberated from the conceptual straightjacket imposed by simplistic application of the social control interpretation of its historic role and the corollary of self-interested motivation. It is no great revelation to learn that sponsors of charity organization or of social settlements were members of the educated elite who acted on the basis of self-interest (class, gender, nationality, professional). The alternative is absurd—that social work or social welfare should function, in contrast to any other sphere of life, as the realm of pure altruism and self-sacrifice.

NOTES

1. Clarke A. Chambers, "Toward a Redefinition of Welfare History," *Journal of American History* 73 (Sept. 1986): 407–09.

2. Sam Bass Warner, *The Private City: Philadelphia in Three Periods of Its Growth* (Philadelphia: University of Pennsylvania Press, 1968), pp. xi, 3.

3. Edward T. Devine, "Results of the Pittsburgh Survey," *American Journal of Sociology* 14 (Mar. 1909): 662.

4. Margaret F. Byington, *Homestead: The Households of a Mill Town* (1910; rpt. Pittsburgh: University of Pittsburgh Press, 1974), pp. 12, 14, 15, 159.

5. Paul U. Kellogg, "The Civic Responsibilities of Democracy in an Industrial District," Conference for Good City Government, *Proceedings* (Pittsburgh, 1908, 399–400; Kellogg, "The Social Engineer in Pittsburgh," *New Outlook* 93 (Sept. 25, 1909), pp. 153–154.

6. John K. Alexander, *Render Them Submissive: Responses to Poverty in Philadelphia, 1760–1800* (Amherst: University of Massachusetts Press, 1980), p. x.

7. Eventually, different elements of the mercantile elite dominated both groups.

8. Cecelia F. Bucki, "The Evolution of Poor Relief Practice in Nineteenth-Century Pittsburgh" (seminar paper, Department of History, University of Pittsburgh, 1977), pp. 4, 2.

9. William E. Leuchtenberg, "The Pertinence of Political History: Reflections on the Significance of the State in America," *Journal of American History* 73 (Dec. 1986): 591, 509. See also Ann S. Orloff and Theda Skocpol, "Why Not Equal Protection? Explaining the Politics of Public Social Spending in Britain, 1900–1911, and the United States, 1880s–1920," *American Sociological Review* 49 (Dec. 1984): 726–50.

10. Bucki, "Evolution of Poor Relief," p. 32.

11. Chambers, "Toward a Redefinition," p. 422.

12. Byington, *Homestead*, p. 113.

13. Samuel P. Hays, "The Politics of Reform in Municipal Government in the Progressive Era," *Pacific Northwest Quarterly* 55 (Oct. 1964): 161. Hays's interpretation of the municipal reform process is applied to aspects of urban reform in Ross Messer, "The Medical Profession and Urban Reform in Pittsburgh, 1890–19920" (seminar paper, Department of History, University of Pittsburgh, 1964); and Janet R. Daly, "The Political Context of Zoning in Pittsburgh, 1900–1923" (seminar paper, Department of History, University of Pittsburgh, 1984).

14. For a biographical sketch of Lucy Dorsey Iams, see my article in *Notable American Women.*

15. Beulah Kennard, "Pittsburgh's Playgrounds," *Survey* 22 (May 1, 1909): 195, 196.

16. Dominick Cavallo, *Muscles and Morals: Organized Playgrounds and Urban Reform, 1880–1920* (Philadelphia: University of Pennsylvania Press, 1981).

17. Ibid., pp. xi, 2.

18. Ibid., p. 58.

19. Ibid., pp. 112–46.

20. Lillian D. Wald, *The House on Henry Street* (New York: H. Holt, 1915), p. 95.

21. Felix Adler, "The Ethics of Neighborhood," *University Settlement Studies* 2 (July 1906): 28.

22. Richard Ely, *The Coming City* (New York: T. Y. Crowell, 1902), p. 71; Josiah Strong, *The Twentieth-Century City* (New York: Baker and Taylor, 1898), pp. 122, 123, 127, 128.

23. The role of women in the early Pittsburgh settlement is explored in Elizabeth Metzger, "A Study of Social Settlement Workers in Pittsburgh, 1893 to 1927" (seminar paper, Department of History, University of Pittsburgh, 1974). The subject is also covered in Raymond McClain, "The Immigrant Years: Irene Kaufmann Settlement, 1895–1915" (seminar paper, Department of History, Carnegie Mellon University, 1969). Other studies of contemporary Pittsburgh settlements include Donald L. Taylor, "The Woods Run Settlement, 1895–1932" (seminar paper, Department of History, Carnegie Mellon University, 1970) and Laurie Billstone, "Soho Community House, 1905–1940" (MS thesis, School of Allied Social Sciences, University of Pittsburgh, 1942).

24. See William H. Matthews, "Lillian Home: Which Affords Its Guests Fresh Air, Farm Life, and Every Country Joy," *Survey* 24 (June 4, 1910): 407–19.

25. Kingsley House Association, *First Annual Report*, June 20, 1894, p. 5.

26. Ibid., pp. 5, 9.

27. William H. Matthews, *Adventures in Giving* (New York: Dodd, Mead, 1939), pp. 48, 52.

28. Ibid., p. 65.

29. *Kingsley House Record,* 7 (Jan. 1903): 1, 2.

30. Kingsley House Association, *Fifteenth Annual Report,* 1909, photographic section, n.p.; William H. Matthews, "A Discussion of Housing Conditions in Pittsburgh," *Kingsley House Record* 10 (Jan.–Feb. 1907): 1.

31. It should be noted that Matthews's successor as head worker, Charles C. Cooper, continued the Kingsley House battle against environmental pathologies. According to Cooper, the settlement's knowledge and influence must be used in interpreting the life of the people to the more fortunate, and particularly "in depicting the ills of child-labor, in showing the damning influence of the social evil and of the saloon, in presenting bad housing and unsanitary conditions, in exemplifying the need of industrial, educational, and recreational work in an over-crowded metropolitan city." Kingsley House Association, *Nineteenth Annual Report,* 1913, p. 16.

32. For the liberal idea of community, see Roy Lubove, "Frederic C. Howe and the Quest for Community in America," *The Historian* 39 (Feb. 1977): 270–91.

33. Andrew Carnegie, "The Gospel of Wealth," in Edward C. Kirkland, ed., *Andrew Carnegie: "The Gospel of Wealth" and Other Timely Essays* (Cambridge, Mass.: Harvard University Press, 1962), pp. 16, 15.

34. Ibid., pp. 14, 23.

35. Ibid., p. 26; Carnegie, "The Best Fields of Philanthropy," in ibid., p. 31.

36. Byington, *Homestead,* p. 178; and "The Family in a Typical Mill Town," *American Journal of Sociology* 14 (Mar. 1909): 658.

37. Stanley Buder, *Pullman: An Experiment in Industrial Order and Community Planning, 1880–1930* (New York: Oxford University Press, 1967).

38. John A. Fitch, "A New Profit-Sharing Plan," *Survey* 26 (April 1, 1911): 31; Fitch, "The Human Side of Large Outputs: Steel and Steel Workers in Six American States. VI. The Labor Policies of Unrestricted Capital," *Survey* 28 (April 6, 1912): 27.

39. The labor policies of the steel industry, including the welfare programs, are examined in David Brody, *Steelworkers in America: The Nonunion Era* (Cambridge, Mass.: Harvard University Press, 1960). Also useful are: Charles A. Gulick, *Labor Policy of the United States Steel Corporation* (New York: Columbia University Press, 1924); and Ida M. Tarbell, *The Life of Elbert H. Gary: The Story of Steel* (New York: D. Appleton, 1925). The welfare programs of U.S. Steel are discussed in Roy Lubove, *The Struggle for Social Security, 1900–1935* (1968; rpt. Pittsburgh: University of Pittsburgh Press, 1986), chap. 1; and Marion Thompson, "Accident Reforms in the United States Steel Corporation" (seminar paper, Department of History, Carnegie Mellon University, 1969).

40. Quoted in John A. Fitch, "Ford of Detroit and His Ten Million Dollar Profit Sharing Plan," *Survey* 31 (Feb. 7, 1914): 550.

41. Charles Dellheim, "The Creation of a Company Culture: Cadburys, 1861–1931," *American Historical Review* 92 (Feb. 1987): 32.

42. Ibid., pp. 14, 42.

43. The impact of the English garden city and suburb, and experiments in American industrial town development, are discussed in Roy Lubove, *The Progressives and the Slums: Tenement House Reform in New York City, 1890–1917* (Pittsburgh: University of Pittsburgh Press, 1962), chap. 8.

44. Annie L. Diggs, "The Garden City Movement," *Arena* 28 (Dec. 1902): 627.

45. These included sick benefits, pensions, medical and dental care, academic and technical education, sports and recreation, profit sharing, and reduced work week. See Dellheim, "Creation of a Company Culture," p. 29.

46. Gerald Zahavi, "Negotiated Loyalty: Welfare Capitalism and the Shoemakers of Endicott Johnson, 1920–1940," *Journal of American History* 70 (Dec. 1983): 607.

47. Ida M. Tarbell, *New Ideals in Business: An Account of Their Practice and Their Effects Upon Men and Profits* (New York: Macmillan, 1916), p. 154.

48. Ray Burkett, "Vandergrift: Model Worker's Community" (seminar paper, Department of History, University of Pittsburgh, 1972). Vandergrift is also discussed and lauded in Eugene J. Buffington, "Making Cities for Workmen," *Harper's Weekly* 53 (May 8, 1909): 16.

49. Albert H. Spahr, "The Town of Midland, Pa.: A New Development in Housing Near Pittsburgh," *Architectural Review* 4 (Mar. 1916): 33. Also, John Ihlder, "Midland," *Survey* 33 (Dec. 12, 1914): 300.

50. Agnes Wilson Mitchell, "The Industrial Backgrounds and Community Problems of a Large Steel Plant: The Jones and Laughlin Steel Corporation, Aliquippa, Pa." (Ph.D. diss., Department of Economics, University of Pittsburgh, 1932) p. 6.

51. *Boot Straps: The Autobiography of Tom M. Girdler* in collaboration with Boyden Sparkes (New York: Scribner's, 1943), pp. 177, 171, 169.

52. Stuart D. Brandes, *American Welfare Capitalism, 1880–1940* (Chicago: University of Chicago Press, 1976), p. 147. Another comprehensive analysis is David Brody, "The Rise and Decline of Welfare Capitalism," in John Braeman et al., *Change and Continuity in Twentieth-Century America: The Twenties* (Columbus: Ohio State University Press, 1968), pp. 147–78.

53. Fitch, "Human Side," 27.

54. Welfare capitalism was also a setting for the development of social work, if the definition is stretched to include the company nurses who performed extramedical welfare. See Brandes, *American Welfare Capitalism*, pp. 111–18.

■ The Soul of the City: A Social History of Religion in Pittsburgh

LINDA K. PRITCHARD

S ECULAR RATHER than religious images usually characterize the Pittsburgh region. Popular images of the city and hinterland have more to do with steelworkers and sports teams than with church steeples. Pittsburgh historians occasionally mention the religious affiliations of founding families and subsequent immigrants or that Western Pennsylvania was the birthplace of several religious movements. But these meager accounts stand in stark contrast to regions where religion is viewed as the cultural core of society, such as Puritan New England, the Bible-belt South, or New Age California. Pittsburgh's religious history has generally been regarded as unrelated to the city, unimportant to regional development, and irrelevant to national patterns.

In the last twenty-five years, however, a number of studies have uncovered evidence of religion's primary importance to the region. Few focus directly on religion, but the social history approach employed by most, stressing the interdependence of social processes and structures, documents the religious development of the region. While analyzing economic growth, ethnicity, community-building, or politics, for example, they have demonstrated how religion was as central to this area as elsewhere. Far from being peripheral to Pittsburghers, religion defined their values, structured their institutions, and organized their environment.

The information about religion in these studies comes largely from congregations and other religious institutions in the region. Although only one of several dimensions of religion, the congregation, or local meeting of like-minded believers, has been the primary outlet for personal religious beliefs in the United States.[1] The total number and size of congregations and their denominational affiliation, membership base, geographical location, and community activism provide a systematic overview of Pittsburgh's changing religious configuration.

Four specific time periods, 1770 to 1830, 1830 to 1880, 1880 to 1920, and 1920 to the present, demarcate the religious growth and development of Western Pennsylvania. Protestants from traditional groups dominated the period of white settlement, especially Presbyterians who were more successful in this region than anywhere else in the country. When commercial development began to reshape the social and economic base of the area after 1830, Protestant denominations maintained their hegemony by rallying around new evangelical ideas, while Irish and German immigrants organized Roman Catholic and Jewish congregations. The transition to industrial manufacturing after 1870 strengthened these non-Protestant institutions, and Protestants were forced to share religion's moral authority with increasing numbers of Catholics and Jews. As smokestack industries declined after 1920, Protestant, Catholic, and Jewish organizations consolidated their respective institutions and together dominated organized religion. Their position was never secure, however, as even newer religious ideas proliferated outside mainline religions.

In terms of religious development, massive industrial expansion led to a variegated religious configuration which historians have mistakenly interpreted as religious decline. But religious diversity and change did not mean secularization. Socioeconomic shifts promoted new ways to integrate a fragmenting religious community. Lay and clerical leaders demonstrated extraordinary success in organizing and maintaining congregations differentiated by class, ethnicity, race, and neighborhood. This essay will illustrate how religion remained a central focus of social life in Pittsburgh.

Religion in Frontier Pittsburgh

Little is known about religion in Western Pennsylvania before 1830, because Pittsburgh historians have devoted their attention to the later periods of industrial development. Fortunately, Thomas Cushing's *History of Allegheny County*, a nineteenth-century county history, recorded detailed descriptions of nearly every congregation founded by whites through its date of publication in 1889.[2] The baseline religious parameters of the region followed two common patterns in American religion. The ethnicity of the immigrants and their premigration religion determined the original white religious imprint on the region, while the building of community shaped the number and location of their congregations.

Colonial explorers brought European religion into the Pittsburgh region. As they passed through the "gateway to the West," travelers held religious services for themselves and the natives. In the 1750s,

Catholics, Quakers, and Moravians all attempted to Christianize the local Delaware, Shawnee, and Iroquois tribes. A Catholic friar stationed at Fort Duquesne in 1754 became the first resident pastor in the region. Five years later, on the Sunday following the French abandonment of Fort Duquesne, the chaplain of the British forces preached the first Protestant sermon in the territory. During the next decade, religious services, probably Baptist and Episcopal, were reported at Brownsville, the most important site for fording the Monongahela River.[3]

Sustained religious organization occurred only after permanent settlers arrived during the early 1770s. The rich lands south of the Monongahela and Ohio rivers were settled first, so two Baptist churches, one in Washington County and one at Library in Allegheny County, were organized in 1772 and 1773 respectively.[4] Permanent congregations did not appear at the Point until the locational advantages of the area bordering the three rivers prompted land developers to lay out a town in 1784. In that year, the original surveyor of Pittsburgh reserved three large lots for churches and donated the land to the Protestant denominations in the town.[5]

As elsewhere, land donations demonstrate the intimate relationship between community development and religious organization. Real estate developers offered gifts of land and money to congregations when they put other town lots up for sale, apparently believing that churches and schools would bring in more land buyers at higher prices. The land was usually in the center of the new town, insuring that churches became landmarks of the business districts dotting the Pittsburgh region. For example, the first McKeesport developers granted four lots for "the use of a place of worship and a seminary of learning" when they laid out the town in 1795.[6] Land speculators and community developers continued to invest in churches throughout the nineteenth century. When Wilkinsburg and Edgewood became railroad suburbs of Pittsburgh in the 1860s, a large original landowner, Squire Kelley, sold off large chunks of his estate to business developers and donated lots to any Protestant congregations wishing to erect a church.[7]

The ethnic composition of the early population determined which religious groups benefited from these economic favors. The first residents of Pittsburgh came predominantly from the former colonies of Virginia and Maryland; the British Isles, especially Scotland, northern Ireland, and England; and Germany. As a result, one of three original downtown church lots was donated to German Protestants, who established Smithfield United Evangelical Church in 1782. Another went to the First Presbyterian congregation, incor-

porated in 1787, and the last to former Anglicans who finally organized Trinity Episcopal Church in 1825.[8]

At the beginning of the nineteenth century, the religious mix of
Pittsburgh, now a town of fifteen hundred people, diversified. A
Methodist class and a Reformed Presbyterian congregation appeared
just before 1800, and a Roman Catholic parish and Baptist congregation incorporated soon thereafter. The 185 free blacks in the city
began Bethel African Methodist Episcopal (AME) Church in 1818,
and a Unitarian congregation was organized ten years later. Evidence
of Jews in Pittsburgh exists by 1786, including Wilkinsburg's first
owners, but no religious services were held until mid–nineteenth
century.[9]

Heavy Scots-Irish immigration after 1800 meant that Presbyterians predominated in Western Pennsylvania. In Pittsburgh, five of the
twelve congregations in 1830 were Presbyterian. These churches represented a full denominational spectrum, from the "uncompromising
rigidity" of the Associate and Associate Reformed Presbyterians to
those with American-born members affiliated with the more moderate Presbyterian, U.S.A. denomination. Both wings were strong
enough to support important Presbyterian institutions in the city,
including seminaries which still exist today. Traditional Presbyterian ministerial candidates attended Allegheny Theological Seminary, opened in 1825. Moderates went to Western Theological Seminary, later Pittsburgh Theological Seminary, founded in 1827 with a
grant of eighteen acres of land and more than $21,000 in loans from
Allegheny County's public funds.[10]

Rural areas surrounding Pittsburgh developed an even stronger
Presbyterian identity. Until after 1830, for example, more than one-
third of the population in Westmoreland County to the east and in
Washington County to the south was born in the Presbyterian
strongholds of Scotland and northern Ireland. The first congregations in these counties, as well as in the towns of McKeesport and
Sewickley, were Presbyterian. In addition, the four counties just
north of Allegheny produced thirty-two rural churches by 1810, and
twenty-eight were traditional Scots Presbyterian. Even though other
ethnic groups also organized congregations in rural locations, including German communitarians in Harmony, the concentration of Presbyterians in Western Pennsylvania qualifies the region for the title,
the "Presbyterian valley."[11]

This first phase of congregation-building in the Pittsburgh area
was distinguished by a relative lack of conflict within the fledgling
churches. Individual members undoubtedly had personal grievances
with each other, but religious disagreements rarely disrupted the

congregation publicly. The case of Smithfield may be instructive. This congregation managed to keep Lutheran and Reformed worshipers under the same roof and pastor for fifty-five years between its founding in 1782 and 1837.[12] Perhaps the financial and emotional demands of a congregation in a new environment initially caused individuals to play down their differences to insure group survival.

Yet two studies of the period before 1830 uncover portents of the noisy controversies which became the hallmark of Pittsburgh's next period of religious development. Both examine the nineteenth-century triumph of evangelical religion, including the optimistic view that men and women, with the help of God, could control their own destiny, over traditional religious forms emphasizing fatalism.

One is set in the rapidly developing commerical agricultural hinterland southwest of Pittsburgh, where the Baptist churches of the Redstone Association battled over missions between 1815 and 1845. Congregations here and throughout the nation parted ways over whether Baptists should sponsor centralized missionary activities or whether the responsibility for propagating the faith rested with individual church members. The evangelicals who advocated organized missionary endeavors eventually triumphed over the "antimission" diehards, but not without shattering Baptist congregations and associations in the region.[13]

Entirely new religious movements were born of this Baptist disharmony. The "Christians," inspired by Thomas and Alexander Campbell, first appeared during the conflict in Washington County. Soon after reorganizing their strict Presbyterian church into an antimission Baptist church, these Campbellites moved across the river into Ohio bearing a new name, the Disciples of Christ. The same controversy also produced an early Mormon convert. Ten years after the First Baptist Church of Pittsburgh was organized in 1812, church members still could not agree on whether to be antimission or evangelical. One of the charter members was Sidney Rigdon, later infamous because of his widely rumored role in publishing the Book of Mormon.[14]

The second study highlights early tension within the First Presbyterian Church of Pittsburgh. By 1800, a faction within the church objected to members of the congregation who belonged to the Masonic Order. Leaders of the Antimason group were early and vocal evangelical converts who formed Pittsburgh's Second Presbyterian Church in 1802.[15]

As the settlement stage came to a close, Pittsburgh took pride in its considerable religious accomplishments. Even though one traveler, Ann Royal, noted disdainfully that the city had more taverns than churches in 1828, Pittsburgh residents at a later date could brag that "although many of the people of the day were noted for their disregard of the discipline laws of the churches to which they belonged, there was not the religious indifference at the time" that they believed their generation suffered.[16] The ratio of congregations to population in the Pittsburgh region was comparable to the national average at the time: slightly more than one church per 1,000 people in 1828, totaling twelve congregations for 10,600 people.[17] Congregations were even more plentiful in rural areas. In 1830, for example, Washington County had twice as many churches per capita (two per 1,000) as Pittsburgh, or one every three miles in heavily settled areas.[18]

Even so, the next fifty years saw phenomenal religious development throughout the region. Between 1830 and 1880, Western Pennsylvanians founded and supported congregations, organized religious agencies, and engaged in religiously based social reform activities more strenuously than they ever had or would again.

Religion in a Developing Region, 1830–1880

Major social and economic forces accounted for the tremendous burst of religious organization during this period in Pittsburgh, as well as in most other parts of the nation. The Second Great Awakening, emanating from fractious controversies over evangelicalism, increased Protestant activity in the region, while Pittsburgh's coming of age as a regional center for commerce and industry promoted new immigrant religions and the building of congregations. My own study of religious change at mid-century in the upper Ohio Valley, whose eastern boundary was Western Pennsylvania, demonstrates that the distribution of evangelicalism, immigrant religions, and traditional denominations was closely related to the presence or absence of economic development.[19]

Pittsburgh's surging regional economy, with attendant population growth and optimism about future expansion, encouraged the development of religious institutions. In 1850, Allegheny County achieved the highest congregation to population ratio ever, 1.3 congregations for every 1,000 people, and Pittsburgh became "a City of Churches."[20] However, as elsewhere in the upper Ohio Valley, the commercial farming hinterland which surrounded the city continued to support an even higher number of churches. The lucrative commercial countryside of Washington County, for example, had nearly three times as many congregations (three per 1,000 population) in 1850.[21]

Despite the general expansion of organized religion throughout the region after 1830, the strength and mix of different Protestant groups remained similar to the prior time period. Religious development between 1830 and 1880 left Protestantism in the same preeminent position in Western Pennsylvania as before, with the rank order of the largest denominations largely unchanged. Between 1830 and 1859, at least forty Presbyterian congregations were founded in the city, and a similar number in proportion to the population appeared in the rich agricultural areas of the region.[22] Between 1850 and 1890, Presbyterians controlled approximately 40 percent of Allegheny County's religious establishments.[23] The mainstream Presbyterian, U.S.A. denomination and the traditional Scottish groups maintained nearly equal numbers of communicants. The importance of Pittsburgh for both types is demonstrated by the fact that each denomination held a major conference in the city at mid-century. In 1858, Associate and Associate Reformed groups organized the United Presbyterian Church, and in 1868, Pittsburgh was the site of the Presbyterian, U.S.A. unification of the thirty-year-old Old School and New School wings.

Even though sustained growth kept Presbyterianism in the numerical lead, other Protestant denominations also grew rapidly between 1830 and 1880. Protestant Episcopal activity increased in the 1830s and again in the 1850s. Both the Methodist and Baptist denominations had their greatest successes in terms of new congregations between 1830 and 1860 when Baptists went from one to twenty-one congregations, for a total of 2,314 members, and the Methodists from one to twenty-six with 3,924 members. Black Methodist denominations also did well. With slightly more than twenty-five hundred blacks in Allegheny County, at least two AME and three African Methodist Episcopal Zion (AMEZ) congregations were organized before 1860.[24]

The fastest growing Protestant denomination between 1830 and 1880 was Lutheranism. After 1837, the Smithfield Church was no longer the only German Protestant church in the city. In that year, Lutherans separated from the congregation and formed two new churches. With the subsequent tidal wave of German immigrants, the Lutherans went from fifth largest to third largest Protestant denomination in the county, trailing only the Presbyterians and Methodists in number of congregations by 1860.[25]

Several studies of farming communities during this period illustrate the central role of Protestantism in rural areas. An ambitious look at 130 rural congregations shows that Presbyterians dominated in all but a few locations from 1800 to 1976. Peters Township in

Washington County was just such a pocket. In 1850, its two churches were Methodist and, since elite families were often instrumental in the religious development of hinterland towns and hamlets, were named after large landowning Methodist families. Children's socialization in rural Washington and Greene counties between 1840 and 1900 was thoroughly Protestant. And these religious views were unlikely to change in townships like Cross Creek which had few connections to neighboring communities.[26]

Yet Western Pennsylvanians surely would have emphasized the upheaval rather than the stability in the region's religious landscape between 1830 and 1880. While institutional growth and development sustained the dominance of Protestantism, the divisive power of evangelicalism became paramount after 1830. Congregational splintering, inter- and intradenominational conflict, and new religious alternatives enlarged and enlivened the mid-nineteenth-century religious milieu of Pittsburgh. At the same time, rapid Roman Catholic expansion exerted a new religious influence.

Most Pittsburgh studies of evangelical conflict focus on mainline Presbyterians. At the congregational level, one study looks at the division of the Sixth Presbyterian Church (U.S.A.) in 1860, when Rev. Samuel Findley resigned under pressure and several prominent families left the church. Evidence is scanty, but the New School leanings of Findley, pastor of a devoutly Old School congregation, undoubtedly contributed to the disagreement.[27] Another study shows that wealthy members who moved from Pittsburgh to Sewickley split the basically Old School Presbyterian church over the purchase of an organ and other innovations.[28]

At the denominational level, career patterns of Old and New School Presbyterian ministerial candidates were different. Those who graduated from Pittsburgh's Western Theological Seminary, which, with Princeton, was an Old School school, were older at ordination, more likely to remain parish ministers throughout their careers, more often served poorer churches, and were involved more in the antislavery movement and less in temperance than were their counterparts from Auburn Theological Seminary in central New York, a major New School training facility.[29] Despite the strong Old School orientation in Western Pennsylvania, prominent Presbyterians helped to initiate the last round of Second Great Awakening revivals in Pittsburgh in 1857–1858.[30] Such an enthusiastic embrace of a decidedly evangelical technique by this Old School stronghold set the stage for the later Presbyterian, U.S.A. reconciliation.

Concurrent with the rise of evangelicalism, Pittsburgh became a Catholic stronghold, largely due to the immigration of large numbers of Irish and Germans from 1830 until the end of the century. Not all immigrants were Catholic: twenty percent of the Irish were Presbyterian Orange men and women from northern Ireland, and over 50 percent of the Germans were Lutheran, Reformed, or Evangelical Protestant.[31] Even so, the immigrants from these two countries assured a sizable increase in the proportion of Roman Catholic congregations in Pittsburgh.

Catholic parish development started slowly. From 1808 to 1829, the entire city had only one parish, Old St. Patrick's. After 1,214 baptisms in ten years, St. Paul's Cathedral was built in 1830, becoming the institutional center for Irish Catholics. In that year, two visiting Roman Catholic bishops estimated a burgeoning Catholic population in the city at around four thousand. Ten years later, only two additional parishes, one using English and one German, had been formed.[32] But by 1880, the Roman Catholic church could boast of two dioceses, thirty five churches, twelve chapels, one college for men and five "young ladies' academies," forty-three parochial schools, two hospitals, eight "homes," over twelve thousand children in parochial schools, and eighy-three thousand Catholics within Allegheny County.[33]

Two approaches typify the study of Roman Catholicism in Pittsburgh during this period. A traditional interest in the development of Catholic hierarchy characterizes a detailed account of the first bishop of Pittsburgh, Michael O'Connor.[34] This young Irish prelate took on the demanding new Diocese of Pittsburgh in 1843. During the next seventeen years, O'Connor presided over a Catholic population subjected to bitter anti-Catholic denunciation by Pittsburgh nativist Joseph Barker and his cronies. O'Connor lost power to clergy who successfully fought his authoritarian leadership for a larger role in diocesan decisions; he called his German constituents "those people" who "do not wish to adapt to the customs of the nations to which they have come"; and he unsuccessfully fought to restrain Boniface Wimmer, the head of St. Vincent's Monastery (later abbey) near Latrobe, from engaging in beer production and running a seminary for German priests. Despite controversy, O'Connor oversaw an impressive increase in the number of parishes and churches, including the dramatic renovation of St. Paul's Cathedral in 1855. The next two bishops, however, struggled to repay church debts and heal the wounds he incurred.[35]

The other approach to Pittsburgh Roman Catholicism comes from the social history commitment to study people instead of institutions. Parish studies of mid-nineteenth-century German and Irish communities explored the relationship between religion and the socioeconomic position of these immigrant groups. Although historians have generally believed that traditional religious affiliations retarded the acculturation of immigrants, the Pittsburgh studies show that immigrant religion provided an important bridge from the Old to the New World. The changing membership of eight Catholic and Protestant German churches in Pittsburgh and Allegheny City between 1850 and 1880, for example, demonstrates that an evolving religious diversity allowed Germans to keep a common ethnic identity, even though the Pittsburgh economy was dividing them along class lines.[36] In the case of the Irish middle class, new industrial values including temperance and lay authority dominated Roman Catholic development.[37]

The mid-nineteenth century also saw the initial religious organization of Pittsburgh's Jews. The first formal services were held in 1842 at a private home. A burial society followed in 1847 and a synagogue in 1848. Religious dissension wracked the original congregation, Shaare Shamayim, because two-thirds of the original twenty to thirty Jewish families came from southern Germany and one-third from Poland. Polish Jews, less Americanized and demanding more traditional religious practices than their German counterparts, organized Beth Israel in 1851, which lasted only two years. Germans organized Rodef Shalom in 1854. When the latter congregation adopted the American Reformed platform ten years later, the Tree of Life congregation was organized as an orthodox Polish alternative. The number of Jews remained small, however, and in 1870, only two thousand lived in the city.[38]

Evangelicalism and the success of non-Protestant groups created a volatile religious environment in Western Pennsylvania between 1830 and 1880. Although religious leaders complained that such religious warfare undermined attempts to expand institutions and serve the community, religious innovation and strife seemed to fuel congregational growth and individual religious commitment.

Political parties and reform movements provide natural arenas for religious conflict. A pioneering attempt to see if Germans voted Republican in 1860, as their leaders encouraged them to do, discovered that German and Irish Protestants in Pittsburgh voted Republican, while German Roman Catholics voted Democratic.[39] Another study showed that this religious split extended into the German elite of the city.[40] Several inquires into Antimasonry during the 1830s illustrate the complex relationship between religion and politics, especially third parties, in the region. Antimasonry was strongest among emerging middle-class Protestants who lived in the presence of a growing

Roman Catholic population, but the denominational breakdown differed by region. In rural Somerset County, southeast of Pittsburgh, antirevival Lutherans, German Reformed, and German sects made up the Whig and Antimason coalition, which opposed the largely Methodist and Disciples Democrats; in Pittsburgh the Antimason supporters were primarily Presbyterian, Methodist, and Disciple evangelicals.[41]

The temperance crusade enlisted the same anti-Catholic and pro-evangelical nativists as Antimasonry. The concentration of the dry vote in evangelical Protestant areas of Western Pennsylvania during the 1854 state temperance referendum and the entrepreneurial and professional class background of temperance activists in Pittsburgh demonstrate this familiar national pattern.[42] However, although temperance was the most important reform movement in the Pittsburgh area, the relatively low voter turnout for the 1854 referendum suggests that this primarily religious conflict was not as salient here as elsewhere. No riots comparable to those in Philadelphia in 1844 occurred, although Joseph Barker was elected mayor of Pittsburgh in 1850 while jailed for inciting anti-Catholic demonstrations in Market Square.[43]

Unlike temperance, antislavery sentiment was not rooted in a nativist constituency of evangelicals and anti-Catholics. In fact, local denominations and churches rarely took a stand on the issue. In Washington County, individual Protestant clergy from several religious orientations organized an underground railroad depot in the 1830s. An active antislavery society, started in 1834, polarized several Methodist congregations in the county. The Wesleyan Methodist and other Protestant denominations gained several congregations when these churches divided over the issue of slavery, but abolition was less prominent than temperance and other social reforms.[44]

The enthusiastic religious growth from 1830 to 1880 blanketed the city of Pittsburgh with a greater number and variety of congregations than ever before. Two demographic measures, however, may have worried church leaders by the end of this period. The central business district, the location of most of Pittsburgh's churches, lost population rapidly after 1860. Furthermore, the overall ratio of congregations to population in Allegheny County had declined slightly by 1870. But these went unnoticed as congregations continued to proliferate. In the downtown area, new religious groups organized congregations (Latter Day Saints, Swedenborgians, and Universalists), denominational schisms produced congregations (Wesleyan Methodists, Congregational Disciples, and German Methodists), and the first YMCA opened between 1850 and 1870.[45] In the long run, however, the changing size and location of population would alter Pittsburgh's expansive religious growth.

Religion and Steel, 1880–1920

Between 1880 and 1920, the booming steel industry affected every aspect of Pittsburgh's religious community. Protestant, Catholic, and Jewish institutions strained to absorb massive numbers of new immigrants and to adjust to residential dispersion and occupational dislocations. The denominational direction of the city shifted dramatically, with new workers more likely to come from Poland, Hungary, and Russia than from the United States, Ireland, or Germany. By 1920, immigrants doubled the Catholic population of Allegheny County and multiplied the number of Jews in the city. The mills and their allied industries mushrooming along the Monongahela and Allegheny riverbanks pulled this growing population into new neighborhoods such as Birmingham, Lawrenceville, and Oakland.

In 1890, organized religion appeared daunted by these tasks, with the number of congregations per capita reaching the lowest point in the history of Allegheny County (0.73 per 1,000). By 1910, however, the downward trend had been reversed (0.81 per 1,000), and the ratio continued to climb throughout the century.[46] Each major religious group was successful in organizing its potential new members. As a result, Catholics closed the gap on Protestants in the number of congregations and surpassed them in the number of communicants by 1910. Even so, Protestants remained more influential in terms of social status and wealth.

As in the previous historical period, doctrinal conflict and denominational fragmentation, as well as cooperation and organization, contributed to religious expansionism. Congregations in all major religious groups, following the earlier Protestant pattern of conflict and division, became more homogeneous, more defined by neighborhood, ethnic group, or occupation, and more detached from the larger environment. Yet expanding city-wide and regional denominational structures enmeshed individual congregations. Diversity and physical dispersion within centralizing bureaucracies became Pittsburgh's religious hallmark in the early twentieth century.

The second wave of Catholic immigrants entered Pittsburgh after the Roman hierarchy had accommodated the Irish and German newcomers of a half-century earlier. In 1889, under Bishop Phelan, the diocese was free enough from noisy tensions and debt to move St. Paul's Cathedral to Oakland. This calm was short-lived, and between

1904 and 1921, a new church was founded every thirty days.[47] Most were ethnic, commonly called national, parishes organized by Poles, Slovaks, Hungarians, Croatians, and Italians. They were following the Germans who had earlier established the right to attend a parish populated by Germans, lead by a German priest, and stocked with German rituals. Ironically, the first American-born Pittsburgh bishop (later archbishop), Regis Canevin, led this period of Roman Catholic diversity.

National parishes, however, did not insure Catholic integration. Between 1880 and 1920, Pittsburgh developed one of the largest aggregations of non-Roman Catholic parishes in the country. The area became a center for Eastern Orthodoxy in the United States after St. Nicholas Carpatho-Russian Greek Church was organized in Duquesne in 1890. Later, Greek, Russian, Serbian, Syrian, and Ukrainian immigrants demanded their own Orthodox jurisdictions. After 1900, at least seven Orthodox congregations existed in the Pittsburgh area.

Pittsburgh also became a major location for the Byzantine Rite Catholic Church after Czech immigrants on the South side organized St. John the Baptist Church in 1890. Although under the Roman pope, Byzantine Catholics in Pittsburgh were granted their own bishop in 1907 and a separate diocese in 1924. Other Catholics who did not recognize the Vatican included Syrian immigrants on the Hill, who organized St. Anne's Maronite Roman Catholic Church. Pittsburgh was never a hotbed for the independent Polish Church, an American-born church, but issues of clerical authority and rituals drove some Pittsburgh Poles into two congregations of the Polish National Catholic Church in the early decades of the century.[48]

Information on Pittsburgh Catholics during this period is largely embedded in accounts of single immigrant groups. Histories have been written for groups which came to Pittsburgh after 1880, including Poles, other Slavs, Hungarians, Greeks, and Syrians.[49] These studies typically document the arrival and subsequent community and religious development of the ethnic group. Most demonstrate the wide religious differences within the immigrant groups and their often bitter struggles to create separate ethnic congregations.

A welcome comparison of Protestants and Catholics within the same immigrant group can be found in a detailed study of Slovak Catholics and Lutherans between 1880 and 1915. Both religious groups consciously organized congregations in Pittsburgh which mirrored the religious institutions and practices of their homelands. Once established, Slovak Catholic and Lutheran congregations were shaped by the differing responses of the respective American denominational hierarchies to their premigration religion. The Roman Catholic national parish system allowed Slovaks easy and uncontroversial entrance into the Pittsburgh diocese, while the doctrinal battles in Lutheranism promoted a fractious and unstable Slovak Lutheran presence in Pittsburgh.[50]

This study calls special attention to the role of ethnic fraternal associations in providing funding for ethnic churches in the city. Although some were secular in outlook and function, most fraternals gave considerable financial support to the first congregation organized by members of their ethnic group, regardless of the denomination, in much the same way that Protestant land developers helped churches earlier in the century.[51]

Several important areas of immigrant religion, however, remain obscure. Since the history of women within the Pittsburgh region has been hidden until recently, it is not surprising that the religious experiences of immigrant women remain virtually unexplored. The few studies that do exist, however, focus on Roman Catholic women's religious orders, especially the Sisters of Mercy. Most simply tell of the struggle to found the orders and subsequent triumphs, but one social history study compares the ethnic dimensions of three Roman Catholic women's religious orders from 1890 to 1940, including the Polish Felicians, the Irish Sisters of Mercy, and the German Sisters of St. Francis.[52]

Neither is the important Jewish expansion well documented for this period. The number of Jews increased to sixty thousand by 1914, and immigration, doctrinal innovations, residential changes, and benevolent organizations redefined Pittsburgh's Jewish community. Eastern European migration brought large numbers of Jews to the city who demanded the Orthodox rituals of their homelands. National congregations did not emerge as often among Jews as Catholics, but after 1885, rabbis schooled in Eastern European Orthodox Judaism were in great demand. Native-born Jews resisted the more traditional religion of immigrant Jews. At a national meeting of rabbis at Rodef Shalom under Dr. Lippman Mayer in 1885, the "Pittsburgh Platform" became the foundation for Reform Judaism which adopted more American-style religious services, including seating by family instead of by gender. A middle ground between the Americanized Reform and traditional Orthodox Judaism was carved out by the turn of the century, and B'Nai Israel, a Conservative congregation, was organized in 1924.[53]

As elsewhere, wealthy, American-born, German Jews created a substantial community infrastructure and practiced "preventive philanthropy" in the form of education and community services in order to integrate new Jewish immigrants into the economic and social structure of Pittsburgh. At the same time, the Jewish population shifted its residences and religious institutions from downtown to the East End and Squirrel Hill. Both Rodef Shalom and Tree of Life moved in 1906. As a result, Jews continued to be one of the most residentially segregated ethnic groups in the region.

Another area in need of study is the relationship between industrialization and religious change from 1880 to 1920. One monograph on working-class life, which does not directly analyze the role of religion or religious groups, shows that an Irish Catholic school committee appointed a priest to head a public school in 1887 and that union leaders supported the Young Men's Temperance Union.[54] Other studies confirm the importance of religion to industrial workers and union organizing, although no consistent role is evident. For example, the late-nineteenth-century union movement was weaker in Johnstown, where iron workers belonged to fraternal lodges, ethnic societies, and church-centered associations, than in Pittsburgh's Woods Run, which did not divide along religious lines. In contrast, Catholic steelworkers organized in the Mon Valley even though the religious leadership spoke loudly against the union.[55]

Unlike immigrant religions, mainline Protestant denominations grew slowly during this period of population increase. A comparison of the five largest Protestant denominations, Presbyterian, Methodist, United Presbyterian, Episcopal, and Baptist, between 1910 and 1929, reveals that none increased its membership by more than 30 percent, and all cut back the number of congregations they maintained. For the first time, the number of Catholic parishes in the city matched Presbyterian congregations and greatly surpassed them in members.[56] Roman Catholicism was the fastest growing religious group in the region's rural areas as well.[57]

Yet the Protestant churches continued to exert primary religious authority over Pittsburgh between 1880 and 1920 because their membership included a disproportionate number of the wealthiest and most powerful families in the community. Protestant denominations simply had much greater access to wealth and political power than other groups. Elite congregations and denominations developed common strategies, including residential segregation, which undercut the immigrant influence. Protestantism also dominated because it promoted the values of the new social order, such as individualism and acquisitiveness, which leading non-Protestants eventually would adopt and propagate.

Class distinctions among Protestants continued to sharpen. Episcopalians tended to be upper class, as they were throughout the

country, and favorite son Andrew Carnegie demonstrates the staying power of Presbyterianism among the city's wealthy industrialists. A study of four German churches between 1850 and 1930 illustrates the class base of religion in the general population as well. The
Smithfield and Grace Reformed congregations had a high proportion
of white-collar and professional members, consistent with findings
for the earlier period, while blue-collar members predominated in St.
Matthew's Lutheran and in Zion Evangelical Association Church.[58]

Pittsburghers who moved up the social ladder, however, rarely
changed religions. The Germans, for example, remained Lutheran,
Evangelical and Reformed, and even Catholic as they became community leaders. By 1880, the social, business, and political elite of the
city included Lutherans and Jews.[59] They did, however, alter the content of their religion. Respectable middle-class German Jews and Irish
Catholics preached the same values of sobriety, upward mobility, and
hard work to their recently arrived Eastern European counterparts as
did Protestants.[60]

The reaction of wealthy Protestant and non-Protestant communicants to the massive influx of immigrants followed a common pattern. A study of the Presbyterian Church, U.S.A. between 1880 and
1906 developed a useful three-part model to characterize elite responses to immigration. The dominant religious groups underwrote
benevolent and reform activities to encourage industrial values, set
up missions to provide alternatives to the immigrants' premigration
religion, and moved into new suburbs to avoid contact.[61]

Pittsburgh was in the forefront of benevolent and urban reform
activities throughout the country. Four years after Jane Addams
founded the prototypical Hull House in 1889, Kingsley House
opened in the Strip district under the direction of George Hodges,
the rector of Calvary Episcopal Church. Originally proposed by Presbyterian, Unitarian, and Roman Catholic pastors, Kingsley House,
named after the well-known British proponent of the Social Gospel,
had impressive backing from an array of Pittsburgh elite.[62] A Jewish
counterpart to this largely Protestant settlement house, the Irene
Kaufmann Center, opened in 1895 to help immigrant Jews who still
lived on the Hill.[63] Both institutions used American money and
values to socialize foreigners.

The studies of settlement houses and benevolent activities between 1880 and 1920 suggest the important role of Protestant and
Jewish women in religion.[64] Although women participated in earlier
Pittsburgh reform movements, they remained in the shadow of male
leaders and male-dominated institutions. By 1920, women had become recognized leaders of benevolent and reform movements in

Pittsburgh and other major cities. For example, the Columbian Council, a middle-class Jewish organization descended from the Ladies' Aid Society founded during the Civil War, was instrumental in establishing the Kaufmann Center.[65] Furthermore, the staff in both the Kaufmann Center and Kingsley House was religiously motivated. Most were unmarried women with some college or theological training who volunteered as part of a family commitment to charity. As these women increasingly became paid staff and even supervisors at the centers, they set the stage for the commanding position of women in the emerging field of social work.

Ministering to the immigrants, whether by men or women, took two directions: attempts to ameliorate poor conditions and to reform the personal habits of immigrants. The first aproach is demonstrated by the Kingsley House staff who helped organize the Pittsburgh Survey, the national study funded by the Russell Sage Foundation to study conditions among immigrants.[66] As early as 1880, however, religious leaders shifted away from improving conditions to inculcating values enabling individuals to adapt to or eliminate their inhuman conditions. Curbing drinking, dancing, and other personal vices took priority over the elimination of crowded and filthy housing, child labor, or low wages. The dominant churches also cooperated in fighting prostitution in the first two decades of the twentieth century.[67]

Most Protestant denominations also made efforts to convert the immigrants to their own brand of Christianity. If "the foreigner" was "mixed up with the problems of the liquor business, Sabbath desecration, and commercialized vice," as one of the several studies undertaken by the Pittsburgh Council of Churches of Christ reported, then missionary activities were necessary. This Protestant-dominated interdenominational agency sponsored three missions to convert the Jews of the city, and they directed several others toward Roman Catholics. Individual denominations also created outreach programs, although none was very effective.[68]

Yet suburban flight, not evangelizing or reforming the new immigrants, was the most significant response of the dominant religious culture to the new arrivals. Among Presbyterians, for example, mission activity did not occur at all until after 1900, and even then, the first targets were not recent immigrants—Poles or Russians—but the older, already Americanizing, French, Belgian, and German immigrants.[69] Presbyterians began moving to suburbs and small towns outside Pittsburgh before the Civil War, but subsequently the rate accelerated.[70] Bitter struggles over who would retain control over church property divided congregations and the presbytery, but the

new suburban churches effectively insulated the Scots-Irish Presbyterians from direct contact with foreigners.

Suburban churches provided Protestants with ethnic islands in a
sea of diversity in the same way that emerging national parishes did
for Catholics and Jews. Episcopal churches concentrated in areas of
the city with the largest English immigrant population. When migration from the South accelerated during the 1910s, black churches,
including AME and AMEZ, Methodist Episcopal, and Baptist congregations, underwent a growth spurt. Germans maintained English
and German Lutheran, Methodist, Reformed, and Evangelical Association, as well as Roman Catholic churches. Magyar Reformed,
Slovak Lutheran, and Swedish Evangelical congregations also were
founded in Pittsburgh during this period.[71]

Less respectable religious responses to the new immigrants also
occurred in the Pittsburgh area. During the 1920s, six local Ku Klux
Klan "Klaverns" were formed in the city of Pittsburgh and in surrounding mill towns where large immigrant populations gave rise to
blatant anti-Catholic activities. Klan efforts to influence local school
boards and borough councils included donating King James Bibles to
schools and opposing the hiring of Roman Catholic teachers. The
Klan was also accused of hangings, floggings, and kidnapping in the
area. At its peak, the Klan engaged in violent encounters with Catholics and others in Carnegie and several nearby small towns.[72]

None of these studies, however, has linked religious activities
directly to the growing fundamentalist-modernist rift within Protestantism between 1880 and 1920. As the new theology of the Social
Gospel justified interdenominational benevolent agencies, such as
Kingsley House, some Protestants felt that the distinctive nature of
their religion was being compromised. Fundamentalists reasserting
the need for individual repentance based on a literal interpretation of
the Bible, and modernists advancing ideas of the Christian social
conscience, precipitated sharp differences within the city. For example, the ministers of the prominent First and Second Presbyterian
churches clashed over the interpretation of the gospel to needy immigrants, and the strong antiprostitution coalition broke apart by 1915
over the issue of who should regulate morality.[73]

Churches, leaders, and agencies represented denominations from
both sides of the theological conflict. Modernism was present in
Western Theological Seminary when several faculty members practiced the "new Biblical scholarship." Clarence Darrow defended evolution in the city about the same time that ministers from the interdenominational Ministerial Union became the Pittsburgh Council
of Churches in 1915. Salvation Army and Christian Science congregations were organized during this period as well.[74]

On the fundamentalist side, Pittsburgh was the center of the new Jehovah's Witnesses until 1909 when the founder, Charles Russell, moved to Brooklyn. Russell, a converted Seventh Day Adventist, organized Bible classes to preach that Christ would come to judge the world. For thirty years, the successful *Zion's Watchtower* was published in Pittsburgh. Billy Sunday also wooed the city with his preaching ability, and in 1904, a "giant city-wide revival" broke out. Before 1920, the fundamentalist Church of God (Anderson, Indiana), Nazarenes, Holiness groups, and the Assemblies of God founded congregations in Pittsburgh.[75]

By 1910, the industrial momentum of Pittsburgh had peaked. Overall population growth began to slow as unemployment accelerated. Pittsburghers were more likely to return to, than come from, Europe. The percent of foreign-born population in Allegheny County declined throughout the twentieth century, from over 25 percent in 1910 to less than 5 percent in 1970. The expansion of Catholic Pittsburgh came to a halt, and all denominations were faced with the likelihood of declining enrollments and failing congregations.

Religion in a Deindustrializing Region

This grim prognosis failed to materialize between 1920 and 1960. New migrants, residential dispersion, denominational consolidation, and the growth of sectarian groups kept organized religion vibrant and expanding within the region. Although few Pittsburgh scholars have undertaken studies of religion during the transition of the Steel City to the Steelers' city, the work that has been done, especially O. M. Walton's bicentennial description of Pittsburgh's religion in 1958, confirms that previous religious patterns extend to the present.

Despite deteriorating demographic and economic conditions, new congregations continued to appear between 1920 and 1960. Until 1940, places of worship increased more than population, even as the number of churches decreased in downtown areas which were losing residents. Between 1940 and 1960, the population of the entire city fell sharply by 10 percent, but the number of churches only by a minimal 1.1 percent. As a result, the number of congregations per one thousand inhabitants was 0.88 in 1958, slightly up from 0.81 in 1910 and substantially higher than 0.73 in 1890.[76]

Blacks were the largest contributors to the per capita growth in congregations between 1930 to 1960. These three decades represented the greatest black religious development in the history of Allegheny County. Migration from the South multiplied the proportion of blacks in the county's population from 3 to 7 percent.[77] Like

the immigrant groups before them, black migrants founded congregations faster than their population grew during the peak migration
years. By 1958, Pittsburgh had two hundred churches "of Negro
origin," compared to fifty-six in 1930 and twenty-seven in 1910.[78]

Just as immigrant parishes had done for Europeans, new black
churches offered homestyle beliefs and services. They protected
families and preserved southern culture; baptized, married, and buried newcomers; provided employment information and welfare assistance; and sponsored a myriad of social functions.[79] An ethnography
of one black congregation on the Hill in 1973, affiliated with a tiny
southern Pentecostal sect, details how "intimate, spontaneous,
imaginative worship" tied the previous rural life of black migrants
to their new urban existence.[80]

Differences between the religious experience of blacks and European immigrants are also evident. A study that compares three
groups in Pittsburgh between 1900 and 1960 suggests that religion
was important to black migrants as individuals, while Polish and
Italian parish churches served as a focal point for the entire immigrant community. The authors speculate that although the church
was the only significant black social institution, the multiplicity of
black religions, often of the sectarian storefront variety, fragmented
the black community. Combined with the economic precariousness
of black workers and the lack of additional fraternal, insurance, and
social agencies, cohesive black neighborhoods remained elusive.[81]

Yet there are many signs that black religious patterns were con-
verging with those of immigrants. By 1958, 75 percent of all black
churches in the county were Baptist, most of them affiliated with
the National Baptist Convention.[82] The few, well-established black
churches, such as Ebenezer Baptist, sponsored the usual array of
social services for southern migrants.[83] Not surprisingly, these theo-
logically moderate congregations were at odds with the strong fundamentalist tenets of those stressing the importance of a sinless life.
But as in the larger Protestant religious community, the modernist-
fundamentalist conflict actually fueled religious enthusiasm and
congregational organization. Black church attendance and membership was greater in some Pittsburgh neighborhoods than in the
United States as a whole and certainly than among many white
populations.[84]

Furthermore, black congregations in Pittsburgh shared with black
religion elsewhere a history of direct political involvement. More
research is needed to fill out the twentieth-century equivalents of
antislavery campaigns in antebellum Pittsburgh, but in one case,
three Moslem sects on the Hill in 1948 believed that Christians in-

vented the idea of the Negro race as a means of discriminating against people with dark skins. These unaffiliated Islamic churches, embracing what they thought was original slave religion, were precursors of the Black Muslims in the 1960s.[85] As urban renewal displaced blacks from the Hill into neighborhoods like Homewood, black congregations catalyzed community organizing. Ministers and churches became the rallying points for political and social causes including civil rights, labor organizing, and anti–Vietnam War protests.[86]

Black congregations were not the only contributors to the per capita increase in religious organizations between 1920 and 1960. With geographical mobility widespread within the region, new congregations were organized and older churches changed locations according to ethnic residential patterns. For example, second- and third-generation Italians, displaced by the Larimer Avenue renewal program, moved into Morningside. The local parish, St. Raphael's Roman Catholic Church, went from 5 to 40 percent Italian during the 1960s, while the Lutheran, Methodist, and Presbyterian congregations lost 75 percent of their members who moved further into the northeastern suburbs between 1930 and 1970.[87]

A study of Episcopal parishes between 1870 and 1975 suggests that most churches eventually abandoned changing neighborhoods. Nevertheless, church closings trailed population loss in transitional locations. A few older congregations, including Calvary Episcopal in Shadyside, adopted programs which drew in new members from the altered communities.[88] Others, especially ethnic congregations, maintained a geographically widening circle of members who returned for services and social gatherings.[89]

Religious expansion was by no means uniform among denominations. In general, Roman Catholic growth leveled off during this period. Although the number of Italian and Orthodox Catholic parishes increased as immigration continued to bring Southern and Eastern Europeans to Pittsburgh, Catholics, with a third of the county's population, maintained about 18 percent of congregations in the city from 1920 to 1960.[90]

Parochial education replaced the goal of creating additional parishes. The number of Catholic schools increased from 390 to 448 between 1921 to 1950 under Bishop Hugh Boyle. This was remarkable, given the sharp opposition from Eastern Europeans who believed their religious culture would be ignored in parish schools. In particular, Poles refused to support Boyle's citywide efforts, because their parishes were burdened with expensive building projects and they feared that the local hierarchy favored Irish Catholics.[91] Church leaders believed that their schools added to neighborhood cohesion and desirability, but by the early 1980s, no statistically significant relationship was present.[92]

Generational conflict was the most important spur for organizing new ethnic parishes. Additional churches appeared when second- and third-generation immigrants could not agree with more recent immigrants on the proper role of their churches. For example, a Greek Orthodox parish was added in 1952 because immigrant Greeks were not satisfied with the ethnic churches that already existed. A member survey of three post–World War II congregations revealed that immigrants believed the mother church of St. Nicholas focused too much on social activities and not enough on religion.[93] A second-generation Syrian expressed the same sentiment in another study: "My father goes to church and pays a lot of attention to it. . . . He meets his friends there and has a good time talking to them. But why should I bother? . . . I go to church to worship, not for social life."[94] New churches also accompanied the relocation of earlier immigrants to the suburbs.

Jewish congregational growth continued unabated after 1920. Even though the number of Jewish immigrants declined, the Jewish share of Pittsburgh congregations doubled from 2.7 to 4.9 percent between 1910 and 1960.[95] No research has tried to account for this increase, but if other ethnic groups are any indication, growing religious divisions could be the cause. By 1978, Reform, Conservative, Orthodox, Reconstructionist, and Hassidic congregations were present in Pittsburgh. Even so, Jews remained in the same neighborhoods they had moved to early in the century, using kinship ties to promote their own upward mobility.[96]

In contrast, a sharp decline in the relative strength of established Protestant denominations marks this period. Although the total number of Presbyterian, Baptist, Methodist, Lutheran, and Episcopal congregations remained the same through the twentieth century, their collective percentage of all congregations in Allegheny County dropped from 73.9 to 54.9 between 1910 and 1960. Since the number of Catholic congregations remained stable and Jews never had more than a small share, new Protestant groups undercut these mainstream denominations. The new groups, largely fundamentalist in theology, sectarian in polity, and often with a black membership base, included Jehovah's Witnesses and Mormons, organized prior to 1920, and more recent ones with names like the Church of God in Christ, Church of Our Lord Jesus Christ of the Apostolic Faith, and Kodesh Immanuel. Most had only one to twenty churches locally, but together they represented 21.7 percent of all congregations in the city in 1960, a fourfold increase since 1910.[97]

Despite the growth of these sects and denominations, the rank order of the major Protestant groups changed little throughout the history of the Pittsburgh region. Presbyterians continued to dominate other denominations in terms of membership and congregations. In 1958, 22 percent of Protestant congregations in Allegheny County belonged to the United Presbyterian church, formed by a merger of the two largest Presbyterian bodies in the Syria Mosque during that year. Presbyterian strength was actually increasing in the rural areas of the county. Depopulation strengthened the ethnic character of rural areas, with many returning to their earlier predominantly Scots-Irish character.[98]

The remaining large Protestant denominations in the county included Baptists, Methodists, Lutherans, and Episcopalians. The only significant shift among them during the twentieth century was the Baptist displacement of the Methodists as the second largest Protestant group, caused in large part by the black preference for Baptists. Although some congregations in these mainline denominations closed or moved out of the city, membership in these denominations remained stable. In the case of the Episcopal churches, total membership stayed the same, while the number of churches dropped from twenty-two to fifteen between 1920 and 1974.[99]

Like Pittsburgh itself, religion did not stagnate or decline during the twentieth century; it only changed locations, leaders, and orientations. As the city moves into the next century, mosques and Hindu temples, EST hotel rooms and Scientology halls, and non-denominational fundamentalist churches most likely will continue to alter the religious contours of Pittsburgh. But the explanation for these changes will remain rooted in the same basic processes of religious change which have governed the region from the earliest days of white settlement, including migration into the city, movement within the city, and religious conflict within and between denominations.

Conclusion

The studies that I have examined provide neither a comprehensive nor an in-depth account of religion in Pittsburgh. In most cases, they did not even focus on religion directly. Nonetheless, the bits of religious information they record about specific periods, places, and religious groups, when connected, provide a useful frame for viewing religion in Western Pennsylvania. More important, this neglected aspect of regional history begins to form a pattern familiar to historians of religion elsewhere. Community development, migration, and economic change intersect with ethnic, racial, gender, class, and political factors to define the soul of every city and region in the nation. Although specific premigration religions, denominational exigencies, and individual leaders and worshipers make each region unique, Pittsburgh illustrates the fundamental patterns of religious development throughout the United States.

The success of the Pittsburgh studies in illuminating basic American religious patterns stems from the social history approach they employ. The major contribution of social historians to the study of religion is the recognition that religion exists in a social context; it cannot be divorced from the rest of society. When Pittsburgh authors searched for meaningful social relations and explanations of social patterns, they discovered that religion was tightly interwoven with other social elements and that religion often offered more interpretive power than did traditional explanations. Religion is not fundamentally different from other social elements, such as occupation or ethnicity, and should be studied with the same tools.

Even so, the exploration of religion and society in Pittsburgh has been one-sided. Pittsburgh historians most consistently have examined the influence of religion on secular society, showing the remarkable degree to which religion shaped Pittsburgh's life from early settlement to the present by influencing politics, ethnicity, moral reform, social welfare, gender relations, and community-building. Less often have studies shown how these same social patterns altered religion. Only a few have illustrated the crucial impact of social change, especially class formation, immigration, and residential dispersion, on religious evolution.

The best-known study of Pittsburgh religion from the point of view of social history comes in the area of politics. Paul Kleppner's 1963 article on the role of Germans in Pittsburgh's 1860 election undercut political rhetoric, ethnic stereotypes, and alternative explanations to illustrate the power of religion to account for how people voted. Lee Benson had already introduced "cultural" issues into voting behavior based on New York state during the same period, so when other historians demonstrated the same religious differences among Germans and other ethnic voters throughout the nation, the ethnocultural explanation of nineteenth-century American politics gained acceptability.[100]

Another major contribution of the Pittsburgh studies is the documentation of the symbiotic relationship between religion and community expansion. Both traditional religious historians, who focus on the internal aspects of religion such as theology and denominational leadership, and social historians, who rarely investigate religion directly, have neglected the environmental context. Religion was not simply an individual act of the spirit, but a public act in which people with diverse motives participated.

The Pittsburgh evidence suggests, for example, that the overall economic condition of the region was more important for advancing the number of churches than denominational or congregational strength. Urban expansion led to institutional religious growth, with city fathers providing economic incentives and moral support to a variety of religions, whether or not they were members of the denominations they were aiding or even believers. Furthermore, congregations in Pittsburgh regularly built churches in the centers of business rather than in residential areas. As a result, the relationship between changing neighborhoods and church relocation was erratic.

Religious institutions eventually followed their parishioners to new locations, but the timing depended on several factors. Congregations usually could not agree on whether or when to build a new church; a division often resulted in which one faction stayed in the same location. When an ethnic group moved into a community, religious organizations were sure to follow. But when the group left, some of their institutions remained to intermingle with the new neighborhood occupants. Hostility and distrust among the diverse groups usually ensued, but occasionally religious institutions successfully tackled common community problems together.

The theme of religion is most prominent in studies of ethnic groups. Pittsburgh's social historians, like their counterparts elsewhere, have been successful in retrieving the religious experiences of incoming minority groups, arguing that they must be seen on their own terms and not through the eyes of the dominant culture. These studies underscore the religious diversity of ethnic groups, the likelihood that religious controversies originating in the homeland will remain salient in the immigrant community, and the importance of religion in setting the ethnic groups apart from native-born Americans, other immigrant groups, and each other.

Their conclusions strongly support the findings of Timothy Smith and others who have countered the notion that immigrant religion is a static holdover from the homeland which eventually fades away. Immigrant and ethnic religion is always in transition, interacting with the new world and trying to accommodate changing constituencies with evolving religious perspectives. The complexity of ethnic communities makes a single religious standard impossible. Each change in the church is seen by some members as too much and others as too little "Americanization," while the introduction of English, musical instruments, or temperance, for example, become symbols of Pittsburgh's secular social structure.

Overall, however, the value of the Pittsburgh studies depends on how well historians can use them to compare religion in Western Pennsylvania with religion elsewhere. Since the focus of most of these studies is not on religion itself, using them to judge the strength and distinctiveness of religious groups, behaviors, and attitudes elsewhere is difficult. In order to enhance Pittsburgh as an important case study of religion from frontier settlement to post-industrialism, religious historians with a social history outlook must put religion at the center, not at the periphery, of new studies. Until then, the Pittsburgh studies will not make a dent in the thorny conceptual problems plaguing religious history.

One intriguing unresolved issue is how individuals, across denominational lines, have made decisions to pursue religious change. A premier feature of American religion has been that new ideas—such as evangelicalism, fundamentalism, and the contemporary New Age—periodically erupt and cause a realignment. Few Pittsburgh authors have tried to separate the innovators from the conservatives. Since the basic religious actor is a person who also must live in the nonreligious world, examining possible secular as well as religious motivations is necessary. The congregation is a useful analytic location, because religious conflict among formerly like-minded people often includes social differences.

Crucial issues of ethnicity and religion also remain unexplored. Most scholars would agree with one Pittsburgh author that "the Church was one of the most pervasive forces in the life of immigrants."[101] Yet the relationship between religion and ethnic identification is not straightforward. Since most ethnic groups have more than one religious tradition, important questions will remain unanswered as long as religion is discussed as either the same as, or derivative of, ethnicity, and until more comparative studies of the role of religion in ethnic groups are undertaken.

Much still needs to be done to uncover the full array of minority religion in the region. Of the major groups, Judaism remains understudied in all time periods. The lack of any studies on the origins of the Jehovah's Witnesses is a notable gap; nor are there many studies of the Disciples of Christ, New School Presbyterians, or Wesleyan Methodists. The divisions of the late-nineteenth-century ethnic groups along distinctive religious lines also need much more attention. Even the well-known Polish National Church and the Lithuanian National Church have not been studied in their crucial Pittsburgh setting.

New research in women's history offers ample direction for the exploration of Pittsburgh women and religion. Thus far, the studies have concentrated on women's work experience in the region rather than on their social and cultural activities. The scant information that does exist, however, is consistent with conclusions drawn from other regions. Women dominated the membership of religious organizations, but lacked the power and authority of their numbers. Initially segregated into their own socieites, women had a separate arena in which to exert religious authority, but they were isolated from centers of power. But when integration occurred, women lost their administrative positions to men.

Furthermore, Pittsburgh studies do not adequately illuminate the role of religion in the life of ordinary workers. Even the many labor history projects from the University of Pittsburgh fail to explore the influence of religion on working-class life or labor organizing. This lack of attention is surprising, since labor historians have uncovered important religious links to industrial environments elsewhere.[102] Many studies pit ethnicity and religious identification against class consciousness and collective action in a mutually exclusive way. Although the issue is still controversial and unresolved among historians, there is no doubt that religion acts in different capacities depending on specific circumstances. In Pittsburgh, the class basis of religious denominations seems clear, although not explored very well.

Religion may have been largely invisible to historians of the City at the Point, but not to the waves of immigrants who settled the region after the French and Indian War. Much about religion has changed in the last 250 years. In particular, religious diversity has replaced established Protestant hegemony; congregations are divided by class, race, ethnicity, and geography; and religious institutions share their members with an increasing variety of nonreligious organizations. Nevertheless, religion continues to provide a focal point for Pittsburgh's families, neighborhoods, and social institutions.

NOTES

I want to thank my former colleague Susan Smulyan, now at Brown University, for her help in revising this essay.

1. George Bedell, Leo Sandon, Jr., and Charles T. Wellborn, *Religion in America* (New York: Macmillan, 1985), pp. 11–12.

2. Thomas Cushing, ed., *History of Allegheny County, Pennsylvania*, vols. 1–2 (Chicago: A. Warner & Co., 1889). County histories are generally reliable when it comes to documenting when and where congregations were organized.

3. Cushing, *Allegheny County*, pp. 12–31, 55, 124; O. M. Walton, *Story of Religion in the Pittsburgh Area* (Pittsburgh: Pittsburgh Bicentennial Association, 1958), p. 27; *Churches of Allegheny County, Pennsylvania* (Chicago: A. Warner and Co., 1889), p. 7.

4. Boyd Crumrine, *History of Washington County, Pennsylvania* (Philadelphia: L. H. Everts & Co., 1882), pp. 668–69; *Churches of Allegheny County*, p. 66.

5. Cushing, *Allegheny County*, p. 109, 450.

6. Donald S. Wood, "A Geographical Study of the Economic and Urban Evolution of McKeesport, Pennsylvania: From Frontier Settlement to Industrial City, 1755–1900" (MA thesis, Department of Geography, University of Pittsburgh, 1950), p. 29.

7. Evan Hammar, "Wilkinsburg and Edgewood: Commuter Suburbs" (seminar paper, Department of History, Carnegie Mellon University, 1972), p. 11.

8. Cushing, *Allegheny County*, p. 491.

9. Mitchell A. Nathan, "The Jewish Community of Pittsburgh: A Beginning" (research paper, Department of History, History Workshop, Carnegie Mellon University, 1982), p. 2.

10. *Churches of Allegheny County*, pp. 7–14, 24–30. Pittsburgh was also the original home of the Presbyterian U.S.A. Western Mission Society, the forerunner of the Board of Foreign Missions, and four major national boards for the Scottish Presbyterian groups.

11. Hal Kimmins, "Westmoreland County, 1783–1790" (seminar paper, Department of History, University of Pittsburgh, 1964), pp. 2–3; Edward H. Hahn, "Social Changes in a Small Community: 1860–1880" (seminar paper, Department of History, University of Pittsburgh, 1974); J. E. Davidson, " 'God Speed the Plough': A View of Agricultural Society" (seminar paper, Department of History, University of Pittsburgh, 1969); Wood, "McKeesport," pp. 29, 39; Fred Wallhausser, "The Upper-Class Society of Sewickley Valley, 1830–1910" (seminar paper, Department of History, University of Pittsburgh, 1964), pp. 9–10; and Thomas J. Hannon, Jr., "The Process of Ethnic Assimilation in Selected Rural Christian Congregations, 1800–1976: A Western Pennsylvania Case Study" (Ph.D. diss., University of Pittsburgh, 1977), p. 38.

12. Nora Faires, "Ethnicity in Evolution: The German Community in Pittsburgh and Allegheny City, Pennsylvania, 1845–1885" (Ph.D. diss., University of Pittsburgh, 1981), pp. 167–68.

13. Harold L. Twiss, "The Development of Missionary Support by Baptist Churches and Associations in Western Pennsylvania, 1815–1845" (seminar paper, Department of History, University of Pittsburgh, 1963).

14. *Churches of Allegheny County*, p. 67.

15. Scott C. Martin, "Fathers Against Sons, Sons Against Fathers: Antimasonry in Pittsburgh" (seminar paper, Department of History, University of Pittsburgh, n.d.).

16. Cushing, *Allegheny County*, pp. 548, 146.

17. Linda K. Pritchard, "The Burned-Over District Reconsidered: A Portent of Evolving Religious Pluralism in the United States," *Social Science History* (Summer 1984): 247.

18. For the city of Pittsburgh, see Cushing, *Allegheny County*, pp. 548–49. For the four rural counties just north of Allegheny, see Hannon, "Process of Ethnic Assimilation," p. 49.

19. Linda K. Pritchard, "Religious Change in a Developing Region: The Social Contexts of Evangelicalism in Western New York and the Upper Ohio Valley During the Mid–Nineteenth Century" (Ph.D. diss., University of Pittsburgh, 1980), pp. 178–213.

20. Walton, *Story of Religion*, p. 1; *Seventh Census of the United States, 1850* (Washington, D.C.: Robert Armstrong, 1853).

21. *Seventh Census, 1850*.

22. *Churches of Allegheny County*, pp. 15–19, 24–29, 32–33.

23. *Seventh Census, 1850; Eighth Census of the United States, 1860: Statistics of the United States* (Washington, D.C.: U.S. Government Printing Office, 1866); *Ninth Census of the United States, 1870: Statistics of Population* (Washington, D.C.: U.S. Government Printing Office, 1872); and *Compendium of the Eleventh Census, 1890* (Washington, D.C.: U.S. Government Printing Office, 1892.

24. Walton, *Story of Religion*, pp. 7, 17, 18–19; *Eighth Census, 1860*.

25. *Seventh Census, 1850; Eighth Census, 1860; Ninth Census, 1870*.

26. Anna-Mary Caffee, "The Socialization of Rural Children in Southwestern Pennsylvania, 1840–1900" (seminar paper, Department of History, University of Pittsburgh, n.d.); Hahn, "Social Changes in a Small Community"; Hannon, "Ethnic Assimilation"; and Davidson, " 'God Speed the Plough'," pp. 38, 43.

27. Jeffrey O. Siemon, "Division Within the Sixth Presbyterian Church, Pittsburgh, Pa., 1850–1862" (research paper, Department of History, Carnegie Mellon University, 1982).

28. Wallhausser, "Upper Class Society," p. 14.

29. Linda K. Pritchard, "Ministers: A Comparative Framework" (seminar paper, Department of History, University of Pittsburgh, 1970).

30. Glenn E. Myers, "Reflections of the Revival of 1857–1858 in Greater Pittsburgh" (research paper, Department of History, Carnegie Mellon University, 1980).

31. Victor Anthony Walsh, "Across 'The Big Wather': The Irish-Catholic Community of Mid-Nineteenth-Century Pittsburgh," *Western Pennsylvania Historical Magazine* 66 (Jan. 1983): 2; and Faires, "Evolution of Ethnicity."

32. Henry A. Szarnicki, *Michael O'Connor, First Catholic Bishop of Pittsburgh . . . 1843–1860: A Story of the Catholic Pioneers of Pittsburgh*

and Western Pennsylvania (Pittsburgh: Wolfson Publishing Co., 1975),
p. 10.

33. *Churches of Allegheny County*, p. 98.

34. Szarnicki, *Michael O'Connor.*

35. Buddy Hobart, "Nineteenth-Century Church Politics" (seminar paper, Department of History, Carnegie Mellon University, n.d.); and Daniel
Donahoe, "The Catholic Community in Pittsburgh: 1865–1890" (research
paper, Department of History, Carnegie Mellon University, n.d.).

36. Faires, "Ethnicity in Evolution."

37. Victor Anthony Walsh, "Across the 'Big Wather': Irish Community
Life in Pittsburgh and Allegheny City, 1850–1885" (Ph.D. diss., University
of Pittsburgh, 1983); and Patricia K. Good "Irish Adjustment to American
Society: Integration or Separation?" *Records of the American Catholic Historical Society of Philadelphia* 86 (Mar.–Dec. 1975): 7–23.

38. Walton, *Story of Religion*, p. 39; Laurie Mizrahi, "The History of
the Jewish Community of Pittsburgh: 1847–1890" (seminar paper, Department of History, Carnegie Mellon University, 1981); and Nathan, "Jewish
Community."

39. Paul J. Kleppner, "Lincoln and the Immigrant Vote: A Case of Religious Polarization," *Mid-America* 48 (July 1966): 175–95.

40. Layne Peiffer, "The German Upper Class in Pittsburgh, 1850–1920"
(seminar paper, Department of History, University of Pittsburgh, 1964),
p. 25.

41. John W. Brant, " 'Those Damn Ignorant Somerset Dutch': An Analysis of Antimasonic and Whig Voting in Somerset County, Pa., from 1828 to
1840" (seminar paper, Department of History, University of Pittsburgh,
1969), pp. 21, 53–59; Duane E. Campbell, "Anti-Masonry in Pittsburgh"
(seminar paper, Department of History, Carnegie Mellon University, n.d.),
p. 7; and Ray Burkett, "The AntiMasonic Party in Pittsburgh" (seminar
paper, Department of History, University of Pittsburgh, 1971), pp. 3, 48.

42. John Dankosky, "The Ante-Bellum Temperance Movement in Pennsylvania" (seminar paper, Department of History, University of Pittsburgh,
1971), pp. 32–33; and Leeann Rosner, "The Pittsburgh Temperance Movement, 1830–1850" (research paper, Department of History, Carnegie Mellon University, 1982).

43. Hal Kimmins, "Joseph Barker, Mayor of Pittsburgh, 1850–51" (seminar paper, Department of History, University of Pittsburgh, 1963); and
Szarnicki, *Michael O'Connor*, pp. 120–24.

44. Edwin Brownlee Spragg, "Antislavery Sentiment in Washington
County in the 1830s" (MA thesis, Department of History, University of
Pittsburgh, 1962), p. 34; and Crumrine, *Washington County*, pp. 510–24,
787, 839.

45. Howard V. Storch, Jr., "Changing Functions of the Center-City: Pittsburgh, 1850–1912" (seminar paper, Department of History, University of
Pittsburgh, 1966), p. 6.

46. Oscar Sloan Whitacre, "A Comparative Study of Five Major Protes-

tant Denominations in Pittsburgh, 1910–1929" (MA thesis, Department of Sociology, University of Pittsburgh, 1973).

47. Walton, *Story of Religion*, p. 30.

48. Ibid., pp. 34–36, 46–47. See also Morris Zelditch, "The Syrians in Pittsburgh" (MA thesis, Department of Social Work, University of Pittsburgh, 1936).

49. Louise Misko, "A Study of Political Activities and Attitudes of Pittsburgh Poles Relative to Achieving the Independence of Poland Through Preservation of Religious, Fraternal, and Cultural Institutions" (seminar paper, Department of History, University of Pittsburgh, 1975); Joseph Kenneth Balogh, "An Analysis of Cultural Organizations of Hungarian-Americans in Pittsburgh and Allegheny County" (Ph.D. diss., University of Pittsburgh, 1945); Howard F. Stein, "An Ethno-Historical Study of Slovak-American Identity" (Ph.D. diss., University of Pittsburgh, 1972); Georgia Katsafanas and Alice Flocos, "The Greek Immigrant and the Greek Orthodox Church in Pittsburgh" (research paper, Department of History, University of Pittsburgh, 1974); and Zelditch, "Syrians."

50. June Granatir Alexander, *The Immigrant Church and Community: Pittsburgh's Slovak Catholics and Lutherans, 1880–1915* (Pittsburgh: University of Pittsburgh Press, 1987).

51. See also Katsafanas and Flocos, "Greek Immigrant," p. 12; and Stein, "Slovak-American Identity," pp. 198–199.

52. Kathleen Healy, "The Early History of the Sisters of Mercy in Western Pennsylvania," *Western Pennsylvania Historical Magazine* 55 (April 1972): 159–70; Patricia McCann, *On the Wing: The Story of the Pittsburgh Sisters of Mercy* (New York: Seabury Press, 1980); and Lois Kalloway, "Sisters All: Polish, Irish, German, and Lithuanian Religious Sisters in Pittsburgh, 1890–1940" (Ph.D. diss., University of Pittsburgh, in progress).

53. Walton, *Story of Religion*, pp. 37–42; Mizrahi, "Jewish Community"; Ida Cohen Selavan, "The Founding of the Columbian Council," *American Jewish Archives* 30 (April 1978): 24–42; and Ida Selavan, ed., *My Voice Was Heard* (Pittsburgh: KTAV Publishing House, Inc., and the National Council of Jewish Women, Pittsburgh Section, 1981).

54. Francis G. Couvares, *The Remaking of Pittsburgh: Class and Culture in an Industrializing City 1877–1919* (Albany: State University of New York Press, 1984), pp. 51–52, 71–72. Shelton Stromquist, "Working-Class Organization and Industrial Change in Pittsburgh, 1860–1890" (seminar paper, Department of History, University of Pittsburgh, 1973), fails to mention religion in a section on neighborhoods.

55. John William Bennett, "Iron Workers in Woods Run and Johnstown: The Union Era, 1865–1895" (Ph.D. diss., University of Pittsburgh, 1977); and Frank H. Serene, "Immigrant Steelworkers in the Monongahela Valley: Their Communities and the Development of a Labor Class Consciousness" (Ph.D. diss., University of Pittsburgh, 1979).

56. Whitacre, "Five Major Protestant Denominations," pp. 14, 16, reported that the two major Presbyterian wings had eighty-one congregations,

while Walter J. Fitzpatrick, "Places of Worship in Pittsburgh: A Study of Changes in Location and Distribution, 1910–1960" (MA thesis, Department of Geography, University of Pittsburgh, 1967), p. 17, reported eighty-one Roman Catholic and Eastern Orthodox congregations in 1910.

57. Hannon, "Ethnic Assimilation," pp. 173–220.

58. G. Dale Greenawald, "Germans in Pittsburgh, 1850, 1880, 1930: Residency, Occupations, and Assimilation" (seminar paper, Department of History, Carnegie Mellon University, n.d.).

59. James C. Holmberg, "The Industrializing Community: Pittsburgh, 1850–1880" (Ph.D. diss., University of Pittsburgh, 1981).

60. Layne Peiffer, "German Upper Class."

61. Tom Callister, "The Reaction of the Presbytery of Pittsburgh to the New Immgrants" (seminar paper, Department of History, University of Pittsburgh, n.d.).

62. Trisha Early, "The Pittsburgh Survey" (Research Seminar, Department of History, University of Pittsburgh, 1972); and Elizabeth A. Metzger, "A Study of Social Settlement Workers in Pittsburgh, 1893 to 1927" (seminar paper, Department of History, University of Pittsburgh, 1974).

63. Metzger, "Social Settlement Workers"; and Selavan, "Columbian Council."

64. Metzger, "Social Settlement Workers"; and Beth Kurtz, "Women's Charity Work: Its Place and Value in America During the Second Half of the Nineteenth Century" (seminar paper, Department of History, Carnegie Mellon University, 1980).

65. Selavan, "Columbian Council," pp. 28–29. Rosner, "Pittsburgh Temperance Movement," pp. 5–6, points out that no women participated in drinking reform in the 1830s, and Edwin Spragg, "Anti-slavery Sentiment in Washington County," p. 17, indicates that the Washington Colonization Society "respectfully" suggested that the "ladies of the town and county" should form an auxiliary.

66. Holmberg, "Industrializing Community," pp. 222–223; Early, "Pittsburgh Survey," pp. 6–7, 18–24; and Callister, "Reaction of the Presbytery," pp. 13–22.

67. Ida Cohen Selavan, "The Social Evil in an Industrial Society: Prostitution in Pittsburgh, 1900–1925" (seminar paper, Department of History, University of Pittsburgh, 1971); and George H. Westergaard, "Prostitution and Reform in Pittsburgh, 1906–1915: A Study in Social Control" (seminar paper, Department of History, Carnegie Mellon University, n.d.).

68. Selavan, "Social Evil," p. 17.

69. Callister, "Reaction of the Presbytery," pp. 23–34. See also Wayne H. Claeren, "The Ministry in Pittsburgh During the Progressive Era" (seminar paper, Department of History, University of Pittsburgh, 1964).

70. Wallhausser, "Upper-Class Society"; Hammar, "Wilkinsburg and Edgewood"; and Jack L. Hiller, Jarrell McCracken, and Ted C. Soens, "Morningside: An Urban Village" (seminar paper, Department of History, Carnegie Mellon University, 1969), pp. 13–14.

71. Callister, "Reaction of the Presbytery," p. 4; Elizabeth J. Masson, "Episcopal Church Development and Church Neighborhood Change, Pittsburgh, Pennsylvania, 1870–1975" (MA thesis, Department of Geography, University of Pittsburgh, 1976); Greenawald, "Germans in Pittsburgh"; and Walton, *Story of Religion*, p. 3.

72. John Barnes, Jr., "The Ku Klux Klan in Western Pennsylvania," (seminar paper, Department of History, Carnegie Mellon University, n.d.).

73. Claeren, "Ministry in Pittsburgh," p. 8; and Westergaard, "Prostitution and Reform," p. 22.

74. Walton, *Story of Religion*; and Claeren, "Ministry in Pittsburgh."

75. Walton, *Story of Religion*.

76. Fitzpatrick, "Places of Worship," pp. 69, 86; and Walton, *Story of Religion*.

77. John Bodnar, Roger Simon, and Michael P. Weber, *Lives of Their Own: Blacks, Italians, and Poles in Pittsburgh, 1900–1950* (Urbana: University of Illinois Press, 1982), pp. 8, 265.

78. Walton, *Story of Religion*, p. 8; and Bodnar et. al., *Lives of Their Own*, p. 199.

79. Bodnar et. al., *Lives of Their Own*, pp. 73–74.

80. Melvin D. Williams, "A Penecostal Congregation in Pittsburgh: A Religious Community in a Black Ghetto" (Ph.D. diss., University of Pittsburgh, 1973), published as *Community in a Black Pentacostal Church: An Anthropological Study* (Pittsburgh: University of Pittsburgh Press, 1974), p. 12.

81. Bodnar et. al., *Lives of Their Own*, pp. 78, 265. See also Michael P. Weber, John Bodnar, and Roger Simon, "Seven Neighborhoods: Stability and Change in Pittsburgh's Ethnic Community, 1930–1960," *Western Pennsylvania Historical Magazine* 64 (April 1981): 121–50.

82. Walton, *Story of Religion*, p. 8.

83. Bodnar et. al., *Lives of Their Own*.

84. M. Ruth Mcintyre, "The Organizational Nature of an Urban Residential Neighborhood in Transition: Homewood-Brushton of Pittsburgh" (Ph.D. diss., University of Pittsburgh, 1963), pp. 47–51, 64–66.

85. Noshir K. Kaikobad, "The Colored Moslems of Pittsburgh" (MA thesis, Department of Social Work, University of Pittsburgh, 1948).

86. Mcintyre, "Organization Nature," pp. 127–129.

87. Hiller et. al., "Morningside."

88. Elizabeth J. Masson, "Episcopal Church Development," pp. 88–91.

89. Greenawald, "Germans in Pittsburgh," p. 24.

90. Walton, *Story of Religion*, p. 65.

91. Daniel S. Buczek, "Polish American Priests and the American Catholic Hierarchy: A View from the Twenties," *Polish American Studies* 3 (Jan. 1976): 36–37.

92. Linda K. Neff, "Parochial Schools and Neighborhood Attractiveness: A Study of Allegheny County, Pennsylvania" (Ph.D. diss., University of Pittsburgh, 1985).

93. Walton, *Story of Religion*, p. 2; Katsafanas and Flocos, "Greek Immigrant"; and Demetrius S. Iatridis, "The Post-War Greek Newcomer in Pittsburgh: A Study for Community Organization" (MA thesis, Department of Social Work, University of Pittsburgh, 1952).

94. Zelditch, "Syrians in Pittsburgh," p. 29.

95. Fitzpatrick, "Places of Worship in Pittsburgh," pp. 17, 32.

96. Myrna Silverman, "Class, Kinship, and Ethnicity: Patterns of Jewish Upward Mobility in Pittsburgh, Pennsylvania," *Urban Anthropologist* 7 (Spring 1978): 25, 28.

97. Fitzpatrick, "Places of Worship"; and Walton, *Story of Religion*.

98. Hannon, "Ethnic Assimilation," p. 220.

99. Masson, "Episcopal Church Development," p. 23.

100. Kleppner, "Lincoln and the Immigrant Vote"; Lee Benson, *The Concept of Jacksonian Democracy* (Princeton, N.J.: Princeton University Press, 1961); Paul Kleppner, *The Cross of Culture* (Glencoe, Ill.: The Free Press, 1970); Richard Jensen, *The Winning of the Midwest: Social and Political Conflict, 1888–1896* (Chicago: University of Chicago Press, 1971); and Philip VanderMeer, *The Hoosier Politician* (Champaign: University of Illinois Press, 1985).

101. Balogh, "Hungarian-Americans," p. 22.

102. For example, see David Montgomery, "The Shuttle and the Cross: Weavers and Artisans in the Kensington Riots of 1844," *Journal of Social History* 5 (Summer 1972): 411–46; Bruce Laurie, *The Working People of Philadelphia* (Philadelphia: Temple University Press, 1980); and Paul Faler, *Mechanics and Manufacturers in the Early Industrial Revolution* (Albany: State University of New York Press, 1981).

Community-Building and Occupational Mobility in Pittsburgh, 1880–1960

MICHAEL P. WEBER

THE PHYSICAL LANDSCAPE of metropolitan Pittsburgh experienced changes of almost revolutionary proportions in the four decades following World War II. Two massive urban redevelopment projects occurring during the mayoral administrations of David L. Lawrence, Joseph Barr, and Richard Caliguiri almost completely restructured the skyline of the central business district. Hotels, railroad stations, skyscraper office buildings, a historic commercial complex, parking lots, and numerous two- and three-story loftbuildings gave way to the Gateway Center and Civic Arena developments, to the corporate headquarters of Mellon Bank, PPG Industries, and United States Steel (USX). Major hotels, commercial office buildings, a mid-sized convention center, and a sports stadium added to the changing panorama of the city.

The structural environment created by Pittsburgh's major heavy industries had also changed dramatically by the late eighties. The banks of the region's three major rivers, once lined for fifteen miles in either direction from the city's apex with factories of international significance were turned over to recreational and commercial uses or became small industrial and research parks. Other areas were in the process of being reclaimed by nature as trees and grass began to grow where once mighty furnaces and forges stood. Huge plants owned by Jones & Laughlin, Wheeling-Pittsburgh, U.S. Steel, Dravo, and Armstrong Cork had been dismantled or stood idle awaiting purchase and conversion to other uses.

Perhaps surprisingly, the most enduring aspect of the Pittsburgh landscape during the twentieth century has been its distinctive residential neighborhoods and communities. Surveys of the region conducted in the late eighties reveal that more than two dozen industrial towns remain from the era when manufacturing dominated the Western Pennsylvania economy. In many of these communities the

361

industry on which they were founded had recently fled but the residential neighborhoods and the institutions which served them were still struggling with an uncertain future. Urban social scientists, historians, and news reporters similarly identified inner city and suburban communities in which particular class, racial, or ethnic groups continued to reside. While few of these areas could be classified as homogeneous, the group which once dominated the neighborhood remained important. Social and religious institutions, cultural and class patterns, a common history, and the region's rugged topography all helped to preserve distinctiveness. Few serious observers could overlook the "Germanness" of Troy Hill or Elliot, the Italian character of Bloomfield and parts of East Liberty, the dominance of Poles and other Eastern Europeans in Polish Hill and parts of the South Side, or the persistent Jewish influence in Squirrel Hill. Upper classes and professional groups, too, maintained their historic hold on parts of Shadyside and Point Breeze in the city and in Sewickley, Fox Chapel, and Mt. Lebanon in the suburbs. Blacks became dominant in the Hill district in the mid-1930s and in Homewood-Brushton in the 1950s.

The persistence of these historic communities into the mid-1980s is somewhat remarkable when one considers the commercial and industrial change which took place in Western Pennsylvania during the last two generations. It is perhaps even more striking when examined in the context of the changes occurring in other urban American centers. With the exception of the persisting black ghettos in the urban northeast, most of the historic neighborhoods had disappeared or been amalgamated to such an extent that they lost their distinctiveness. Once easily identifiable ethnic and social class neighborhoods in Philadelphia, Chicago, Cleveland, New York, and St. Louis had ceased to exist. While a few remained in each city, the formerly dominant neighborhood mosaic was no longer a key spatial characteristic of urban America.

Pre–World War II Research

Researchers in Pittsburgh have long recognized the importance of community and neighborhood life in this industrial region. A substantial portion of the Pittsburgh Survey (1909–1914) was devoted to life in Pittsburgh's black and immigrant enclaves such as Painters Row, Skunk Hollow, Tammany Hall, and the Slavic quarters in Homestead. The topic became a favorite with reporters from the Pittsburgh *Leader* and Pittsburgh *Press* during the decade before

World War I. Each newspaper ran several series on life among the immigrants prior to the war. Most were curiosity pieces describing the peculiar habits and activities of the Hun, the Negro, the Slav, and the Italian. A brief work published by the Irene Kaufmann Settlement House in 1916 identified the location of racial and ethnic clusters in Pittsburgh's Hill district and described the social and physical deterioration of the area. None of the articles recognized the areas they described as neighborhoods or communities in which the residents shared common concerns, problems, or solutions. Neighborhood institutions such as churches, schools, business districts, and fraternal associations were generally ignored. At that time, only the presence of extensive problems or cultural anomalies made these areas worth writing and reading about.

The first scholarly concern with community life in the Pittsburgh region was fostered by several social science departments and the Department of Social Work at the University of Pittsburgh and by the Pittsburgh Urban League. Between 1918 and 1940 nearly two dozen articles, master's theses, or doctoral dissertations were produced which exhibited a significant interest in community life in Pittsburgh. Each study reflected the particular emphasis of the discipline or agency for whom it was produced. Several of the works, particularly those by the Urban League, were primarily concerned with identifying the social and economic problems faced by minority groups in their attempt to adjust to urban-industrial life. A number of theses defined the spatial patterns of ethnic and racial groups residing in the Pittsburgh industrial area. Others, more sociological in nature, examined family work patterns, neighborhood development and change, and community relationships. Those produced by economics students were particularly concerned with the occupational opportunities of various groups. Several attempted to measure the impact of education on occupational choice.

While a number of the works, particularly those discussed below, presented important evidence to accompany their discussion, none are analytical and they all lack a longitudinal perspective. Each study examined existing problems in a current context. They also failed to present any comparisons with conditions in other cities, much in vogue today, to enable the reader to draw any cross-urban or national conclusions. Surprisingly, however, while a number show the effects of limited training and most are primarily descriptive, they did not rely merely on anecdotes and impressions to tell their stories. Data appear to be carefully assembled, sources have been evaluated, and the quantitative manipulation, while simple, is revealing. Historians or others interested in learning about community and occupational patterns during the first third of the twentieth century will find these studies useful.

Perhaps reflecting their concern with important social problems, half of the studies produced during the interwar period focus on social and occupational patterns in Pittsburgh's black communities. Nearly all deal with the same issues: migration patterns, job procurement and stability, education, homeownership, and social and health conditions in the black neighborhoods. Fortunately for the contemporary researcher, although the studies were written by different authors at different periods of time, these same questions are considered at frequent intervals over a twenty-year period. One can easily examine the same concerns to determine the extent of stability or change over a period of time.

The earliest serious work dealing with residential and occupational patterns of blacks in Pittsburgh is Abraham Epstein's *The Negro Migrant in Pittsburgh*. Epstein, who was employed by the Irene Kaufmann Settlement in the Hill district, produced his work under the supervision of the University of Pittsburgh's School of Economics. His primary concern was the social condition of more than twenty thousand black workers who migrated to Pittsburgh after World War I. The study, one of the most often cited in recent works on blacks in Pittsburgh, is rich in economic and social data. Because it provides a valuable benchmark from which one can measure change in the black community, a brief discussion of its contents is appropriate.

Typical of students of the period, Epstein was motivated by two concerns. "The primary purpose of the study was to learn the facts about black Pittsburghers who participated in the Great Migration." Secondly, he hoped "that the data presented might lead to the amelioration of certain existing evils."[1] Relying primarily on personal knowledge and field inspections, Epstein identified six black neighborhoods in Pittsburgh in 1918. They included several sections in the Hill district; Penn Avenue in Lawrenceville between Twenty-eighth and Thirty-fourth streets; Beaver and Fulton streets on the North Side; East Liberty near Mignonette and Shakespeare streets; and a newly developing section in the central business district on Second Avenue, along Ross and Water streets. Through the use of graphs, tables, and photographs he documented the deteriorated condition of each area. High density and transiency rates, low homeownership, a young median age, juvenile delinquency, infant mortality, and disease rates of almost epidemic proportions characterized these neighborhoods. The common accommodations were cellar apartments and tenement houses. While it was the author's intention to examine conditions in the neighborhoods of the new black migrant, other sources indicate that, with the exception of a small middle-class neighborhood in Homewood-Brushton near Tioga and Susquehanna streets, Epstein's neighborhoods were also the residences of the older, more established black population. What is clear from his data is that the neighborhood institutions and any sense of community established by the earlier black population were overwhelmed by the post-1916 migration. Churches, fraternal associations, and welfare organizations were totally unequiped to deal with the massive influx.

Epstein also documented, through 505 interviews, that the recent arrivals to the city were intent on becoming permanent family members of the community. Many migrated as a family or sent for wives and children within three months of their arrival. Others sent money home to help care for family members until they could be reunited in Pittsburgh. Although nearly half responded that they intended to return home or move elsewhere, most indicated that inadequate housing was the reason for their planned departure. In sum, Pittsburgh's black population, whether settled in the city for a full generation or more, or recently arrived, lived in unorganized transient slums. A community where individuals shared a sense of belonging and mutual responsibility simply did not exist. Institutions which could provide services and foster community-building were absent or inadequate.[2]

Epstein also provided a useful measurement from which to examine black occupational mobility during the first two decades of the twentieth century. Turn-of-the-century manuscript census data reveal that blacks were scattered about the city in a wide variety of unskilled and semiskilled occupations. Unlike their immigrant counterparts, they were unable to establish an occupational beachhead in the city's metal industries.[3] Epstein showed, however, that by 1918 blacks had become a significant force in the mills of the region, principally with Carnegie Steel and Jones & Laughlin. Nearly six thousand blacks worked for these two firms alone. As might be expected, 95 percent of all blacks in industrial firms were in unskilled positions. Only those in service industries, self-employed, or working for small firms held jobs at higher levels. Almost none were in white-collar positions. While a more recent study data indicates that some mild upward mobility did occur for blacks remaining in Pittsburgh for two full decades, 1900–1920, Epstein's work makes it clear that as a group the newer migrants began their career in Pittsburgh at the bottom of the economic ladder.[4] Both studies conclude that the lack of economic opportunity, along with poor housing conditions, contributed to the high rate of intra- and intercity transiency among the city's black population.

Three master's theses produced by University of Pittsburgh economics students added to the literature on the black community and mobility. Produced during the mid-1920s, all three focused primarily on occupational opportunities for blacks in Pittsburgh's major industries.[5] One of the works also provided a vivid account of conditions in the black community in 1924. One has only to compare it with Epstein's study to conclude that physical conditions grew worse between 1918 and 1924. In the Hill district site of one of Epstein's neighborhoods, Abram Harris selected a sample of housing units at random. They contained an average of 2.5 rooms per unit with approximately four persons occupying each room. Half were in deteriorated condition. A number had no sewerage, lighting, or water. Tables showing the black-white differences in rates of disease and mortality, serious and petty crimes, and public disturbances such as drinking and fighting demonstrate that regardless of the issue identified, it was still much less healthy to live in the black areas than in white ones. While community development was not a central issue in any of these studies, the evidence suggests that the process had still not begun in the black neighborhood.

All three studies are rich in data regarding the economic plight of the black worker. Generally they mirror the earlier findings of Epstein in suggesting that overall conditions had remained static or deteriorated. An analysis of the data presented by Harris and by Ira Reid approximately fifteen months later suggests how volatile the black employment situation could be. In August 1923 Harris found that twenty-seven major industries in Western Pennsylvania were employing 17,224 black workers. Less than a year and a half later, in the midst of a serious economic downturn in Pittsburgh, the number had dropped to 7,636.[6] Black employment in Pittsburgh was clearly at the mercy of the city's economic condition. The occupational data presented by all three authors indicates that unskilled and semi-skilled work dominated the employment profile of blacks throughout the decade. Fewer than one of every ten black workers held skilled or white-collar positions. The studies do not provide an accurate picture of occupational mobility, but it is not difficult to impute rough rates from the data presented. Nearly every black worker in Pittsburgh in 1918 (Epstein), 1924 (Harris), 1925 (Reid), and 1928 (Covington) were in the bottom two occupational classes. While a few perhaps managed to slip through the barriers holding them back, the number or percentage was almost too small to record.

Additional information provided by Reid documents the low occupational status of Pittsburgh's black workers. High worker turnover rates, difficulties with foreign-born foremen, and racial discrimination all contributed to the inability of blacks to secure better jobs. Where they were able to acquire skilled work, as in the building trades, blacks were paid approximately 75 percent of union scale. More than half of the building trade unions, however, prohibited blacks from joining.[7] Other unions were even more restrictive. Blacks constituted less than 6 percent of the union membership in 1920, and their membership was halved during the economic downturn of 1924. More revealing, four-fifths of those who did gain union entry belonged to the Hod Carriers, Building and Common Laborers' Union. All were members of the black local.[8]

The inability of blacks to find suitable employment, however, did not deter their children's aspirations. Floyd Covington's study of occupational goals of 434 black high-school students in Pittsburgh in 1928 reveals that nearly two-thirds aspired to white-collar professional or semiprofessional occupations. Surprisingly, their choices closely match those of a similar study of white students in Madison, Wisconsin. That these students desired to move up on the economic scale can be seen when their aspirations are matched with the occupations of their parents. Excluding ministers, less than 3 percent of their fathers and 6 percent of their mothers held white-collar occupations. Their parents' work history, moreover, failed to dampen the goals of Pittsburgh's young blacks. Roughly 40 percent of the fathers and 20 percent of the mothers had experienced downward occupational mobility during their lifetimes.[9]

While Covington suggests that black children held unrealistic views of their employment possibilities, his examination of enrollment in the Pittsburgh trade schools presents some idea of the dilemma they faced. Aware that few trades would employ blacks, the city school district followed a policy of discouraging their attendance in the vocational schools. Blacks were permitted to constitute no more than 8 percent of the vocational population. That they learned the lesson well is demonstrated by the fact that only fourteen black boys (less than 1 percent) were so enrolled in 1927.[10] Thus, eliminated from the skilled occupations, young blacks were forced to choose between unskilled and semiskilled labor and the white-collar professions. Perhaps not realizing that professional occupations were closed to them, they chose the latter.

Covington provides no longitudinal analysis to measure how well these optimistic students fared when they graduated two years later. Several other studies, however, confirm the suspicion that whatever difficulties these young people faced, the going would be even more difficult in the midst of the Great Depression.[11]

The unemployment rate among Pittsburgh's black adult population, which ranged between 33 and 40 percent during most of the Depression, wiped out even the modest occupational gains blacks enjoyed during the boom times of the twenties. By mid-1930, 41 percent of a sample of 2,700 black families were classified as destitute while another 33 percent fell into the poverty class. Only 11 percent lived in "comfortable circumstances." Conditions were even worse among female-headed black families.[12] Among those blacks able to find work, as a group they were marginally worse off than the 1920 cohort. Nearly all black male and female employees during the Depression held jobs at the lowest occupational classifications. The nature of most of the occupations, moreover, suggests that few provided long-term, steady employment.[13] Clearly the black high-school youths studied by Covington must have been highly disillusioned by their employment experience during the 1930s. Upward occupational mobility was an unattainable illusion for them.

The studies of the black occupational experience carried out during the interwar period provide no historical examinations and most eschew analysis. The somewhat crude data also fail to consider the possibility of spurious evidence and none attempt to determine which of the multitude of variables, if any, contributed significantly to the absence of black mobility. Nevertheless, when taken in total, they provide important data regarding the occupational status of blacks during this volatile industrial period. High job aspirations, educational training, persistence in the community, or a willingness to work hard were clearly no match for the weak economic conditions and strong racist attitudes of the Pittsburgh industrial region.

No doubt due to their economic training, most early observers of the black experience confined their efforts to chronicling the dismal black work patterns. A few, however, provide an important look at residential conditions and give strong clues to the causes of the absence of community development among Pittsburgh's black residents. Again, the studies are descriptive, painting an accurate picture of rootlessness and powerlessness. They are also an important source of data for students attempting to understand the social history of Pittsburgh's black population.

Sociological studies by Alex Pittler, Alonzo Moron, and Delmer Seawright document the tremendous dispersion of the black population across the entire Pittsburgh landscape during the early 1930s. Blacks inhabited every area of the city, living in all but 16 of the 174

Pittsburgh census tracts, although strong concentrations existed in 9 tracts. Seven of these were located in the Third and Fifth wards (Hill district); the other two were in the Tenth and Twelfth wards respectively. What is most striking, however, is that these 9 areas combined made up only 36 percent of the city's black population. The overwhelming majority lived in tracts containing less than six hundred blacks. Interestingly, particularly in the face of the massive black in-migration which occurred during the period, the rate of black dispersion increased slightly between 1910 and 1930. The proportion of blacks in the 9 major tracts actually declined from 39 to 36 percent during the period.[14] All three studies, as well as several others cited above, document the exceptionally high rates of migration into and out of Pittsburgh after World War I.[15] Seawright and Moron, in addition, provide a careful exposition of intracity residential movement. Both show that the typical Pittsburgh black who remained in the city for any length of time likely moved five or six times during a decade and that most of the moves were from one neighborhood to another. Moron also notes significant black movement into and out of Homewood-Brushton during the two-year period after the Seawright study. A useful study by Moron and a colleague for the Bureau of Social Research confirmed these findings.[16] While the studies also found significant black concentrations in the sections of the city noted above, they all concluded that the transiency of the population prevented the development of any meaningful community life. "The absence of a solidly Negro community in Pittsburgh," Moron and Stephan concluded, "reduces very materially the power of the Negro population to compel retail dealers to employ Negro clerks in Negro residential districts and to secure Negro political representation as in some Northern cities."[17]

Without labeling it as such, Moron correctly speculated that the journey-to-work was a key factor in most black residential movement. He noted, for example, the rapid increase in the black population of the Greenfield-Hazelwood-Glenfield section of the city between 1910 and 1930. He then identified the expansion of the manufacturing firms in the area for which these individuals worked. While later studies have verified the Moron hypothesis, the data he presented was mostly speculative, as was the relationship he attempted to establish. Actual disaggregate information on employment and residence was not presented.[18]

While all three authors suggest that the constant movement of the black population within Pittsburgh prevented the development of a community focus in the black neighborhoods, one must turn to a later work by Ira Reid for a comprehensive examination of black life in Pittsburgh. Shortly after completion of his graduate work at the University of Pittsburgh, Reid became director of the Pittsburgh branch of the National Urban League. He organized a committee, which he chaired, to study social conditions among blacks in the Hill district. In 1930 the committee published its 117-page report written primarily by Reid. Topics examined in the report included environment and housing, health conditions, industry and employment, delinquency and crime, recreation, education, Negro churches, and social agencies of the community. The report concluded with recommendations to the city, the black community, and various social welfare groups for means to improve living conditions on the Hill.[19]

The study was clearly the most detailed of life in any Pittsburgh neighborhood since the publication of Margaret Byington's major work on Homestead.[20] The exceptional data documented the transition of the Hill district from an area of first settlement by Pittsburgh's ethnic and black in-migrants to a primarily black ghetto. Although the process would not be completed for another twenty years, the direction was unmistakable and probably irreversible. The black population increased by several hundred percent between 1900 and 1930 while white residency began to decline. Overcrowding, deteriorated housing, lack of sanitation, high disease and mortality rates, rampant vice, and a marked increase in crimes against person and property are all identified by Reid's committee. Several women's organizations, Reid pointed out, gave aid to needy black families, but they were simply too few to alleviate the problems. Other neighborhood organizations, such as fraternal associations and welfare groups, Reid concluded, lacked the power and the finances to have a significant impact. Constant in-migration of blacks from the deep South and the outward migration of established black and white residents seeking work or better living conditions robbed the community of much of its potential leadership. While the churches were viewed as "the strongest institution among Negroes," the presence of forty-five different congregations in the two wards prevented any from playing the leadership role found among the nationality churches in the city's ethnic communities.[21] The study by Reid and his associates is not without its limitations. The authors are content to describe a variety of existing social and economic conditions. No attempts are made to determine causes or provide any long-range perspective. Secondly, in their zeal to describe the Hill, they overlook the dynamics of its residents. Daily life, social interaction among blacks and between black and white ethnic residents, and family or neighborhood networks are all ignored. Finally, their recommendations—enforce existing housing and criminal codes, provide additional health care personnel, and establish programs to train female domestics—are modest, and one can easily detect a patronizing tone: "There should be a general campaign urging upon the colored people the desirability of keeping their properties in an attractive condition."[22]

These criticisms notwithstanding, the study presents a convincing case that social disorganization characterized the Third and Fifth Ward black neighborhoods. Population turnover, occupational patterns which kept blacks moving throughout the city, inability to purchase housing, and the impotence of religious and social welfare institutions prevented the development of community networks in Pittsburgh's black neighborhoods before World War II.[23]

The apparent interest in black occupational patterns and neighborhood life among researchers during the interwar period was not matched by an equal curiosity with life among Pittsburgh's foreign-born immigrants. Two master's theses provide only a glimpse of life among the Italian communities during the first third of the twentieth century.[24] The descriptive and numerical data they provide point to the establishment of strong community feeling in the neighborhoods of Bloomfield and East Liberty between Larimer and Lincoln avenues. Ethnic homogeneity, evidence of considerable intraresidential interaction, religious and fraternal institutions active in providing community social and welfare services, the strong visible presence of ethnic business districts, and community celebrations and activities all attest to the existence of a strong sense of neighborhood identity. While neither author offers documentation on homeownership or intragenerational residential persistence, their data and descriptions suggest that a hypothesis of residential and community stability among Italians would not be far off the mark.

By providing information on the occupations and occupational backgrounds of some Italian businessmen, Migliore suggests some limited upward occupational mobility. He identified numerous merchants and a small number of physicians (23), dentists (14), attorneys (16), and teachers (31) from among the first- or second-generation Italian immigrants and more than one hundred fifty Italian immigrants who graduated from the University of Pittsburgh between 1911 and 1927.[25] While the numbers are not overwhelming, when compared with the data previously presented on black migrants, they strongly suggest that the Italian experience in Pittsburgh was significantly different from the black one. In addition to information on community and occupation, the student of the Italian experience in the Pittsburgh region will find the study by Migliore intriguing for

other reasons. Migliore, a United Presbyterian minister sent to Pittsburgh to "evangelize the Italians in Allegheny County," offers a
useful account of the growth and services of the Italian Catholic
church in Pittsburgh, 1900–1925, and provides a brief but interesting account of a seventeen-year attempt by the Presbyterian church
to win Italian converts. The effort, by Migliore's own admission,
failed and was evidently abandoned in 1927.[26]

Two somewhat related possibilities may explain the absence of
pre–World War II studies on Pittsburgh's other ethnic groups. They
reveal perhaps as much about the social and economic condition of
the various groups as they do about the state of scholarship during
the era. Many of the studies examined above were conducted by
members of the group being examined. Blacks such as Reid, Moron,
and Covington studied black economic and social conditions; an
Italian, Migliore, is the major recorder of the Italian experience. That
most works were carried out as part of the requirements for a graduate degree indicates that some blacks and Italians succeeded in attaining such high educational prizes. While the data presented by
the studies above indicated that such achievements were rare, they
obviously did occur. In the absence of educational data on other
immigrant groups, particularly Eastern Europeans, one might speculate that few reached graduate school, and thus none were available
to study their own history.[27]

Another possible explanation lies in the economic and social
conditions of those being studied. In nearly every study of blacks,
while not always explicitly stated, the authors expected that revealing the degrading and often dangerous conditions would lead to
some reform. In fact, most of their conclusions called for specific
reforms. Migliore's work is a thinly veiled attempt to understand the
failure of the Presbyterian evangelical movement among Pittsburgh's Italians. While conditions among the other ethnic groups in
the city were never luxurious, they were perhaps never bad enough
to spur aspiring scholars—even co-ethnics—to study them. Modern
scholars wishing to understand the history of Eastern Europeans in
Pittsburgh will need to turn to the works of the Pittsburgh Survey or
Thomas Bell's novel, *Out of This Furnace*, both written before
World War II, or to the postwar studies discussed below.[28]

Research Since World War II

Interest among local scholars in the community and/or occupational experiences of Pittsburgh's social, ethnic, and racial groups
ceased between 1940 and 1960. The reasons are not apparent, but

neither published nor unpublished authors devoted much effort to either topic during the two decades. Three papers, each with only a tangential interest in either of the topics of central importance to this account were produced during the period.[29] Beginning in the 1960s, however, a flood of student papers dealing with a variety of social and urban history topics—including immigrants, blacks, elites, neighborhood life, mobility, the influence of technology, and gender history—were produced.

While the papers vary considerably in quality and breadth and depth of treatment, they have a number of features in common. First, all attempt to determine change or stability in the area of primary interest over time. Second, the authors are not content with describing existing conditions or with simply demonstrating that change had or had not taken place; they are interested in causation. Third, the use of different types of sources and a more complex methodology is immediately apparent. The authors examine a wide variety of data (maps, company work records, real estate title transfers, mortgage records, building, sewer, and road permits) and conduct oral interviews in seeking answers to their questions. Fourth, a number of the papers struggle with attempts to place their work into a broader historical context by providing comparisons with other, mostly published, works on the same topic. Finally, many of the works are characterized by a sophisticated conceptualization of the topic at hand, and they rely heavily on quantitative information. Not all the recent unpublished works on community or occupational mobility (and few of the published works) will be examined in the remainder of this essay; but the findings and methodology of those discussed are representative of the important works produced during the last decade.

Three papers written in 1964 examined community life among the region's elite. The questions asked by the authors are as instructive as the conclusions they reach. Each attempts to determine the origin, heritage, and organization of the upper class in Sewickley Valley, Sewickley Heights, and in Shadyside.[30] Using records from the Pittsburgh Blue Book, the Social Register, church and club rosters and the *Index*, a society newspaper published weekly through the first two decades of the century, all three papers demonstrate that the city's elite came from wealthy families, belonged to the Presbyterian or Episcopal church, attended the same schools, shared exclusive recreational interests and resorts, and gave their sons and daughters in marriage to each other's offspring. In each community these families formed an interlocking economic elite. Half a dozen men in Sewickley Heights, for example, exercised control in some twenty-five banks and at least fifty corporations in Pittsburgh. They also took matters into their own hands to maintain and extend their exclusiveness. They joined restrictive organizations such as the Duquesne Club or the more exclusive Pittsburgh Club. They formed country clubs for the men, exclusive social organizations for their wives, and private academies for their children. The upper class in Sewickley also formed the Sewickley Heights Protective Association and petitioned the state legislature to form a Sewickley Heights Borough with power to set restrictive zoning ordinances which would prevent encroachment from the middle classes.[31]

Many of the questions raised by these student authors have become commonplace among social historians examining elite networks today. Excellent studies by Edward Pessen, John N. Ingham, and others have convincingly demonstrated the restrictive social and family networks which enabled the elite to retain their exclusiveness.[32] The works discussed here provide ample evidence to demonstrate that the Pittsburgh elite community was no different.

Curiously, recent students of Pittsburgh history have exhibited less interest in the experiences of the city's lowest economic class, the black worker, than their contemporaries during the post–World War I period. Peter Gottleib's work on migration patterns of Southern blacks to Pittsburgh; a comparative study of community, family, and work experiences among two generations of blacks; and two studies of changing spatial patterns among blacks are the lone works produced during the past two decades.[33]

The studies by Gottleib and John Bodnar and his associates differ in a number of respects. Gottleib is concerned primarily with the experiences of the blacks who arrived in Pittsburgh during the Great Migration, while Bodnar et al. compare the experiences of the old and new generations of blacks in the city. The latter study, moreover, attempts to analyze the black experience in the light of parallel but different experiences of Poles and Italians in the city. Thus the two studies ask somewhat different questions and arrive at mildly different conclusions. Gottleib sees more limited occupational goals among those he studies while the Bodnar analysis of two generations shows that blacks held greater aspirations than their immigrant counterparts. The two studies generally agree that both generations were systematically excluded from all but the most menial, low-paying occupations. The result, Bodnar et al. contend, influenced the development of the black family, contributed to exceptionally high rates of occupational and residential transiency, and inhibited the development of stable black communities for most of the twentieth century. The primary reason for this negative experience, the three authors conclude, was a pervasive racism which permeated Pittsburgh society but which was most evident among the group they call the "gate-keepers." The gatekeepers, or industrial foremen, prevented many blacks from gaining a foothold in any permanent industrial work, thus forcing them to accept unstable work among the city's smaller industrial firms or in low-paying service jobs. Even during the 1920s, when the doors to the mills were opened to blacks, they were always among the first fired during economic slowdowns. Unlike the immigrants who arrived in Pittsburgh at the same time, the great majority of even the earliest black migrants to the city remained at the bottom of the occupational and social class structure.

Peter Gottlieb, who conducted some of the oral interviews for the Bodner study—the two works shared some of the same sources—arrives at similar substantive conclusions but his interpretations differ somewhat. He combines a mass of quantitative data and oral interviews to compare the pre- and postmigration experiences of southern blacks who traveled to Pittsburgh during the period of the Great Migration. By identifying not only the jobs men and women held before and after migration, but their work conditions, level of skill, and daily and seasonal work rhythms, Gottlieb concludes that the "similarity between black migrants' jobs in the South and in Pittsburgh as well as the proximity of their rural homes afforded them a relatively easy transition to steel mill work."[34] The casual approach to work which blacks brought with them, rather than the prejudices of the gatekeepers and the low pay resulted in an occupational transiency which inhibited both occupational mobility and community development. Moreover, Gottlieb argues that while some black communities did exist, they were separated by class and origin. Older, established, upper-class blacks lived in small clusters isolated from their less affluent counterparts. Like the white residents of Sewickley Heights and Shadyside, they followed a separate social life in exclusive black men's and women's clubs and at upper-class churches. Middle-class blacks, mostly northern-born or among the second generation in Pittsburgh, inhabited the central city or parts of the North Side, mainly away from the recent southern migrants. Most of the workers among this older group exhibited hostile feelings toward the recent southern arrivals. Northern black women spurned attempts at affection by southern migrants, and some northern blacks actually moved from their old neighborhoods to new areas of black settlement in the city.[35] Recent southern migrants, reflecting their condition as temporary industrial laborers were thus segregated in temporary quarters hard-by the mill sites.

Blacks themselves, Gottlieb argues, intensified the divisions within their own neighborhoods, thus contributing to their inability to create stable communities in which residents could aid one another and develop a mutual sense of belonging.[36] Many southern black migrants, he concludes, maintained southern customs and patterns throughout their lives. "Dwelling, laboring, and worshiping among other blacks from the South, their social and geographic orientation remained the basis of their identity even after many years of residence in the North."[37]

Following in the wake of Stephan Thernstrom's pathbreaking work on Newburyport, Oscar Handlin's work on Boston's immigrants, and a volume of essays on occupational mobility in nineteenth-century American cities, immigrant and neighborhood studies and concerns about occupational mobility proliferated in the urban field.[38] A full generation of historians, or so it seemed, produced dissertations which examined social class, occupational procurement and mobility, and residential persistence among various ethnic groupings. Students of Pittsburgh history, perhaps motivated by the lack of serious work on the numerous ethnic groups which continued to inhabit neighborhoods throughout the region, joined the movement.

A seminar paper by Carolyn Schumacher in 1970 examined social mobility among school-age students during the nineteenth century. One year later Nora Faires produced a detailed study of the changing occupational patterns among the Germans in Allegheny City in 1850 and 1860. Both works contained numerous quantitative data and raised significant questions about the ethnic experience in Pittsburgh.[39] My own work compared many of the conditions Stephan Thernstrom discovered in Newburyport with those in a small northwestern Pennsylvania oil town.[40] All three studies suggested that, unlike conditions which Thernstrom found in economically stagnant Newburyport, limited occupational mobility was likely for certain ethnic groups in a growing or rapidly developing local economy. The first two works, particularly Faires's study, demonstrated the importance of ethnic differences in family patterns, occupational experiences, and community development. Even within nationality groups, strong differences could be detected among the first generation of German immigrants based on province of origin. Religion, too, proved to be a significant factor. German Catholics were not only different from their Protestant counterparts in their political preferences, as suggested by Paul Kleppner, they exhibited different occupational, social, and residential characteristics. They were also less successful in achieving upward occupational mobility.[41]

A number of the studies which followed those mentioned

above, while not duplicating their methodology, confirmed the suggestion that the experiences of ethnic groups in Pittsburgh differed greatly in both occupational structure and mobility and in community formation.

Studies of the two major groups which characterized the first large migration to Pittsburgh, Germans and Irish, suggest that both established neighborhoods throughout the city, that the neighborhoods were organized according to province or place of origin, and that both groups worked in manufacturing, particularly in the city's rapidly expanding metal industries. Within the largest German neighborhood in 1860, Dutchtown in Allegheny City, residential clusters from Hesse, Bavaria, Wurtemburg, and to a lesser degree Baden, Hanover, and Prussia could be detected. Members of each provincial neighborhood apparently interacted with each other rather than with Germans from other provinces. Approximately 50 percent of the men and 65 percent of the women found a spouse from the same province.[42] The Irish also established neighborhoods segregated by place of origin although to a lesser degree than did the Germans. Victor Walsh found three major Irish ghettos in Pittsburgh in 1850, each with distinct provincial origins: the Point Irish came mainly from Connaught, while those in West Pittsburgh and Monongahela on the South Side came from Connaught, and counties Donegal, Cork, Kerry, and Clare. The Irish, who were more likely to labor at unskilled occupations than the Germans, 70 percent to 30 percent respectively, could also be found in mixed neighborhoods in the Strip district, along Second Avenue, in Oakland, and in the Hill district.[43] Their search for work in the iron factories of the city evidently carried them beyond the confines of the segregated Irish neighborhood.

Within these homogeneous neighborhoods, the authors identified networks and institutions which helped to maintain the solidarity of the provincial group. To what extent this provincial loyalty prevented the Irish or Germans from acting as unified nationality groups to deal with issues of common interest to all remains unexplored. Their ability to put aside Old World differences and form larger nationality clusters may have contributed to their rapid demise as ethnic communities.

Between 1890 and 1910 Pittsburgh was the recipient of the entire spectrum of nationality groups which characterized the "new immigration." Jews, Poles, Italians, Hungarians, Greeks, Ruthenians and Lithuanians, and a variety of others arrived in the city seeking work in steel and other manufacturing enterprises. Most of these groups have not yet found their historian, but the studies which do exist confirm the earlier claims of diversity. Where groups were numerically large enough, such as the Poles, Italians, and Jews, they quickly formed independent ethnic neighborhoods with a complex network of support groups and institutions. The land areas they inhabited became more than geographical neighborhoods. A sense of community based on shared experiences and a sophisticated organizational network enabled them to establish ethnic villages which lasted two and three generations. Like the Germans and Irish who preceded them, the Italians retained some distinctions based on province of origin, but the differences were primarily confined to recreational associations or fraternal groups. Jews and Poles from Russia, Prussia, and Austria, in contrast, quickly subverted premigration differences to form unified communities. For the largest of the new ethnic groups arriving at the turn of the century, common religions enabled them to overcome Old World ethnic differences.[44]

Little is known about some of the numerically smaller groups which arrived in Western Pennsylvania during the twentieth century. Small ethnic clusters have been identified in the Hill district, in Oakland, and on the North and South sides in the city and in steel towns such as Braddock, Homestead, McKeesport, and Aliquippa. Too small to support their own institutions, most of these groups shared churches, schools, and other neighborhood organizations. While ethnic purity was retained within fraternal organizations, by the second generation most such associations were primarily recreational and apparently played a limited role in community formation.[45]

Studies produced thus far have provided a sufficiently detailed analysis of the occupational mobility experiences of Pittsburgh's immigrant population. Different rates of occupational success among specific groups have been established with native-born Americans, Germans, and Jews near the top and Poles and blacks at the bottom. Plausible explanations for such differences have been offered. Numerous other groups have been ignored but it appears that, with the possible exception of some work on the most recent migrants to Pittsburgh—Vietnamese, Indians, and so forth—little would be gained by further mobility studies.

In addition, the oral interviews provided by Bodnar and his associates and by Peter Gottlieb offered an important new dimension to studies of this type. First, one of the motivating drives of the social history movement was to uncover meanings and patterns in the lives of the great mass of nonliterate American workers and families. Yet the very method which served these ends, the ability to analyze large amounts of disaggregate data, also tended to depersonalize and dehumanize them. They became statistics to be manipulated, numbers in a table. Neither their struggles and goals, nor their place in a community ever came to life. Secondly, and relatedly, while we were able to measure their success, or lack thereof, on an objective and often elaborate statistical scale, we overlooked the worker's own measurement of the sum total of his or her life. It was only when I began to consider the perceptions of my own father—a lifelong unskilled worker—and of Dobie, one of the heroes in Thomas Bell's novel, *Out of This Furnace,* who was a failure on all of our measurement scales, did the inadequacy of our attempts become clear. It made no difference how we judged these workers. What mattered were their own assessments. A man who provides food and shelter for his family, sees them grow up in a psychologically healthy and happy environment, and observes similar success in his offspring may feel quite satisfied. A woman who provides support and love for her family might also derive a strong measure of satisfaction. Moreover, their behavior often reflected their feelings. In our attempts to determine the success or failure of individuals, and often to account for the absence of a labor mentality in American society, the personal dimension has too long been ignored. The oral interviews have returned some of this to recent social history.

The intricate process of community development and its significance in the lives of the residents of a geographical area, however, is just beginning to be understood. That significant differences in the developmental process and organization of communities exist seems certain from the available studies. Ethnicity, social class, group size, religion, and even geographic location all play important roles. Additional studies of other, as yet ignored groups, may help determine the strength of each variable. More importantly, a number of questions have been unasked. What has been the role of leadership and of political affiliation and participation in community-building? What factors contribute to the persistence or demise of an ethnic or social class community? How has community-building differed in the region's smaller industrial towns such as Turtle Creek, Wilmerding, or Duquesne? What has been the role of the company in one-industry towns such as East Pittsburgh or Ambridge? Has it inhibited or fostered community development? Has ethnicity in these towns played a less significant role than common work and economic experiences?[46] We also need to know more about the development of community feelings in situations where members do not inhabit contiguous land areas and its persistence in areas also inhabited by others (the contemporary Jewish community in Squirrel Hill, for example). We also know almost nothing about attempts to develop community spirit in suburban neighborhoods. What institutions, for example, exist to create a sense of belonging in transient, middle- and upper middle-class communities? Can organizations and activities such as neighborhood associations, men's and women's bowling, bridge, or garden clubs, or progressive dinners substitute for ethnic identity, church membership, or fraternal associations? Do community efforts succeed in entire towns, Mt. Lebanon or Upper St. Clair, for example, or can one find small neighborhoods—subdivisions—within these suburban towns which parallel the conditions found in nineteenth- and early-twentieth-century ethnic communities? Finally, one must also ask if the presence or absence of community identity or spirit makes any significant difference in the lives of modern urban dwellers. What difference does it all make anyway? The possibilities for meaningful research by students and professional historians in the area of community development in Western Pennsylvania seem endless.

NOTES

1. Abraham Epstein, *The Negro Migrant in Pittsburgh* (Pittsburgh, 1919), p. 8.

2. Ibid., pp. 27, 28. Cf. John Bodnar, Roger Simon, and Michael P. Weber, *Lives of Their Own: Blacks, Italians, and Poles in Pittsburgh, 1900–1960* (Urbana: University of Illinois Press, 1982), esp. chap. 3.

3. Bodnar et al., *Lives of Their Own*, pp. 61, 64.

4. Epstein, *Negro Migrant*, p. 32; Bodnar et al., *Lives of Their Own*, p. 132.

5. Abram Harris, "New Negro Worker in Pittsburgh" (MA thesis, Department of Economics, University of Pittsburgh, 1924); Ira Reid, "The Negro in Major Industries and Building Trades of Pittsburgh" (MA thesis, Department of Economics, University of Pittsburgh, 1925); Floyd C. Covington, "Occupational Choices in Relation to Economic Opportunities of Negro Youth in Pittsburgh" (MA thesis, Department of Economics, University of Pittsburgh, 1928).

6. Harris, "New Negro Worker," p. 45; Reid, "Negro in Major Industries," p. 10.

7. Reid, "Negro in Major Industries," p. 37.

8. Ibid., pp. 35–43.

9. Covington, "Occupational Choices," pp. 31, 52–54. Twenty-three of the mothers (5.3 percent) held jobs as secretaries (p. 60).

10. Ibid., p. 68.

11. Ira De A. Reid, *Social Conditions of the Negro in the Hill District of Pittsburgh* (Pittsburgh: General Committee on the Hill Survey, 1930); Howard Gould, "An Analysis of the Occupational Opportunities for Negroes in Allegheny County" (Ph.D. diss., University of Pittsburgh, 1934);

John N. Rathmell, "Status of Pittsburgh Negroes in Regard to Origin, Length of Residence, and Economic Aspects of Their Life" (MA thesis, Department of Sociology, University of Pittsburgh, 1935).

12. Rathmell, "Status of Pittsburgh Negroes," p. 30.

13. Gould, "Occupational Opportunities," pp. 25–41.

14. Alex Pittler, "The Hill District in Succession" (MA thesis, Department of Sociology, University of Pittsburgh, 1930); Delmer Seawright, "Effect of City Growth on the Homewood-Brushton District of Pittsburgh" (MA thesis, Department of Sociology, University of Pittsburgh, 1932); Alonzo Moron, "Distribution of the Negro Population in Pittsburgh, 1910–1930" (MA thesis, Department of Sociology, University of Pittsburgh, 1933). While comparisons are inexact, the concentration of Pittsburgh's blacks during the period appears to be less than in Chicago, New York, Cleveland, Buffalo, Milwaukee, Detroit, and Boston.

15. For a vivid description of the Hill district's transition from an elite residential area at mid-nineteenth century to an immigrant, black section at the turn of the twentieth century, see Pittler, "Hill District," pp. 17–60.

16. Moron, "Distribution," p. 33; Alonzo Moron and F. F. Stephen, "The Negro Population and Negro Families in Pittsburgh and Allegheny County," *Social Research Bulletin* (Pittsburgh) 1 (April 20, 1933):1–7.

17. Moron and Stephan, "Negro Population," p. 4.

18. Moron, "Distribution," pp. 33–34.

19. Reid, *Social Conditions.*

20. Margaret Byington, *Homestead: The Households of a Mill Town* (1910; rpt. Pittsburgh: University of Pittsburgh Press, 1974).

21. Moron, "Distribution," pp. 40–47, also describes the proliferation of black churches in Pittsburgh's Hill district. In addition, Moron presents useful data indicating that the churches were not only small and ineffective in community-building but that many were also transient. Within a twenty-year period, five of twenty-seven "organized" congregations ceased to exist, ten changed location, and a dozen new ones were formed.

22. Reid, *Social Conditions,* pp. 11–19. A reading of Epstein, "Negro Migrant," and Harris, "New Negro Worker," provides some indication that the conditions Reid et al. describe in the Hill in 1930 were persisting problems. A comparison of the data presented in the three works, however, makes it obvious that the overall condition of the Hill as a residential section has deteriorated.

23. See Bodnar et al., *Lives of Their Own,* esp. chaps. 3 and 7 for a comparison of community-building in the black and ethnic communities. These authors also argue that the development of a job procurement strategy which fostered individualism among black offspring resulted in somewhat loosened family ties. Blacks were more likely to live in nuclear families than their European immigrant counterparts. Thus, intergenerational immigrant families were more likely to concentrate in one geographic location whereas second-generation blacks were likely to live in a different neighborhood than their parents.

24. Ella Burns Meyers, "Some Italian Groups in Pittsburgh" (MS thesis, Department of Sociology, Carnegie Institute of Technology, 1920); Salvadore A. Migliore, "Half a Century of Italian Immigration into Pittsburgh and Allegheny County" (MA thesis, the Graduate School, University of Pittsburgh, 1928).

25. Migliore, "Half a Century," pp. 72–74.

26. Ibid., chap. 4. Other interesting, if somewhat thin, chapters include one describing attempts on the parts of the Catholic and Protestant churches to Americanize the Italian immigrant and one on the role of the Italian press in Pittsburgh.

27. This suggestion is admittedly speculative. Later studies, for example, indicate that some Eastern European Jews did attend college in the twenties and thirties, and certainly a substantial number of descendents of the "old immigration" must have populated the city's colleges and universities. Yet no studies of Germans, Irish, Jews, French, or other groups were done prior to World War II. The educational history of these groups remain fertile fields for exploration by future researchers.

28. Works of fiction provide interesting and surprisingly accurate accounts of the lives of numerous social and ethnic groups who lived in Pittsburgh. David P. Demarest has identified more than one hundred and fifty novels in which life in the Pittsburgh region plays a significant part. For an excellent sampling of such works, see David P. Demarest, *From These Hills, From These Valleys* (Pittsburgh: University of Pittsburgh Press, 1976).

29. Thomas Augustine, "The Negro Steelworkers in Pittsburgh and the Unions" (MA thesis, Department of Sociology, University of Pittsburgh, 1948); Ruth Simmons, "The Negro in Recent Pittsburgh Politics" (MA thesis, Department of Political Science, University of Pittsburgh, 1945); Robert Sullivan, "Some Economic Aspects of the Development of a Neighborhood Shopping District: Squirrel Hill" (MA thesis, Department of Economics, University of Pittsburgh, 1949).

30. Fred Wallhausser, "The Upper-Class Society of Sewickley Valley, 1830–1910" (seminar paper, Department of History, University of Pittsburgh, 1964); Stephen J. Schuchman, "The Elite at Sewickley Heights, 1900–1940" (seminar paper, Department of History, University of Pittsburgh, 1964); Renee Reitman, "The Elite Community in Shadyside, 1880 to 1920" (seminar paper, Department of History, University of Pittsburgh, 1964).

31. Schuchman, "Elite at Sewickley Heights," pp. 8, 14.

32. Edward Pessen, *Riches, Class, and Power Before the Civil War* (Lexington, Mass.: D. C. Heath, 1973); John N. Ingham, *The Iron Barons: A Social Analysis of an American Urban Elite, 1874–1965* (Westport, Conn.: Greenwood Press, 1978); Peter R. Decker, *Fortunes and Failures: White Collar Mobility in Nineteenth-Century San Francisco* (Cambridge, Mass.: Harvard University Press, 1978).

33. Peter Gottlieb, "Making Their Own Way: Southern Blacks' Migration to Pittsburgh, 1916–1930" (Ph.D. diss., University of Pittsburgh, 1977),

published under the same title by the University of Illinois Press, 1987; Bodnar et al., *Lives of Their Own;* Joseph Darden, "The Spatial Dynamics of Residential Segregation of Afro-Americans in Pittsburgh" (Ph.D. diss., University of Pittsburgh, 1972); Jacqueline Wolfe, "The Changing Spatial Patterns of Residence of the Negro in Pittsburgh, Pa." (MA thesis, Department of Geography, University of Pittsburgh, 1964). The latter two studies document the almost constant residential movement among blacks through the decade of the 1950s.

34. Gottlieb, "Making Their Own Way," p. 204.

35. Ibid., p. 260.

36. The argument is similar to one which Kenneth Kusmer offers for the disintegration of the black community in Cleveland. See *A Ghetto Takes Shape: Black Cleveland, 1880–1930* (Urbana: University of Illinois Press, 1986).

37. Gottlieb, "Making Their Own Way," p. 275.

38. Stephan Thernstrom, *Poverty and Progress: Social Mobility in a Nineteenth-Century City* (Cambridge, Mass.: Harvard University Press, 1964); Oscar Handlin, *Boston's Immigrants, 1790–1880* (New York: Atheneum, 1970); Stephan Thernstrom and Richard Sennett, *Nineteenth-Century Cities: Essays in the New Urban History* (New Haven, Conn.: Yale University Press, 1969). Works on immigrants and mobility became so numerous that more than one historian called for a cease-fire.

39. Carolyn S. Schumacher, "Education and Social Mobility: Class and Occupation of Nineteenth-Century High School Students" (seminar paper, Department of History, University of Pittsburgh, 1970); Nora Faires, "The Germans in Allegheny, 1850–1860" (seminar paper, Department of History, University of Pittsburgh, 1971); Nora Faires, "The Germans in Pittsburgh, 1860" (seminar paper, Department of History, University of Pittsburgh, 1972). The works by Schumacher and Faires were expanded into dissertations in 1977 and 1983 respectively.

40. Michael P. Weber, *Social Change in an Industrial Town: Patterns of Progress in Warren, Pennsylvania, from Civil War to World War I* (University Park: Pennsylvania State University Press, 1976).

41. Faires, "The Germans in Allegheny," pp. 36, 43–44; Paul Kleppner, *The Cross of Culture: A Social Analysis of Midwestern Politics, 1850–1900* (New York: The Free Press, 1970).

42. Faires, "The Germans in Allegheny," p. 36.

43. Victor Anthony Walsh, "Across the 'Big Wather': Irish Community Life in Pittsburgh and Allegheny City, 1850–1885" (Ph.D. diss., University of Pittsburgh, 1983), pp. 184–93.

44. Bruce Skud, "Ethnicity and Residence Within the Jewish Immigrant Community" (seminar paper, Department of History, University of Pittsburgh, 1975); Myrna Silverman, "Jewish Family and Kinship in Pittsburgh: An Exploration Into the Significance of Kinship, Ethnicity, and Social Class Mobility" (Ph.D. diss., University of Pittsburgh, 1973); Bodnar et al., *Lives of Their Own.*

45. John W. Larner, Jr., "A Community in Transition: Pittsburgh's South Side, 1880–1920" (seminar paper, Department of History, University of Pittsburgh, 1961); Joseph Johnston, "National Origins and Ethnic Groups of the People of Allegheny, Pennsylvania" (seminar paper, Department of History, University of Pittsburgh, n.d.); Noel Gray, "East Liberty—Portrait of a Changing Community" (seminar paper, Department of History, Carnegie Mellon University, n.d.); Jack L. Hiller, Jarrell McCracken, and Ted C. Soens, "Morningside: An Urban Village" (seminar paper, Department of History, Carnegie Mellon University, 1969); Josephine McIlvaine, "Twelve Blocks: A Study of One Segment of the South Side of Pittsburgh, 1880–1915" (seminar paper, Department of History, Carnegie Mellon University, 1972).

46. Studies by John Bennett, "Iron Workers in Woods Run and Johnstown: The Union Era, 1865–1895" (Ph.D. diss., University of Pittsburgh, 1977); Frank Serene, "Immigrant Steelworkers in the Monongahela Valley: Their Communities and the Development of a Labor Class Consciousness" (Ph.D. diss., University of Pittsburgh, 1979); and Michael Santos, "Skilled Ironworkers in the Era of Big Steel" (Doctor of Arts diss., Carnegie Mellon University, 1984) would seem to suggest that under certain circumstances, common work experiences may supplant ethnicity as a key element in the development of community.

▪ *Pittsburgh: How Typical?*

SAMUEL P. HAYS

T HE FOREGOING CHAPTERS provide state-of-the-art studies of a number of aspects of the history of one city—Pittsburgh. But historians will want to place this in a larger context: How representative was Pittsburgh of cities in the United States, and where does it fit in urban history as a whole? Is Pittsburgh typical or is it in some way distinctive? Amid the vast outpouring of research about American cities over the past three decades, some general notions about urban history ought to emerge. What insight does the research of almost three decades into the history of Pittsburgh contribute to American history in general as a field of study?

As a point of departure we could look upon America as a radical society in which over the course of several centuries fundamental and historically rapid changes have taken place in science and technology, the organization of production and distribution, levels of consumption, education, personal values, the role of women, and in the involvement of diverse cultures, religions, and races. These changes in the larger society were interwoven into a fabric of constant change. Cities were at the center of this process, bringing these various elements together in particular places, each place a vortex of rapid change—and response to change; and the network of cities gave rise to a new national society and culture that transformed America in a short span of time. Cities are, therefore, one of the keys to understanding processes of change within the entire nation.

The city especially pinpoints, in a precise setting, two elements of change: the centralization of institutions and the expanding range of human choice and aspiration. Both are key aspects of American history. A massive process of organizational centralization took place in which economic, social, and political institutions became increasingly larger in size and scale and generated an upward flow of initiative and autonomy from smaller to larger units. Human values changed persistently from more limited patterns of thought, customs, and action to more varied and diverse options. The range of acceptable choice as to what one could think, be, and do steadily expanded, enhancing patterns of aspiration for material living, quality of family life, and personal achievement.

Cities were the centers of transformation from decentralized to centralized institutions. Here were located the people who organized the larger economy, the larger institutions of government, the larger educational and religious establishments, and voluntary associations. They initiated both the centralization of institutions within the city and the more extensive centralizing tendencies among networks of cities within the nation as a whole. Here also were the opportunities for more varied forms of work and leisure, expanded choices as to what people could legitimately think and do in their lives. The availability of such choices attracted people from smaller towns and the rural countryside and challenged those with more limited perspectives and patterns of life to more varied thought and action. In cities, therefore, one can identify in a relatively limited context the fundamental processes of social change occurring in the wider society.

The most dramatic response to these massive changes in institutions and values came in the nation's less settled areas. Here, smaller forms of organization in the economy, government, and society persisted to voice opposition against larger; here also was prolonged institutional and personal support for older patterns of culture which emphasized the importance of maintaining more limited ways of thought and action. But such influences were not absent in the city. There too, older institutional forms long persisted, and migrants brought older patterns of culture and values which they long defended. The ability to withstand change was far more limited in the city than in rural areas. Yet the drama of the tension between change and response took place within the city as well as between the city and the countryside, and urban history provides an especially useful context in which to observe it.

How are these patterns of change and response manifested in the history of Pittsburgh? I suggest that in both Pittsburgh displays distinctive characteristics. With respect to organizational tendencies the city was a more extreme case. Based on the high degree of centralization in iron and steel manufacturing, the entire city displayed a similarly persistent tendency toward centralization in government and institutions. This is not to say that Pittsburgh is an entirely unique case, but that amid institutional centralization it provides an example of a greater rather than a lesser degree of that process.

With respect to changes in cultural values, however, the opposite was the case: the city reflects a more extreme example of cultural conservatism. Value changes in such matters as the role of women, the style of religion, or the level of education came more slowly in Pittsburgh than elsewhere. These conservative tendencies can be associated most readily with the ethnic cultures from Europe that brought into the city a strong sense of the worth of traditional patterns of family and personal values which had significant religious elements and marked effects on the role of women and involvement in education. Among the various ethnic cultures there were variations in this role, with some more involved in value change and others in the retention of older values. But on the whole, the extensive role of these cultures in Pittsburgh made it more conservative with respect to similar changes in values throughout the nation.

This is not to say that other cities did not display similar contradictory patterns. Attempts to generalize are vulnerable simply because of the limited systematic analyses which enable one to distinguish one city from another. But it is reasonable to argue that this dual and contradictory role of Pittsburgh is one of the major implications of these essays. In this chapter I explore this argument more fully; I refrain from summarizing the individual chapters, but, with appropriate reference to other urban studies and general theory, develop their larger meaning for historical patterns of urban structure and change.

The Process of Centralization

The chapters by Paul Kleppner, Roy Lubove, Edward K. Muller, and Joel A. Tarr outline most fully the processes of centralization in Pittsburgh. Over the two hundred years of the city's history a persistent but long-term change took place in the organization of its economy, society, and government. For many decades the major focus of the city's organization was the community. But in the last third of the nineteenth century changes were well under way toward the integration of many aspects of urban life on a larger scale. As these human networks grew in extent, initiative in the city's affairs shifted from the community to the city at large and to the entire region.

During the nineteenth century new places of work were located in areas lying near to but beyond old ones, and around those places new residential areas were formed. Here were schools, churches and shops, neighborhood doctors, funeral directors, saloons, and groceries which served nearby residents. Most daily life took place in this smaller context, as one walked from home to work, shopped among neighbors, and talked with friends across the street from one door stoop to another. Workers went on foot from the mill to home and stopped at the saloon on the way. Each community developed its own network of friendships and leaders who were known and whose opinions were respected. New communities that arose on the city's periphery took their place alongside older ones, each with its own locally controlled schools and its ward representatives in the city councils. Many of these communities had distinctive ethnic and racial characteristics, Irish or German, black, Italian, Polish, or Jewish.

As the city grew in size and geographical extent, however, new types of human contacts developed which crossed community lines. Economic relationships of buying and selling brought retailers together with other retailers elsewhere; people from one community made purchases in another; a small but significant number of young people in one community married others from another. The baseball park on the North Side, then the Kennywood amusement park and other recreational opportunities, attracted people from many places. Doctors, schoolteachers, real estate developers, and bankers in one section of the city began to form associations with those in another and to think not in terms of community alone but of specialized parts of the larger city.[1]

These new networks and connections began to weave an urbanwide fabric that constituted a level of social and economic organization beyond the community and led to ideas and action to integrate the various parts of the city rather than express the interests of any one section. Newspapers and magazines wrote in terms of the city as a whole, and these newer groups talked of the civic life of the entire area. The city became a symbol in which growth in population and area was an overarching sign of vitality and strength and a source of pride.

The physical representation of these innovations was the city center. Here new buildings arose which served as places from which the larger city and its region were organized. Here were the central bankers who financed large-scale projects in contrast with the small community banks that catered to a neighborhood. Here were the offices of doctors who specialized in surgery or eye disease in contrast with the general family physician located in the neighborhood. Here were the headquarters of the larger builders and contractors, the insurance firms, the real estate agents who dealt in larger property in contrast with community home-builders. As the new tall buildings were constructed in the late nineteenth and early twentieth centuries in Pittsburgh's center, they provided the core of the organizational drive within the city.[2]

One could think of these changes as ways in which human relationships became integrated in an ever more complex economy. Business managers once had been located in their factories, but now they chose to be near services which they required, such as banking and insurance. At first economic contacts were maintained through messages conveyed by messenger boys, but in the late nineteenth century the telegraph and telephone came to shape the larger network. For community connections, personal contacts were sufficient, but larger networks required more rapid and flexible means of communication.

These processes of centralization, which became especially visible in the 1890s, continued as the city established its larger network of economic and social relationships in the county outside the city, a topic which Edward K. Muller has outlined. A new wave of decentralization of work and home led to a new wave of centralization to tie it together. Factories spread up and down the rivers, giving rise to new communities, but at the same time to centralizing functions in the center city. Communities beyond Mount Washington and in the East Hills led to new commercial centers and often new office buildings; central city firms provided financial and real estate services for them. While the old decentralization of the nineteenth century had arisen from many smaller and relatively autonomous centers, the new of the twentieth took place within an umbrella of centralized economic life.

New citywide and regionwide networks were not confined to economic affairs. There was the organization of charity and welfare into more centralized patterns, outlined in detail by Roy Lubove, as varied organizations, some local and others emphasizing services to a particular ethnocultural group, became joined together into citywide activities.[3] The United Jewish Federation and the United Fund brought together many organizations throughout the city. Charitable funding became highly centralized so that it was often difficult for smaller and newer activities to persist. In Pittsburgh and Cleveland, the Community Chest, as Judith Trolander has observed, exercised more direct control over settlement house activities than in Chicago and New York, and greatly restricted their functions.[4]

And so also it was with art and culture, which have received far less attention from historians, as the smaller scale activities of an earlier day were superseded by citywide affairs. Varied ethnic cultures were brought together after World War II into an annual folk festival, and the work of local artists was organized into an annual art festival. Those who organized these affairs often exercised considerable influence over the expression of art and culture because of the difficulties experienced by smaller groups in succeeding on their own. Theaters of earlier years now were replaced by new centers of high culture as the citywide symphony orchestra and ballet provided a central cultural focus.

The mass culture that developed with radio and television under the direction of a few dominant stations supplemented the mass communications represented by the city's newpapers. Public radio received community financial support and brought together in one listening audience groups in diverse sections of the city who had a common cultural interest. In similar fashion specialized interests in entertainment, news, or religion, gave rise to specialized radio and television programs that reached people throughout the city to create a set of linkages far beyond community.

The centralization of organization in Pittsburgh was most visible in government (see the chapter by Paul Kleppner in this volume). In the nineteenth century the city council was based upon ward representation. Each ward elected members to both the Select and Common Councils. Emphasizing the political autonomy of each ward, this method of representation was characteristic of the ward focus of many aspects of government: school boards, constables as law officers, justices of the peace, overseers of the poor, and the regulation of commercial weights and measures.

Elected council members often mirrored the characteristics of ward voters in ethnicity and religion, occupation and income. Hence they varied as the composition of the community varied, German or Irish, skilled workers, professional or managerial. Each community developed its own type of leader—lawyers and manufacturers, for example, in the more affluent wards; grocers, saloonkeepers, and funeral directors in the working-class wards. They were people whom the community knew and for whom they spoke. The issues that the council debated were often of interest to the local ward: danger to citizens from the new railroads that ran through neighborhood streets, introduction of gas lights on the street corners, the freedom of citizens to frequent saloons or engage in recreation or amusements on Sunday.

By the late nineteenth century a new tendency emerged toward a more citywide focus and more centralized patterns of decision making, divorced from the elected council. Problems which required day-to-day attention were delegated to special administrative agencies with considerable autonomy and responsibility. As early as the 1837 depression, for example, the ward overseers of the poor were replaced with a citywide board of charity. In later years citywide fire and police departments were formed to supersede similar ward functions. A public health department was established to administer programs for the city as a whole, and commissioners were appointed to organize streets and traffic on a citywide rather than a local basis.

Changes such as these gradually led to the growth of an administrative system divorced from the legislative activities of the council and with a citywide rather than a ward perspective. By the early twentieth century, a drive arose from the industrial, professional, and administrative leaders of the city to modify the government more fundamentally. One proposed innovation was to give the mayor more power, especially over the development and presentation of the budget. Another was to reduce the size of the council to nine and to elect all members from the city at large rather than from the wards, a change made in 1911. And still a third was to shift school administration from the wards to a central school board, a change evolving gradually over the years, culminating in 1913 when the state legislature instituted a central school board appointed by the judges of the Court of Common Pleas.

This growing centralization of power and authority in urban governance took place in most of the nation's cities. But in Pittsburgh it seemed to go further. One feature in such a comparison is the degree of centralization in political representation. While some large cities, such as Chicago, continued the old ward system of selecting city council members, and others, such as Buffalo, combined ward with citywide representation, Pittsburgh was one of the few in which citywide elections fully replaced the ward system.

One may also observe this expression of institutional centralization in the success in Pittsburgh of urban redevelopment in the mid-twentieth century. Beginning in the 1940s Pittsburgh undertook a massive physical reconstruction of selected parts of the city, most notably in its older, central section, under the leadership of the Allegheny Conference for Community Development, a body composed of the leaders of the city's largest and most powerful corporations and the elected officials of the city and the county. This was carried out under the direction of a redevelopment authority, independent of the elected bodies in city and county, with considerable autonomy and the power of eminent domain as well as subsidies to those who would carry out the new construction. This program was widely considered to be a pioneering and unusually successful type of urban reconstruction compared with that in most of the nation's cities; many attributed this to a unique degree of cooperation between the city's corporate elite and government officials.

Two decentralizing tendencies appeared in the twentieth century to counteract some elements of centralization (see the chapter above by Edward K. Muller.) One arose from the vigorous resistance of outlying communities to consolidation with the city in a larger, metropolitan form of government. For many years in the nineteenth century new communities had been absorbed into the city's governing system through consolidation. But as time passed, communities on the city's periphery were less anxious to join in and began to resist consolidation. Pittsburgh's leaders, however, continued their drive for "bigger is better" and in the 1920s proposed a larger metropolitan government. It was rejected, and subsequent proposals for metropolitan government were shot down promptly by the surrounding areas.

Outlying boroughs and townships derived their authority and autonomy from state government which provided them political independence from the city and enabled them to resist Pittsburgh's expansive tendencies. Hence, despite the continual charge that a multitude of autonomous governments in the region was highly inefficient, metropolitan consolidation did not emerge as a serious issue. What did evolve, however, was a piecemeal shift in government from a city to a countywide focus as larger scale administration was established for public health, sewage disposal, water supply, mass transit, redevelopment, and airport management. Just as was the case in earlier years in the smaller city, these new administrative entities of centralized jurisdiction and relatively independent authority shaped the longer term pattern of centralization of governance as a major theme in the city's and the region's political history.

After World War II a move arose to obtain a larger role in public affairs for the city's urban communities. This was fostered by the rise of political consciousness in the black community in the 1960s, particularly as that consciousness was shaped by the destruction of black residential and commercial areas amid redevelopment. Older ideas about local community autonomy received renewed attention. This led to a new policy in which school board members were elected from the various sections of the city instead of being appointed. Similar demands for more community autonomy in City Council fell on deaf ears until by 1986 the council contained no black at all and voters early in 1987 decided to adopt a ward system of election.

The Enhancement of Human Choice

Changes in the scale of social, economic, and political organization were matched by changes in the experience, values, and aspirations of the city's residents. Cities were places of more varied and extended human choice. To those living in rural areas, cities provided greater economic opportunities and sustained more varied options for thought and behavior. People in cities came to want new things, to create a future different from their past, to realize new human aspirations. The drive for change which came from this personal experience was as significant for the history of cities as was participation in the centralization of its institutions.

In addition to fostering variety and differentiation, the city also provided an opportunity for the perpetuation of older values that were widely described as traditional. These were often brought to the city by migrants from rural areas in the United States and abroad where traditional values were more fully rooted, and were often sustained by the first-generation urban resident. Yet the city provided a context in which these older values were eroded, and with new generations came new patterns of life, of what was acceptable to think, be, and do. Older generations did not accept lightly this challenge to their values, and the city, therefore, displayed a persistent blending of old and new, often in sharp conflict but equally often creating impacts which were cushioned by the presence of the old.

The most convenient way of thinking about these inner changes is to focus on the enhancement of human choice. At one time the range of things that one could acceptably think, be, and do was limited; over time that range and variety widened. Modes of behavior of men, women, and children, formerly rejected, now were tolerated and often prized; one thought about appropriate vocations not in terms of those of one's parents but in terms of one's own individual potential and preference. Increases in the range of individual choice can be observed especially in three aspects of the city's history: education, religion, and women. The chapters by Linda K. Pritchard and Maurine Weiner Greenwald highlight some of the contributions in the Pittsburgh research to these features of urban life.

The direction of educational change was toward schooling for more young people for a greater number of years. Over time, elementary schools were supplemented by secondary schools; the first Pittsburgh public high school, Central High School, was established in 1855 to be followed a half century later by high schools scattered throughout the city. Colleges emerged in the twentieth century from their earlier limited beginnings to expand their clientele, often providing education in evening schools as well as full-time instruction. After World War II, junior colleges were added to four-year colleges and universities.

Students attended high school not to prepare for college but to acquire the skills for clerical and other white-collar work. Pittsburgh's blue-collar image has obscured the fact that the evolution of managerial organization provided many white-collar jobs—by 1910, 9.8 percent of the city's jobs—which enabled some to advance in income and occupation. Even more significant was the social context of the high school, a single, central school to which students were drawn from elementary schools throughout the city. Here those from one community met those from another. New ideas and customs were encountered, new friendships made, perhaps new eligible marriage partners found. The high school was a cosmopolitanizing experience.

College carried such tendencies further. Here one met a still wider range of people from varied ethnic and religious backgrounds, from various communities within the city and beyond. Even more important, one encountered diverse subjects and ideas; knowledge was not restricted by the inherited wisdom of the past. Beliefs held to be appropriate according to one's religious traditions or the cherished ideas of one's family or community could be questioned simply because of the tolerance for a wider spectrum of ideas. The range of potential occupations or potential marriage partners was far greater than one had known in secondary school. One could think differently and be a different kind of person.

Changes in religion provided similar occasions for enhancing the range of belief and behavior (see the chapter by Linda K. Pritchard). Many of the initial settlers of Western Pennsylvania were Scots-Irish. Their Presbyterian religion required a strict form of belief and behavior which did not include a varied range of practices. For many years Scots-Irish culture played an influential role in the life of the city and the region. When German and Irish Catholics came to the city, and still later migrants from Southern and Eastern Europe, they provoked a sharp reaction from those who maintained the earlier culture. Customs and belief separated the old from the new. Only slowly and over the years did the older culture mellow sufficiently to allow a legitimate place for newer customs and values.

Many of the city's newcomers brought their own traditional religious beliefs and practices which they sought to sustain. But even among them, change toward greater choice and option came persistently. The younger generation of German Lutherans, for example, wished to have services in English and to join the Masonic Lodge, both practices which required major changes in traditional custom. Among all the groups, the idea of evolution presented a major challenge, as it advocated a more flexible view of the way human life had come to the earth. And especially among Protestants, new ideas emerged which questioned the authenticity of a literal reading of the Bible; biblical accounts should be tested against other historical evidence.

The expansion of thought and practice within the church came first to Protestant groups in the years prior to World War II. For Catholics a similar process of liberalizing came after that war. The English mass was substituted for the Latin, but not without controversy. The priest turned to speak to the congregation during the service rather than facing the altar. In Pittsburgh as well as elsewhere, women demanded a larger role within the church. All this meant simply that while at one time religion had played a major role in limiting what one could think, be, and do, that influence now was diminishing as appropriate standards of behavior came from wider and more diverse sources.

In these changes in both education and religion the shifting role of women played a distinctive part (see the chapter by Maurine Weiner Greenwald). Earlier in the nineteenth century it was thought that the only appropriate role for women was to raise a family and manage the home. While it had been acceptable for women to work outside the home before they married, and later if they were widowed, it was not appropriate for married women to do so. Pittsburgh was distinctive in the small percentage of married women who worked. At the same time, women were subordinated to men in almost every religion; while some changes had taken place, for example the elimination of separate sections in the congregation for men and women, a woman could not become a rabbi, a minister, or a priest. Even with change, the more traditional branches of each religion continued to closely circumscribe the role of women in church and synagogue.

Gradually, however, women moved beyond these narrow limits to increase steadily the range of their choices. At one time most married within their class, ethnic group, or religion. Among Germans in the nineteenth century, 50 percent of the men and 65 percent of the women found a spouse from the same German province as their own.[5] But over the years choices widened. Among young women from blue-collar backgrounds attending high school in the late nineteenth and early twentieth century, 65 percent married husbands with white-collar occupations.[6] Over the years women also became more involved in life outside the home. Upper-class women formed literary clubs which in the late nineteenth century became active in such public affairs as the conditions of work, the life of women and children, and the public schools. Women shaped the settlement house movement, such as the Columbian Settlement in the Jewish community in the Hill district, and played an important role in the development of the new field of social work (see the chapters by Roy Lubove and Maurine Weiner Greenwald).

Changes in the lives and roles of women could be charted through their changing role in the work force. In the nineteenth century they became schoolteachers and contributed to the professionalization of teaching.[7] They began to participate more fully in high-school education and to take up work in the city's growing clerical force. Their role, however, was confined by male supervisors as well as by male workers to the more routine and lower paying jobs.[8] After World War II, the rising number of white-collar jobs and the equally rising level of education enabled women to enter work outside the home in far greater numbers and gradually to take on positions of greater skill, authority, and responsibility.

Education was central in the changing role of women; yet only a few studies chart these changes for Pittsburgh. Some stress ethnic participation; in the mid-nineteenth century, for example, young Irish and German women attended Central High School in significant numbers.[9] Among the post-1890 immigrants, young Jewish women participated in college far more extensively than did those of Slavic origins. These changes took place slowly, and it was not until after World War II that the drive for higher education for women moved into full tide.

Education fostered changes in the values of women. Women who went to college and those who did not displayed marked differences in such diverse matters as work outside the home while married and raising a family, breast feeding rather than bottle feeding of babies, cigarette smoking, professional careers, and more liberal sexual behavior and abortion. No wonder that for those who sought to hold on to the older ways college education was subversive. Here were opportunities for greater choices in what one could think, be, and do that were vastly different from those of one's parents and grandparents.

Ethnic life in Pittsburgh was a major focus of this drama of the enhancement of human choice. Groups from abroad transferred directly to the new country traditions expressed through family and community patterns of life. Some have argued that the strangeness of the new land, in fact, reinforced the desire of newcomers to maintain traditional ways of living as a source of meaning in the midst of the heightened new experience of variety and change.[10] Patterns of life among ethnic groups in Pittsburgh reinforced their desire to sustain tradition in such circumstances; details of this are provided in the chapters by Nora Faires and Michael P. Weber.

The larger cultural climate of the city, which fostered greater options in choice, had an impact on each ethnic group. Some individuals became involved in the new cultural climate and chose to abandon their ethnic roots by choosing different marriage partners, accepting religious practices at variance with tradition, or moving out of ethnic neighborhoods into more mixed communities. Often the key choice was to become involved in education, first to go beyond elementary school to high school, and later beyond that to college. But within each group, as well, many chose not to make such choices and preferred to work out their lives within the more traditional culture.

The ethnic aspect of these processes of value change provides some useful comparisons. Among Jews, for example, values seemed to change more rapidly. Especially with the arrival of Jews from Eastern Europe, programs arose to foster education both among adults and children. The drive for evening high schools had a distinctive Jewish leadership and clientele. Moreover, this process included both men and women. Jewish women developed many voluntary organizations to improve the civic life of the community and, through that, the personal lives of more recent Jewish migrants.[11]

The Slavic community, on the other hand—and Polish people in particular—retained their traditions longer than others. There was a markedly lower interest among Slavic families for their children to achieve higher levels of education; it was not needed for work in the mills which many of their young men preferred. Italian and Polish women, as Maurine Weiner Greenwald has noted above, were far less inclined to pursue employment outside the home, in contrast with the way in which young Irish women sought domestic service in the nineteenth century. These ethnic distinctions in the changing choices of women about family size, education, work, and religion provide a special opportunity to observe the impact of the more general climate of enhanced option and choice on traditional patterns of life.

For some groups the patterns of change were more mixed: higher levels of education were combined with enhanced emphasis on traditional values. A seminar paper comparing Serbs and Greeks in Aliquippa emphasizes differences between them on this score. The study is especially valuable for its focus on three generations of Serbs and Greeks from original migrants in the early twentieth century through their grandchildren in the 1960s. Both were of the Orthodox faith, but while Serbs were inclined to take a relaxed view toward marriage with other ethnic and religious groups they were also less involved in education. Greeks were more strict in their attempt to maintain traditions and much more highly involved in schooling.[12]

Among blacks opportunities were far more restricted. The role of education in enhancing their aspirations was clear enough. In his chapter in this volume, Michael P. Weber cites the study in 1928 of 434 black high-school students in the city in which two-thirds aspired to white-collar and semiprofessional occupations. Such evidence reflects the role of high school in the changing values of younger people. But for blacks, aspirations were frustrated by segregation in jobs and communities. This gap between heightened levels of aspiration and restricted opportunity lay behind the tension within the nation's urban black communities.

Several factors about Pittsburgh have been noted which might well reinforce the persistence of tradition through the persistence of community. One was the city's topography of hills and valleys which tends to separate physically one community from another. Another was the formation of separate new iron and steel towns up and down the rivers, each one becoming a distinct community, each separated from the larger metropolitan area, a feature reinforced by the domination of those towns by steel management. The greatest possibilities for observing these differences in changes in values and ideas about human aspiration, however, are in the context of family, and this, as noted in the introduction, is one of the unexplored aspects of the city's history. A model to follow might well be the work of John Bodnar which focuses especially on the immigrant family.[13]

Inequality and Class

In sharp contrast to the changes brought by centralization and the expansion of human choice are the continuous and persistent social patterns arising from inequality and class. While there were some variations in the outlines of American inequality over the last two centuries, even more striking was its relative stability. The shares of total property owned and income earned by those of the upper, middle, and lower sectors of the social order were quite similar throughout these years. While the average level of family wealth rose, and while individuals gained or lost in relative economic position, the overall pattern of inequality remained with only limited change from one decade to the next.

Just as the city constitutes a sharpened locale of the broad social processes of centralization and enhancement of personal choice, so it also serves as a focus for the persistence of inequality and its self-conscious expression. Inequality and its role in human perception and understanding were pervasive in rural areas, but they were given a special visibility in the city. Here, living in close proximity, people could readily observe differences in economic livelihood; inequality in housing was particularly visible. In the city one could easily form notions of the workings of economic inequality and its close relationship to social and political inequality. Class perception as well as inequality of condition became a sharply delineated aspect of urban life.

Despite the significance of these patterns, a surprisingly limited amount of historical research systematically analyzes inequality and class in the American city to indicate the ways in which they persisted or changed. Little investigation has been undertaken of the middle class; a bit more, but limited, work has been conducted for the upper class; and in labor history major events such as strikes and violence or organized social movements such as labor unions or political parties have often overshadowed systematic class description.

While this is as true of the history of Pittsburgh as for other cities, there are some notable exceptions. No focused work has been conducted on the middle class. Several dissertations and books have outlined important aspects of working-class life such as culture, households, education, and community. And a number of seminar papers have dealt with the city's upper class and its communities. The chapters in this volume by Maurine Weiner Greenwald, John N. Ingham, and Richard Oestreicher deal with some of these issues. Together with portions of the seminar papers and dissertations, they provide some clues about inequality in Pittsburgh's history.

The city's upper class, John N. Ingham argues, was more preoccupied with its role in the city and less involved with national upper-class institutions than was the case elsewhere. His observation is based on comparisons of six cities. This distinctiveness might well have been derived from the peculiar Scots-Irish heritage of the city's elite, as Ingham suggests, which was evident in the city's institutions until World War II. Or it might have been derived from the special role of iron and steel that sustained much of the economic basis of the city's upper class.

As Ingham also observes, most historians have thought about class consciousness largely in terms of the working class. But one can explore the subject also with respect to upper and middle classes. The Pittsburgh upper class, so the research indicates, devoted considerable time and energy to maintaining the stability of its institutions, its family ties through marriage, and its group cohesiveness through exclusive clubs and residential areas. It does not appear that either the middle or working class devoted similar attention to the maintenance of class continuity.

While it is difficult at this juncture to compare upper-class consciousness in Pittsburgh with other cities, what does seem distinctive is the very high degree of fear the elite displayed toward the city's urban, ethnic, working class. Centralization of the city's government, as described earlier, took place partly because urban leaders felt threatened by the city's immigrant workers. In the late nineteenth and early twentieth centuries reform groups representing the upper levels of the city's social order demanded that political power and authority be shifted toward them. They chose to bring about fundamental changes in municipal government not by public referendum but by action of the state legislature imposed upon the city. Here they enjoyed allies from rural areas who were as fearful as they of the urban electorate. The preference for centralized governing institutions was closely related to their own sense of class consciousness and class fear.

A significant aspect of that reaction was the large role which women from the upper levels of Pittsburgh's social order played in urban reform movements around the turn of the century. As in many cities in the nation, after 1890 women became active in public affairs, even before they had the franchise, and were crucial to many social and political reforms. Pittsburgh was no exception; the Twentieth Century Club, an upper-class women's organization, and the Civic Club, in which many women were leaders, spearheaded reform drives. Their notions of desirable elite leadership in public affairs constituted one of the major steps in the changing roles of women in Pittsburgh.

The most intriguing aspect of the evolution of class among the workers of Pittsburgh has been the interconnection between ethnoreligious culture and work.[14] Ethnic and religious ties often separated one group of workers from another. Cultural heterogeneity among Pittsburgh's workers led to a far weaker sense of common interest than was displayed by the Scots-Irish elite.

This was not merely a matter of the conflicting pulls of ethnicity and religion on the one hand and working-class loyalties on the other. Even more significant was the way in which traditional values resisted the pull of the wider society. Workers seemed to be held firmly by a sense of family and community, bound together by tradition and by a common task of sustaining their economic livelihood. A world of larger ideas about class, which required them to think in terms of affairs beyond their immediate experience, came into their lives only slowly.[15]

Involvement of steelworkers in union organization reflected these tendencies. That involvement was retarded prior to the 1930s by a fear that it would undermine family livelihood, by the force of more limited ethnic and religious loyalties, and by the strangeness of abstract ideas about working-class solidarity. By the 1930s, however, a new generation of steelworkers was emerging, with ideas different from their parents, who found working-class action more attractive. They could visualize more clearly the common interest of workers across divided lines of traditional cultures. A shift was apparent when ethnic fraternal societies became more amenable to the use of their halls for union organizing.

By the 1930s a greater sense of class consciousness had arisen that was shaped heavily by inherited cultural values. Appropriate relationships between employer and employee were thought about as much in terms of fairness and justice as in terms of common class interest. And a new working-class loyalty was shaped also by ethnocultural loyalty, by a sense of solidarity among immigrants who faced a dominant Pittsburgh middle and upper class that regarded them as ethnic and religious outsiders.

The objectives of workers in Pittsburgh were shaped also by the emerging desire to acquire the material benefits of expanding consumer goods and services. When workers secured political power in the steel towns around Pittsburgh in the 1930s they used it to carry out a program based not on class theory but on improvement in daily life: the elimination of the hazards of railroad grade crossings or more effective garbage collection.[16]

The "sense of class" which came to Pittsburgh workers in the 1930s emerged at a time when many strata of society were seeking to improve standards of living. The growth of a mass consumption society, with which human betterment was associated and which can be first observed in the 1920s, came to shape the wants and desires of workers as well as middle-class Americans. The class consciousness and class action of workers in Pittsburgh became a part of the striving for improvement in personal and family standards of living and was somewhat divorced from the more abstract notions of class consciousness.[17]

From several quarters, therefore, the power of traditional values helped to shape class consciousness in Pittsburgh both from the upper and lower sectors of the social order. And after World War II the decline of those ethnocultural commitments, as all sectors of society strived to participate in the benefits of a growing economy, tended to mute the older antagonisms. Among workers especially, younger generations reduced their loyalty to inherited cultural traditions and their close association with worker solidarity to seek their future as individuals in a society of expanding benefits.

Amid these changes inequality did not abate; it persisted in much the same degree as before. Despite rising levels of real income for the society as a whole and for individuals within it, despite the upward mobility of many, the overall pattern of inequality remained. It became transformed from unequal rewards for work in a manufacturing economy to unequal rewards for work in a white-collar society marked as much by inequality in levels of education as by inequality in occupation and housing. The class patterns of contemporary Pittsburgh, as measured for example by the inequality in real estate values from one municipality to another, are not less marked than was the case a century ago. But the intensity of class consciousness as displayed in public affairs that was so marked during the 1880s and again in the 1930s has declined.

We can say even less about the Pittsburgh middle class than about either the upper or working class. It may well be that the lack of such studies stems from the difficulties in defining middle-class life and institutions; it also arises from the myth that there was no significant middle class in this iron and steel city. Such a myth is highly suspect as an accurate description of the city's class development. The iron and steel managerial order created a considerable number of white-collar jobs, more than the average for the nation as a whole, and the burgeoning Pittsburgh East End of the late nineteenth and early twentieth centuries contained a middle-class as much as an upper-class community.

There are only a few studies of vertical mobility in Pittsburgh, one seminar paper for a part of Allegheny City in the nineteenth century, and a significant comparison of mobility among black, Italian, and Polish workers.[18] These stress mobility into higher status occupations. But education is equally an aspect of mobility and for this detailed work on individuals in education by Carolyn Shumacher and Ilene DeVault are particularly useful.[19] Several community studies identify important middle-class as well as working-class and upper-class residential areas. But there is little description of middle-class institutions, such as those in John Gilkeson's work on Providence, Rhode Island, ranging from voluntary associations to tax-supported civic institutions and political involvement.[20]

The strong ethnic component in Pittsburgh's population provides an especially useful arena in which to analyze a middle class in the making. Victor Walsh's study of the Irish makes clear a distinction between the middle-class and the working-class Irish communities in the city, and Nora Faires's study of the Germans in the nineteenth century provides a useful analysis of the way in which Germans, as they moved upward in occupation and income, retained for some time, rather than lost quickly, their cultural heritage.[21] Such changes as these provide a peculiar opportunity to investigate the differential way in which strong cultural traditions among ethnic workers became absorbed in a common culture of middle-class consumers as one generation succeeded another.[22]

It may not be too farfetched to argue that the sense of class among the middle class was stronger than for either the upper class or the working class. The middle class did not manifest the tenacity of kinship relationships as a factor in class continuity that the upper class displayed, or a joint sense of economic dependence as was the case with blue-collar wage earners. But it did express common values of individual and family aspiration and reflected decisions to take advantage of opportunities to improve one's livelihood which created a more subtle but still powerful sense of a common role in society and politics.

In this most recent process—in which a new middle class arose with a focus on individual and family improvement, a higher standard of living, and the "good life"—Pittsburgh is becoming more like other cities. The more distinctive elements of Pittsburgh in the earlier years, that stemmed from its advanced role in the centralization of modern management and the strong traditional focus of its ethnic culture, have begun to give way to a prevailing urban middle-class culture, strongly focused on rising standards of living and quality of life, and often content to accept the centralization of power and authority on the part of the city's dominant classes so long as the expected benefits are forthcoming.

These essays on Pittsburgh help to highlight broad social processes which identify a future agenda for urban history. They suggest the way that urban history can move beyond description to concepts of change and stability. In the present state of urban history it is difficult to deal comparatively with such problems as the evolution of organization and human choice, persistence and change in inequality, and class consciousness among the various strata of urban society. But these essays help to define the way in which thought about urban history in the future might well proceed.

NOTES

1. The dissertation by James C. Holmberg, "The Industrializing Community: Pittsburgh, 1850–1880" (University of Pittsburgh, 1981), is especially useful in outlining these changes.

2. Howard V. Storch, Jr., "Changing Functions of the Center-City: Pitts-

burgh, 1850–1912" (seminar paper, Department of History, University of Pittsburgh, 1966).

3. Roy Lubove, *Twentieth-Century Pittsburgh: Government, Business, and Environmental Change* (New York: John Wiley, 1969).

4. Judith Trolander, *Settlement Houses and the Great Depression* (Detroit: Wayne State University Press, 1975).

5. Nora Faires, "The Germans in Pittsburgh in 1860" (seminar paper, Department of History, University of Pittsburgh, 1972). She reports that 77.4 percent of all German family heads had spouses from the same province.

6. See Ileen DeVault, "Sons and Daughters of Labor: Class and Clerical Work in Pittsburgh, 1870s–1910s" (Ph.D. diss., Yale University, 1985), pp. 329–31, for marriage data.

7. Marguerite Renner, "Who Will Teach? Changing Job Opportunity and Roles for Women in the Evolution of the Pittsburgh Public Schools, 1830–1900" (Ph.D. diss., University of Pittsburgh, 1981).

8. Ileen DeVault describes the different male and female role patterns which the commercial department of the Pittsburgh High School stressed in its instruction for women and men. See "The Students of the Pittsburgh High school Commercial Department: 'From Inclination or Necessity' " (seminar paper, Department of History, University of Pittsburgh, 1979).

9. Carolyn Sutcher Schumacher, "School Attendance in Nineteenth-Century Pittsburgh: Wealth, Ethnicity, and Occupational Mobility of School-Age Children, 1855–1865" (Ph.D. diss., University of Pittsburgh, 1977), pp. 28–38, 145–49.

10. Oscar Handlin, *The Uprooted: The Epic Story of the Great Migration That Made the American People,* 2d ed. (Boston: Little Brown, 1973).

11. Ida Selavan, "Immigrant Education in Pittsburgh: An Analysis of Some Roll Call Books from Grant School 'Foreign' Classes" (seminar paper, Department of History, University of Pittsburgh, n.d.).

12. Marcia Chamovitz, "The Persistence of Ethnic Identity in Two Nationality Groups in a Steel Mill Community" (seminar paper, Department of History, University of Pittsburgh, 1976).

13. John Bodnar, *Immigration and Industrialization: Ethnicity in an American Mill Town, 1870–1940* (Pittsburgh: University of Pittsburgh Press, 1977); *The Transplanted: A History of Immigrants in Urban America* (Bloomington: Indiana University Press, 1985).

14. See Richard Oestreicher's chapter in this volume; and Frank Serene, "Immigrant Steelworkers in the Monongahela Valley: Their Communities and the Development of a Labor Class Consciousness" (Ph.D. diss., University of Pittsburgh, 1979).

15. John Bodnar, *Workers' World: Kinship, Community, and Protest in an Industrial Society, 1900–1940* (Baltimore, Md.: The Johns Hopkins University Press, 1982).

16. I am indebted for these observations to Joel Sabadaz who is undertaking a study of the economic and political role of workers in Pittsburgh iron and steel communities during the 1930s.

17. Cf. Lizabeth Cohen, "Learning to Live in the Welfare State; Industrial Workers in Chicago Between the Wars, 1919–1939" (Ph.D. diss., University of California at Berkeley, 1986).

18. John Bodnar, Roger D. Simon, and Michael P. Weber, *Lives of Their Own: Blacks, Italians, and Poles in Pittsburgh, 1910–1960* (Urbana: University of Illinois Press, 1982); Fred Siegel, "Selective Out-Migration of the Fourth Ward of Allegheny City, 1850–1860" (seminar paper, Department of History, University of Pittsburgh, n.d.).

19. Schumacher, "School Attendance"; DeVault, "Sons and Daughters."

20. John S. Gilkeson, Jr., *Middle-Class Providence, 1820–1840* (Princeton, N.J.: Princeton University Press, 1986).

21. Victor Anthony Walsh, "Across the 'Big Wather': Irish Community Life in Pittsburgh and Allegheny City, 1850–1855" (Ph.D. diss., University of Pittsburgh, 1983); Nora Faires, "Ethnicity in Evolution: The German Community in Pittsburgh and Allegheny City, Pennsylvania, 1845–1895" (Ph.D. diss., University of Pittsburgh, 1983).

22. This theme is taken up by Ronald Edsforth in *Class Conflict and Cultural Consumer: The Making of a Mass Consumer Society in Flint* (New Brunswick, N.J.: Rutgers University Press, 1987).

▪ *Pittsburgh and Europe's Metallurgical Cities: A Comparison*

HERRICK CHAPMAN

W HAT HAS BEEN distinctive about the development of Pittsburgh and the experience of its people? One way to answer this question is to compare Pittsburgh's history to that of other American cities. Chicago, for example, had a more diversified manufacturing economy, and its flat topography made for greater residential mobility than in Pittsburgh, with its steep hills and deep ravines. Or take Detroit. It differed, too, from Pittsburgh in its geography, and like Chicago its economy served as a more powerful magnet for black migrants from the South during and after the First World War. Moreover, Detroit's dependence on a burgeoning automobile industry, rather than an established sector like steel, made the city attractive to newcomers well into the 1960s. Cincinnati, to take yet a third example, offers historians of nineteenth-century Pittsburgh an intriguing set of comparisons—two major cities in the upper Ohio Valley with important differences in ethnic composition and economic structure. By investigating these and other comparisons, historians can determine how particular features in the social and physical landscape helped make Pittsburgh distinctive as an American city.

Another approach is to look at Pittsburgh in comparison to similar iron- and steelmaking cities in Europe, since many of the economic changes that lay behind Pittsburgh's development also played a decisive role in European urbanization. There is nothing, of course, quite like Pittsburgh across the Atlantic, that is, a regional center which grew rapidly into the nation's leading producer of steel. In Germany and France, steel towns either remained towns, like Le Creusot, or they became one of several important nodes in a highly developed industrial region, like Essen in the Ruhr or Saint-Etienne in the Stéphanois region of France. Britain's leading metalworking cities, Birmingham and Sheffield, come closer to resembling Pittsburgh, but even Sheffield, so like Pittsburgh with its hills and river location, never achieved Pittsburgh's regional importance, and its economy was built more on fine cutlery and specialty steels than on bulk steel production. Still, despite differences, Europe's metallurgical cities have enough in common with Pittsburgh to warrant comparative analysis. Pittsburgh and Europe's metallurgical cities all grew rapidly after 1850, became pivotal arenas of industrial conflict, and in many cases set the pace for municipal reform by the turn of the century. Likewise, they all shared a common fate in the twentieth century—the struggle to adapt to a changing world economy which diminished the relative strength of their major industry.

Moreover, as visitors often remark, Pittsburgh has a European feel to it, largely, I think, on account of its enduring ethnic neighborhoods. The rugged terrain, the scale of overseas immigration into Pittsburgh, the clustering of ethnic, working-class neighborhoods near the mills along the rivers, the relatively small number of superhighways slicing up the city, and several decades of demographic stagnation (and decline) have all contributed to this pattern of neighborhood stability. Despite the conventional movement of middle-class families into suburbs and the continuing decay of some inner city slums, ethnic enclaves in Pittsburgh have continued to thrive, especially for the descendents of Germans, Italians, Poles, and a variety of other Eastern and Southern European peoples. In neighborhoods such as Troy Hill, Polish Hill, Bloomfield, and Lawrenceville it is commonplace to keep houses in the family from one generation to another. The extended family remains close by. And occupational patterns reinforced these patterns: Poles, Italians, and blacks in Pittsburgh between 1930 and 1960 found it difficult to break out of the entrenched positions they had acquired in the labor market.[1] Poles tended to stick with the steel industry, Italians with construction and steel, and blacks with low-paying work in the service sector. Although the collapse of the steel industry forced young people in industrial suburbs like Homestead and Clairton to abandon their towns, and sometimes even the region, the social geography of Pittsburgh proper has remained much the same.

This pattern of stability makes Pittsburgh like Europe in two senses. First, many Pittsburghers have continued to keep a cultural toehold in Europe by participating in ethnic churches and organizations. To be sure, the cultures which European immigrants created in such neighborhoods as Bloomfield and Polish Hill differed in a multitude of ways from the worlds they left behind. Moreover, today's street fairs, ethnic grocery stores, holiday rituals, folk dancing troops, and church-sponsored summer tours to the old country offer only attenuated connections to the European past. Still, roots die slowly in a town as demographically stable as Pittsburgh has been since the 1920s. The enduring ethnic character of the city has kept reminders of the Old World alive.

Second, neighborhood stability and a continuing dependence on the extended family are features of urban life more common to European cities than to cities in the United States. European cities have their suburbs, of course, but central city neighborhoods have over the past century been much more stable in Europe than in the United States. Many Pittsburghers, then, share not only an ethnic affinity with people in Europe but also an "un-American" proclivity for keeping close to home.

In comparing Pittsburgh to Europe, however, I wish to dwell not so much on these similarities as on differences in historical development that make Pittsburgh a distinctive member in a transatlantic family of iron and steel cities. Although Pittsburgh shares a number of features in common with Sheffield, Birmingham, Essen, Saint-Etienne, and the metallurgical towns in Lorraine and the Ruhr, its evolution as a city bears the stamp of its American context. Peculiarly American features in the evolution of class relations, combined with the enduring effects of America's political culture and state structure, have made Pittsburgh distinct from its European counterparts. To see this, I will compare Pittsburgh with cities in Europe along several lines: the role of the local elite, working-class formation, immigrant experience, and the politics of municipal reform and urban redevelopment in the twentieth century.

The Elite

In the second half of the nineteenth century metal manufacturers became an inordinately powerful segment of the Pittsburgh elite, and they remained so well into the second half of the twentieth century. Their power stemmed in part from the sheer scale of heavy industry. As early as 1850, when Pittsburgh was already being called "the Birmingham of America," a larger portion of Pittsburgh's population was employed in manufacturing than in any other American city, and metal products made up a sizeable share of the city's production.[2] By the 1870s, when the steel barons built the region's first integrated mills, Pittsburgh was well established as the leading center for bulk steel. And by 1900, when the region produced 30 percent of the nation's steel, Pittsburgh's leading metal manufacturers had become not just a local elite but important power brokers at the national level. Although the Pittsburgh steel industry would eventually decline and ultimately collapse, metal manufacturing remained fundamental to the local economy during the first half of the twentieth century, accounting for between 70 and 75 percent of all manufacturing throughout the period.[3] Accordingly, the industrialists, bankers, and lawyers who sat on the boards of the major manufacturing firms of the region remained, as John N. Ingham puts it, "steel city aristocrats."

Ingham's work makes clear that from the late nineteenth century on, Pittsburgh's leading manufacturers, and especially the industrialists and financiers in the iron and steel business, maintained preeminence in an exclusive, tightly knit, upper class. Like the elite in other American cities these men and their wives used posh neighborhoods, social registers, luncheon and country clubs, and Presbyterian and Episcopalian parishes to define themselves as an urban gentry. Prep schools, ivy league colleges, and marital endogamy enabled their children to take their places in the parlors of Shadyside and Sewickley and in the boardrooms of the city's corporations. Most of these people shared similar roots in upper- or upper-middle-class family lineages; Andrew Carnegie's odyssey from rags to riches was the exception, not the rule.[4] Pittsburgh's elite did stand out from that of other cities for being heavily Scots-Irish in origin. As a new American city expanding rapidly, nineteenth-century Pittsburgh offered Scots-Irish entrepreneurs greater access to the top social ranks than did Philadelphia, Boston, or New York, with their older, established families. Moreover, in the twentieth century, Pittsburgh's upper class was more inclined than its counterparts in Cleveland and Philadelphia to stay in the city rather than move to the suburbs, and to remain locally oriented in its business, political, and social commitments.[5] Pittsburgh's elite, then, was a product of the region's growth as a steel center, and it evolved into a powerful, cohesive social force.

Historians still have much to learn about how these businessmen influenced Pittsburgh's political life over the past century, though it is clear that they figured mightily ever since the 1840s, when manufacturers challenged merchants for political control of the city by promoting the Republican party. From the Civil War on, bankers and industrialists enjoyed an overwhelming position of power in Pittsburgh. The city was a Republican stronghold from the 1860s to 1932. And even under Democratic control since that time, Pittsburgh's biggest financial and manufacturing executives played a decisive role in urban redevelopment. Although we still need studies which compare Pittsburgh's elite with its counterparts in other major American cities, it seems safe to say that in a city which grew so rapidly on the basis of a single industry, and where manufacturers rose to dominate the city's upper class, the corporate elite became unusually influential in the public life of the city.

Were manufacturers as tightly integrated a group and as powerful at the local level in Europe's metalworking cities? Industrialists in company towns, of course, like the Schneiders in Le Creusot and the de Wendel family in the Lorraine region, enjoyed enormous local influence, as did some employers in the sprouting towns of the Ruhr district at least until the late 1890s.[6] But in Birmingham and Sheffield manufacturers had a tougher time winning power and status. For one thing, landed aristocrats in England in the nineteenth century still owned a good deal of property in these long-established cities and exerted enormous influence at the county and city level. Only after 1835 could English cities acquire autonomous fiscal and political status within a county, and hence it was only in the mid-Victorian period that commercial and manufacturing interests wrested firm control over local government from a landed gentry.[7] Even then, however, big estate owners continued to serve as important local patrons of churches, schools, and charitable societies. They remained a force for conservative politics in the city. In addition, Birmingham's economy was more diversified than was Pittsburgh's. As a regional hub, Birmingham had a large commercial and professional sector, and its industry rested mainly on a vast network of small metalworking firms making everything from jewelry to guns. To be sure, by the 1860s, manufacturers had became powerful in the city's political life; Joseph Chamberlain's meteoric rise from screw manufacturer to mayor in the 1870s and his subsequent prominence in national politics are emblematic of the success of Birmingham's industrialists. But the Birmingham elite was more diversified—with Liberals and Tories, religious Nonconformists and Anglicans, merchants, bankers, solicitors and industrialists—than was upper-class Pittsburgh.[8]

Sheffield's manufacturing elite, on the other hand, had by the early twentieth century come to resemble Pittsburgh's more closely, at least in its social composition. After 1850, and especially after the introduction of the Bessemer steelmaking process in 1859, several firms emerged to dominate the Sheffield metal industry—Cammell, John Brown, Vickers, and Firth. By the turn of the century Sheffield could boast of national, even international, prominence in special steels, metal castings, and armaments. Vickers was second only to Krupp as an arms producer. Not surprisingly, then, Sheffield's major steel and armaments builders had emerged as the dominant figures in the city, especially since the only other important economic sector, the old cutlery and light metal business, had been greatly overshadowed by the new heavy metal firms that had grown up along the river Don. Sheffield, moreover, lacked the large middle class and commercial sector that made Birmingham's social structure so complex. But in contrast to Pittsburgh, Sheffield's steelmakers did not invest wholeheartedly in controlling the development of the Sheffield region. True, major industrialists won seats on the town council, and John Brown himself served as mayor in 1861 and 1862. But as their businesses grew, Sheffield's metal magnates focused their attention on power at the national level. As arms builders, they sought to influence decisions in Whitehall and the City of London. They socialized at the Carlton Club in London to build a place for themselves in the business and political circles of the capital.[9] Though executives at Vickers and John Brown were clearly men to be reckoned with in Sheffield, they took less of a lead in city-building and civic reform than did their counterparts in Pittsburgh and Birmingham. By looking outward to the ministries in London, the banks of Manchester and the City, and the arms markets of the world, Sheffield's manufacturing elite inadvertently made it more possible for the Labour party to take control of the town council in 1926 and to keep it, almost without interruption, ever since.

When compared to its English counterparts, then, Pittsburgh's manufacturing elite stands out in the extent to which it dominated the upper class and invested in the cultural and political life of the region. In a sense, Pittsburgh's elite, from the 1880s on, has looked more like Sheffield's in its occupational composition but has acted more like Birmingham's, where the elite did so much to promote local institutions and to shape local policy. Pittsburgh's manufacturing elite proved able to assert itself in this unusually forceful way not only because of the pace and scale of industrial change after 1870, but also because it faced relatively little competition from a landed gentry or a well-entrenched commercial elite. Furthermore, as bulk steel manufacturers, rather than arms builders, Pittsburgh's steel barons did not feel the need to cultivate the kinds of political and financial relationships that drew Sheffield's manufacturers out of their region. Pittsburgh's size and regional importance made a difference as well. It was big enough to provide its elite, and especially the Scots-Irish, with a promising arena for building new institutions in its own image. The creation of a civic center in Oakland in the early twentieth century, patronage of universities and the arts, and above all the Pittsburgh Renaissance of the postwar years rested in large part on the eagerness of this elite to enhance its economic investment in the region and to make Pittsburgh a sophisticated city worthy of the lofty ambitions of a prospering upper class.

When compared with other east coast American cities and the metal cities of Britain, Pittsburgh stands out for the way in which it offered a new manufacturing elite an open terrain for asserting itself and consolidating its power over the local region.

Workers and Political Culture

In these cities, the elite operated in a complex, conflict-ridden environment, since the industrial growth from which it derived its power also gave rise to a large working class often at odds with employers. As Richard Oestreicher's contribution to this volume makes clear, Pittsburgh's workers over the past century have taken part in some of the American labor movement's most important triumphs and defeats. When industrial workers made headway in the 1870s and 1880s by organizing trade unions—the Knights of Labor, and eventually the American Federation of Labor (AFL)—Pittsburgh's skilled workers were at the forefront of these efforts. In the 1930s Pittsburgh's steelworkers made a pivotal contribution to the rise of the Congress of Industrial Organization (CIO). By the same token, workers in the Pittsburgh region also endured two periods of collective weakness. After losing the bitterly fought Homestead strike of 1892, steelworkers in the region lived through forty years of employer autocracy which set a discouraging tone for labor relations in Pittsburgh well beyond the boundaries of the steel industry. The second difficult period began in the 1970s and continued through the 1980s, as Pittsburgh's industrial workers struggled in vain to stem the tide of deindustrialization. In short, the ebb and flow of labor's fortunes have washed through America's quintessential industrial city with tremendous force.

Much the same story can be told about Europe's iron and steel towns, since the rise and fall of the steel industry there roughly parallels the American experience. Throughout Europe, trade-union organizers had little luck in the iron and steel industry until the twentieth century. In France, as in Pittsburgh, it was only in the mid-1930s that mass trade unionism finally broke down the thick walls of employer paternalism in heavy metallurgy. Like their Pittsburgh counterparts, working-class families in Lorraine, for example, witnessed a similar period of employer autocracy early in the century, followed by unionization in the late 1930s, and the same recent nightmare of industrial decline. In the Ruhr Valley and in Sheffield, the same basic trends in the steel industry set the tone for working-class militancy in the first decades of the century and struggles to mollify the effects of layoffs and plant closings in our own time. If a family from Homestead or Pittsburgh's South Side could have had the fortune to visit Europe's steel towns in the 1920s or the 1950s, much of what they would have seen would have been familiar.

Despite this similar pattern of economic change, however, two factors have helped to distinguish Pittsburgh's working class from its counterparts in Europe: namely, immigration and political culture. To take the latter first, consider how government structure and political developments in the nineteenth century influenced the way workers learned to defend their interests. In America, democratization preceded industrialization. By the 1840s, middle-class political parties were working hard to win the support of a growing body of working-class voters, and many white male workers, in turn, came to identify with the mainstream parties. Moreover, the decentralized structure of the American state encouraged workers to focus their political energies on voting at the local level, since local rather than national officials decided most aspects of public policy—taxation, policing, poor relief, public schooling, transportation, and sanitation. Under these circumstances workers in Pittsburgh, as in other American cities, came to depend on local party machines which built alliances between working- and middle-class voters.[10] As trans-class organizations, the Democratic and Republican parties made enough headway in working-class communities to make it difficult for third parties, and especially class-based parties, to gain ground.[11] In the meantime, this style of political development encouraged the early trade unionists in Pittsburgh to focus their efforts on workplace issues, rather than to tie their fortunes directly to political parties. Just as workers in big American cities witnessed a growing separation between workplace and neighborhood, so too their organizations made distinctions between workplace and polity, trade unions and parties. Though this pattern of working-class development by no means made workers passive—on the contrary, American urban workers were as prone to strike in the late nineteenth century as workers anywhere in Europe—it did hamper the emergence of a social-democratic movement in Pittsburgh, to say nothing of more revolutionary alternatives.

This pattern of working-class political development stands in sharp contrast to what happened in Europe. There industrialization preceded democratization, making it more difficult for workers to separate their collective struggles at the workplace from the effort to secure a voice in the larger polity. The clearest contrast with the American pattern was in Germany, where workers did not win full-fledged voting rights until the revolution of 1918. Until then, a rapidly growing working class lived in bitter antagonism to an imperial state which deprived workers of the political rights that the middle classes enjoyed. As a result, it was far more difficult to forge trans-class alliances; many workers found it more compelling to link their trade unions to a socialist movement and to a Marxist ideology which viewed the struggle against capitalism and the pursuit of political liberty as one and same.

Similarly, in Britain and France, workers had to fight for the franchise through much of the nineteenth century—the French through a series of revolutionary upheavals, the British through more moderate battles for electoral reform. In both countries, more-over, a unitary, centralized state meant that crucial policy decisions that affected workers' lives were decided at the national level. French and British workers, then, like their German counterparts, tended to focus on the struggle for political change at the national center, and they were less hesitant than American workers to view trade-union organizing as part of a larger political strategy. These basic trends, of course, took different forms on each side of the English Channel: in Britain the Chartist movement of the 1840s and the emergence of the Labour party after 1903 reflected a willingness among British workers to organize politically as a class but to work within the long-standing framework of the parliamentary system; in France a revolutionary tradition, a memory of betrayal by moderate republicans in 1848 and 1871, and an ambivalence about the state tended to fragment the labor movement into several competing fac-tions. Despite these differences in national patterns, however, a great many workers in Britain and France by the early twentieth century, in contrast to their American counterparts, placed their hopes for change on class-based parties designed to struggle for na-tional political power.[12]

In characterizing Pittsburgh's workers in the nineteenth century as locally oriented, tied to mainstream political parties, and willing to separate trade unions from politics, and to contrast this pattern with the more national, class-based political strategies of workers across the Atlantic, we run the risk of oversimplifying the past. Pittsburgh, like other American cities, had many artisans and skilled workers who were eager to promote a more European style of working-class radicalism. And Germany, Britain, and France had plenty of workers who stayed clear of the dominant working-class organizations that set the tone for labor radicalism in these coun-tries. Alternative approaches competed everywhere, and internal dif-ferences between a city like Sheffield, where the Labour party was making a strong showing by 1914 and Birmingham, where it did not, could be almost as great as international differences in working-class politics.[13] Nevertheless, workers in cities like Pittsburgh, Sheffield, Essen, and Saint-Etienne made strategic choices in the course of the nineteenth century based on the state structures and political circumstances that they encountered. And by the turn of the century national patterns had emerged which would continue to have a powerful effect on working-class politics in each of these countries during the twentieth century, despite the consequences of the world wars and Great Depression.

For workers in Pittsburgh, it is hard to overestimate the importance of patterns set by the 1890s. The bread-and-butter unionism of craftsmen in the AFL, working-class loyalty to local political machines, and the failure of semiskilled and unskilled workers in manufacturing to unionize themselves all contributed to an urban political culture that made Pittsburgh's workers hostage to the local elite. To be sure, the working-class families that populated neighborhoods like Lawrenceville and the South Side enjoyed an important degree of autonomy. Compared to steelworking families in Le Creusot or small towns in the Ruhr where employers controlled everything from grocery stores to housing, Pittsburgh offered workers much more room to make lives of their own. Craft unions did have clout in the labor market, and ward politicians did feel the need to grease the wheels of their neighborhood operations by delivering municipal services. Schools, fraternal organizations, sports activites, and local parishes all bore the class and ethnic stamp of the neighborhood. But between the defeat of the Homestead strike and the elections of 1932, working-class self-expression was largely restricted to neighborhood culture, ethnic institutions, exclusionary, nonpolitical unions, and the individualistic struggle to make life better for one's children.

Even with the sea change in Pittsburgh's politics during the New Deal, the patterns of the past exerted their gravitational pull. True, the emergence of the local Democratic party gave Pittsburgh's workers a better channel to city hall, and Roosevelt's New Deal and the rise of the CIO made working-class politics, in Pittsburgh as elsewhere, more nationally oriented. The Steel Workers' Organizing Committee, by finally restoring trade unionism to the mill towns for the first time since the 1880s, transformed local labor relations. Still, the "New Deal compromise," as Richard Oestreicher calls it in his essay in this volume, that emerged by 1946 perpetuated the nineteenth-century pattern of American working-class politics. Trade unionists accepted an arrangement which restricted labor negotiations to the bread-and-butter issues: wages, benefits, and seniority. The unions stayed clear of questions about job control, technological change, investment strategy, and other matters which employers regarded as management's domain. Trade unionists, moreover, preserved the distinction between workplace issues and politics. Though the AFL and CIO became tacit allies with the Democratic party, most trade unionists in Pittsburgh, as elsewhere, had no intention of recasting the traditional relationship between parties and unions into something akin to the European Labour or Social Democratic parties. On the contrary, Pittsburgh's workers still worked within the traditional framework of the local party machine, albeit a more powerful Democratic party machine than before, and as a result their voices still blended into the larger chorus of a trans-class organization. When a Democratic mayor and political boss, David L. Lawrence, joined forces with Pittsburgh's business leaders to chart the city's postwar redevelopment, neither the labor unions nor the party offered workers a vehicle to take part in the process. In contrast to Britain and France, where the Labour party and the French Communist party played decisive roles after the war in urban redevelopment, in American cities workers continued to rely on a system of unions and parties which prevented them from speaking in a separate, collective voice.

Immigration and Ethnicity

Immigration, alongside state structure and political culture, has also had an impact on the formation of Pittsburgh's working class. In contrast to Sheffield and Birmingham, which expanded mainly through natural population increase and rural-urban migration, Pittsburgh relied on immigrants from abroad to enlarge its work force. First came the Germans and Irish in the mid-nineteenth century, who helped fuel Pittsburgh's burgeoning manufacturing sector. Then when the iron and steel business mushroomed in the late nineteenth century, a massive wave of immigrants from Eastern and Southern Europe provided an army of mill hands. As Nora Faires points out in her essay in this volume, from 1880 to 1930 a half to two-thirds of Pittsburgh's population were either foreign born or had foreign-born parents. One more major wave of newcomers was to follow—black migrants from the rural South, who came to Pittsburgh in significant numbers from 1916 to the 1950s. These three waves of in-migration between the 1840s and the 1950s were typical of cities in the industrial Northeast. But the explosive expansion of Pittsburgh's steel industry between 1880 and 1920 and its subsequent stagnation gave Pittsburgh a higher proportion of Eastern Europeans and a lower proportion of blacks than has been the case in cities such as Cleveland,

Baltimore, Philadelphia, and Detroit. Moreover, large-scale immigration and the endurance of strong ethnic communities made Pittsburgh's working class different in important respects from its counterparts in the manufacturing cities of Western Europe.

Not that immigration was unique to American working-class formation. On the contrary, Europeans migrated on an impressive scale within Europe. Italians, for example, migrated to other European countries between 1876 and 1976 in slightly greater numbers than to destinations overseas.[14] In fact, the two largest immigrant groups in Pittsburgh, the Italians and the Poles, served as a crucial source of labor for the mines, smelters, and steel mills of Germany and France. Starting in the 1890s, Polish migrants, mostly from poor, rural areas in the German-controlled regions of Poland, were recruited to work as miners in the Ruhr. By 1914, 2 million Poles lived in the region. Similarly, at the turn of the century, when the Briey basin opened up as the dynamic center for iron mining and smelting in Lorraine, Italian workers became the predominant source of labor there. These two cases of large-scale migration offer a useful contrast to the immigration experience in Pittsburgh.

Turning first to Lorraine, what is striking is the extent to which Italians assimilated into French society and into the French working class; they did not maintain a strong sense of ethnic distinctiveness, at least by Italian-American standards. A number of factors help account for this. For one thing, the rates of return migration were extremely high, as was typical of much intracontinental European migration. High rates of labor turnover in the mines and smelters, and high levels of mobility in and out of towns like Briey and Longwy, made it difficult for workers to establish a stable Italian "colony" in Lorraine.[15] Though many Italian immigrants returned to Italy from the United States as well (1.2 million from 1905 to 1915), it was far more common to do so from manufacturing towns of Germany and France. Of course, some workers stayed in the Briey basin, sent for their families, and lived out their lives there. By the 1920s a small nucleus of settled immigrants provided a stable sense of community in a number of industrial towns. Through them, a second factor came into play—namely, the relative ease with which Italian immigrants in France eventually assimilated into two important institutions, the Communist party and the Catholic church. In this respect, their assimilation into France had a lot to do with their Italian origins. Most migrants to the Briey basin came from northern, industrialized cities in Italy where many of them had become involved in the socialist, anarchist, and syndicalist movements. Some of them, in fact, came to France as refugees from political persecution, and some of those who stayed, especially after 1922, did so as exiles from Mussolini's regime. When trade unionism and the Communist party finally made inroads in the Lorraine region in the 1930s and 1940s, many of these already politicized Italian workers, or their offspring, were quick to support the party. In the postwar period, the presence of workers of Italian origin proved decisive in giving the Communist party dominant voting strength in a number of towns in the Briey basin. By the same token, as trade unionists and Communist voters, they were more inclined than they might otherwise have been to identify with their French working-class comrades. As for the Catholic church, it too served as a cultural bridge between Italy and France, especially in a clerical region like Lorraine and especially for the wives and children of working-class migrants. Although some workers were anticlerical, most maintained a loose affiliation with the church and did not see it in conflict with their support for the Communist party.[16]

A third factor which encouraged Italian workers to assimilate was the cultural homogeneity of France and the way in which a centralized French state tried to enforce it. Unlike the United States, with its tradition of cultural pluralism, France had a long history of monarchs who sought to stamp out local customs, languages, and loyalties in an effort to mold a unified national culture with Paris as its center. Under the Third Republic, government officials perpetuated this tradition, albeit within the framework of a democratic polity, through schools and military service. At the same time, the government sought to lower the barriers of entry for foreign workers badly needed in an underpopulated labor market. In 1889 the government made naturalization easier and declared that birth on French soil conferred nationality (and hence eligibility for military service).[17] In the universalistic spirit of 1789, the French government espoused a policy of assimilation without going so far as to suppress the church parishes and voluntary organizations that immigrants established on French soil.[18] Of course, this enthusiasm for acculturating the outsider went hand-in-hand with an intolerance for the "other"; the French had no special immunity to xenophobia, and the country had its share of anti-immigrant violence. Cultural uniformity, moreover, was only a myth in a nation that still bore the stamp of its multicultural and multiethnic past. But it was a powerful myth in France, and the nationalist passions of the twentieth century made it stronger. Left-wing spokesmen like Jean Jaurès embraced internationalism, but not cultural particularism within France. Not surprisingly, then, all the major immigrant groups in France before the Second World War—Italians, Poles, and to some extent even Eastern European Jews—

found little support in the culture for cultivating distinctive ethnic enclaves. These pressures drove them, and especially their children, toward French institutions—schools, trade unions, political parties— and toward a French identity. As a result, by the 1930s and certainly by the postwar era, class loomed larger on the map of social distinctions than did nationality for Italian immigrant families in Lorraine.

For Polish miners in the Ruhr, acculturation proved more elusive, even though, as migrants from German-controlled Poland, they were considered citizens of the Imperial Reich. For one thing, their living and working conditions tended to isolate them from German workers. Employers recruited them for work in the mines and housed them near the pitheads in company "colonies," a housing arrangement which native German workers tried to avoid. For many Polish migrants, mining offered a step up from an impoverished life as a farm worker east of the Elbe, but once they had settled in the Ruhr they found little chance for further upward mobility. They were stuck in the mining colonies and isolated. The German government, moreover, promoted an anti-Polish assimilationist policy, that is, a policy of Germanization that went far beyond its French equivalent in the effort to undermine a nascent immigrant culture. State officials tried to ban Polish flags and emblems at religious processions, to prohibit Polish translations of mine safety regulations, and to prevent Polish priests from settling in the area. Right-wing nationalists trumpeted in behalf of these measures in the Pan-German press. German trade unions, unfortunately, failed to provide Polish workers with an adequate means of self-defense. The Catholic unions were tied to the Catholic Center party which objected to Polish priests, while Socialist unions had trouble accommodating the workers' religious devotion and Polish nationalism.[19]

As a result, Polish miners in the Ruhr relied on their own cultural resources, and in the years before the First World War they managed rather well to build an ethnic enclave. They created parishes, associations, newspapers, and even a Polish Miners Union (the ZZP). As in America, these kinds of efforts had the paradoxical effect of both insulating them from the majority culture and integrating them institutionally into German society.[20] By the 1920s, however, these efforts became more and more difficult to sustain as some immigrants returned to the new Poland that had emerged after the First World War and others moved on to France to find work in mining and metallurgy. The Polish minority in the Ruhr shrank to less than half what it had been before the war.[21] Those who stayed managed to preserve the Polish language, but much of the organizational foundation of their community eroded beneath them. In 1934 the ZZP dissolved. Fascism, a second world war and massive population resettlements in Central Europe after the war further undermined any prospects for ethnic pluralism in the Ruhr.

When examined alongside these European cases, Pittsburgh's immigrant experience looks distinctive in three respects. The first is the scale and geographical concentration of immigration to Pittsburgh. Poles in the Ruhr and Italians in Lorraine were scattered across dozens of cities, towns, and small communities, whereas in the Pittsburgh region, new immigrants settled into a more concentrated area which centered on a single large city. Demographic concentration made it easier to create an ethnic community with a wider range of talents, resources, and human choices. Moreover, by the turn of the century Pittsburgh was a city of immigrants; even if newcomers had little political and economic power in the region they nevertheless lived in an environment where most people frequented ethnic churches, shops, and voluntary associations. Pittsburgh had plenty of nativism, to be sure. And there were lots of well-meaning reformers eager to make immigrants "Americans" in a hurry. But it was easier to cope with these pressures in an immigrant city than in the more culturally exposed terrain of the Ruhr and Lorraine.

The second distinctive feature of Pittsburgh's immigration is that immigrants were able not only to create but also to sustain ethnic communities more effectively than their European counterparts. In Pittsburgh as in other American cities with large ethnic "colonies," Poles, Slovaks, Ukrainians, Slovenes, and other groups became more, rather than less, involved as time went on in ethnic parishes, fraternal organizations, neighborhood businesses and social clubs. Ewa Morawska calls this process an "ethnicization of consciousness," as immigrants and their offspring became increasingly ensconced in their ethnic communities and increasingly aware of their ethnic separateness.[22] Even in the interwar years, many immigrants were slow to become naturalized citizens, and when they did so they still identified with their homeland. Second- and third-generation offspring, of course, felt little pull to the Old World, but they nonetheless identified as ethnics, as distinctively hyphenated Americans. Though the Depression, the Second World War, and postwar trends toward suburbanization did some damage to the cohesiveness of many an ethnic neighborhood, the crucial pillars of the ethnic enclave—marriage and courtship, ethnic parishes, fraternal organizations, neighborhood segregation—have survived remarkably well in many districts of the Pittsburgh region.

This persistence of the ethnic community in Pittsburgh stems not only from the size and concentration of the initial wave of immigration. It stems too from that peculiar blend of hostile pressures and supportive conditions that encouraged immigrants to cultivate their own ethnic colonies. As in the Ruhr, so too in Pittsburgh, Eastern Europeans encountered hostility—nativists who hated foreigners, Protestants who disliked Catholics, Catholics who objected to ethnic parishes, an upper class that feared the "dangerous" classes. Likewise, immigrants in Pittsburgh, as in the Ruhr, relied on their own collective resources in response. But Pittsburgh offered much more freedom than did the Ruhr for immigrants to build institutions of their own. The state made no effort to stamp out voluntary organizations and public rituals. "Americanization" programs, though culturally coercive, did not carry the same freight of state repression and racial homogenization that characterized Germanization. Above all, a system of legal protections and a tradition of cultural and religious pluralism, while no guarantee against ethnic prejudice, did provide enough shelter to allow Pittsburgh's immigrants to sustain their own organizations. Neither France nor Germany offered immigrants the same level of cultural tolerance and insulation from the state.

The third feature that makes Pittsburgh's immigration distinctive, especially in comparison to the Italian experience in Lorraine, is the extent to which ethnicity interfered with working-class solidarity. To be sure, as Richard Oestreicher points out, ethnic and class solidarities could sometimes be complementary in Pittsburgh. Chain migration and company hiring practices in the late nineteenth century often brought workers of the same ethnicity together into the same job categories and factories. For them a sense of class sprang directly from a common identification as an ethnic minority. Defending one's Catholic, ethnic community against a Protestant, middle-class, native establishment gave immigrant workers of many nationalities a common struggle. By the mid-1930s, moreover, the ethnic workers of different nationalities had come to share enough of a common cultural experience in America to make it easier than before to join forces in the labor movement.[23]

Still, despite the ways in which workers could either use or transcend their ethnic particularities to establish a sense of class identity, ethnicity also undermined efforts to build class solidarity. Barriers of language, custom, religious practice, residential insularity, and labor market segmentation made it difficult for people to cross the boundaries of ethnic nationality, especially from the 1880s through the 1920s. As John Bodnar, Roger Simon, and Michael P. Weber have shown, even cultural differences as subtle as ones that distinguished German, Russian, and Austrian Poles loomed large in Pittsburgh's Polish neighborhoods. Moreover, ethnic differences—and the more basic difference between native and immigrant workers—made it more difficult in Pittsburgh than in many European societies to build political bridges between skilled and unskilled workers. As Michael Hanagan's research on workers in the Saint-Etienne region has suggested, alliances between skilled and unskilled workers became crucial in France for extending a nineteenth-century tradition of working-class radicalism into the manufacturing settings of the twentieth century.[24] But the ethnic and racial segmentation of Pittsburgh's work force, the extent, that is, to which ethnicity and race separated workers residentially and by skill level, impeded this kind of communication. To be sure, workers occasionally proved able to bridge these gaps, as, for example, during the Homestead strike. But overall, employers were able to take advantage of cultural divisions in the work force, and workers found it hard to overcome them, until the second third of the twentieth century.

Municipal Reform and Postwar Redevelopment

If Pittsburgh's takeoff as a major manufacturing city occurred in the late nineteenth century, some of its most dramatic institutional changes still lay ahead—in the reform of municipal government in the early twentieth century, in the Pittsburgh Renaissance of the 1940s and 1950s, and in the struggle of the 1980s to adjust to the collapse of heavy industry. All three episodes in Pittsburgh's history have their parallels in Western Europe. Sheffield, Birmingham, and the cities of the Ruhr all had their own versions of municipal reform, postwar redevelopment, and painful adjustment to industrial decline. And in each case, as in Pittsburgh, nineteenth-century developments in state structure, political culture, and class formation had an enormous impact on how people handled these episodes of urban change in the twentieth century.

Municipal reform in early twentieth-century Pittsburgh, for example, had its roots in social changes underway since the 1880s. As Samuel P. Hays has argued, the upper-class business and professional elite that rose to prominence in the last decades of the nineteenth century became the major force for governmental reform after 1900.[25] Like business leaders in other American cities, Pittsburgh's most powerful and wealthy citizens put their weight behind the effort to centralize and professionalize city government. Through the Civic Club and the Voters' League, Pittsburgh's elite rallied to the cause of eradicating the ward-based political system which, in its view, had made city government too vulnerable to corruption and too responsive to the particularistic interests of lower- and lower-middle-class constituencies. In the case of Pittsburgh, these elite reformers triumphed. By winning approval in the Pennsylvania legislature for a change in the city charter, they abolished the ward system in 1911 and replaced it with a small City Council whose members were to be elected citywide. They also strengthened the executive authority of the mayor. Moreover, they pursuaded the state legislature in 1913 to replace the ward-based school boards with a central board whose members were to be appointed by the judges of the Court of Common Pleas. "Efficiency" became the watchword of civic reform, as business leaders and professional groups applied the ideals of bureaucratic rationalization to city government.[26] In this political struggle between working- and lower-middle-class groups, on the one hand, and business and professional leaders, on the other, the outcome reflected how thoroughly Pittsburgh's tight, cohesive manufacturing elite had come to dominate the city and how poorly equipped Pittsburgh's ethnically mixed working class was to fight municipal reform.

If Pittsburgh's elite succeeded in transforming the political structure, it failed to achieve much in the way of social reform. Not that it ignored problems of housing, health, welfare, and education altogether. The Pittsburgh Survey of 1909–1914 had done a great deal to make the city's business and professional leaders more aware of the wretched conditions in which many people lived in the mill towns and working-class communities of the region. The Civic Club, the Civic Commission, and the Chamber of Commerce—the same crowd that pushed through changes in city government—took an interest in charitable philanthropy, sanitary reform, and model housing programs which aimed to improve the city's ills. Pittsburgh, after all, had come to symbolize not only America's new industrial might but also its filth, its river pollution, its punishing pace of work, the grim conditions under which many new immigrant workers tried to make ends meet. But model housing and tax reforms designed to stimulate real estate development did little to expand the supply of low-income housing. Water and air pollution continued unabated until the 1940s. Even highway construction and city planning languished in the interwar years for lack of public and private leadership.[27] Pittsburgh's elite used municipal reform to strengthen political control in the city, but it was not until the Pittsburgh Renaissance that a new generation of business leaders used their power to reshape the city itself.

This episode in political reform and policy inertia differs in important respects from what happened in German and British manufacturing cities in the first decades of the twentieth century. In the Ruhr, municipal leaders generally went further to try to improve urban services and local housing conditions than was the case in Pittsburgh. There were several reasons for this. For one thing, big industrialists took greater initiative to house their own workers. By 1901, for example, the Krupp firm built over four thousand housing units in Essen to house nearly twenty-seven thousand people. The Pittsburgh region had company housing, too, but not to the extent and to the standard of quality that the Krupps made famous in their housing colonies. In addition to housing, Krupp spent a good deal of money subsidizing schools, clubs, churches, hospitals, and philanthropic organizations. The rapid pace of industrial growth, and the desire to undercut the emergence of a socialist workers' movement, gave German industrialists like Krupp and Thyssen a strong incentive to implement what American companies in the 1920s would call "welfare capitalism." More important, cities in Imperial Germany had professional administrators who wielded considerable influence over city planning and local policy. City officials were well paid, highly trained in public law and finance, and they commanded an army of local civil servants who, as inspectors, welfare workers, educators, and planners, made German city government the envy of American municipal reformers like Frederic Howe.[28] In Essen, the city administration not only promoted improvements in sanitation, transportation, and other city services, but bought a good deal of land for housing construction—an important action since land costs were much higher in Germany than in the United States or Britain. Though conditions in the cities of the Ruhr were far from ideal, in the years before the First World War even socialists from time to time had to admit that city bureaucrats were making improvements.[29]

This administrative achievement in German cities, however, came at a high political price, which American admirers preferred to overlook. Since the Ruhr was part of Prussia, its citizens were subject to the three-tiered Prussian electoral law which weighted a citizen's vote in proportion to his property holdings. The upper class almost always prevailed in elections. In Essen the Krupp family practically owned the city council. This political structure reproduced at the city level the same tight partnership of business leaders and state officials which characterized German political rule at the national level. And this arrangement made it possible for city government to become professionalized and remain autonomous from all but the most powerful political pressures. Under these circumstances city officials could, in the conservative tradition of Bismarck's social welfare provisions, address problems of sanitation, transportation, and housing, and in the process try to check the growth of the socialist movement.

In English cities, such as Birmingham and Sheffield, the initiative for social improvement came more from "below" than "above" as in Germany. The roots of popular pressure for municipal reform go back to the 1830s and the Chartist movement of the 1840s. But it was only in the the last third of the nineteenth century that social reform became a prominent objective of city government. Birmingham led the way. In the 1860s and 1870s, two Nonconformist ministers in the city, George Dawson and Robert William Dale, preached a gospel of civic activism, an ethos of social responsibility which soon became the hallmark of Birmingham's Liberal party. When Joseph Chamberlain became mayor in 1873, he put the "civic gospel" into practice. Under his leadership the city took over the gas system and the waterworks and bought up large tracts of center-city land to launch what we would now call urban renewal. He had old buildings razed and new ones constructed to give the city greater cultural and commercial vitality. He also put a premium on improving city services and professionalizing the civil service. By the 1880s Birmingham had won a reputation as "the best governed city in the world" and a model for "municipal socialism."[30]

Municipal socialism came later to Sheffield, but by 1898 Sheffield's city council had municipalized water, street transport, and electricity. By 1914 city authorities had cleared some of its worst slums and built rental housing both in the center of the city and in the suburbs. After the First World War the city's public housing effort accelerated, especially under the aegis of the Labour party, which won control of the city council in 1926. By 1939 municipal housing estates covered the hillsides north and south of the city. Sheffield's precociousness in the field of public housing, especially in comparison to Pittsburgh, derived in part from the growing political influence of the city's working-class voters, and in part from a national political structure which gave city governments the fiscal tools and governmental authority to take bold measures. The housing acts of 1890 and 1919 made it possible for city councils to clear sums and build new housing, and the latter act in particular provided state subsidies for municipal housing.[31] In contrast to Pittsburgh, where reformers had to rely on weak fiscal incentives to encourage housing construction, in Sheffield the rise of the Labour party and the intervention of the national government in the housing market gave reformers the authority and the resources they needed to expand the stock of working-class housing. In short, from the 1890s through the interwar years, Pittsburgh lacked the key ingredients that made municipal improvements more substantial across the Atlantic: America's steel city had neither a vigorous city bureaucracy under economic and political pressure to address the "social question," as in prewar Germany, nor a working-class party and an interventionist state, as in interwar Britain, which enabled city officials to engage directly in the business of housing reform.

When officials in Pittsburgh finally launched a serious program of urban redevelopment in the 1940s, it was not the pressure of an insurgent social democratic movement or Labour party which provided the impetus for change. Instead, it was a change within the business elite itself which paved the way for the Pittsburgh Renaissance. By the late 1930s the corrosive effects of the Depression had exacerbated what had already been discernible in the 1920s—Pittsburgh's slow, but steady decline as a manufacturing city. To a new generation of businessmen it became obvious that if the city was to have a future as a competitive center for corporate administration, finance, and commerce, and if it was to stem the tide of manufacturing decline, something would finally have to be done about smoke pollution, congested roadways, and a deteriorating downtown business district. Generational change, then, and a pervasive sense of crisis in the business community set the stage for the Pittsburgh Renaissance.

Business leaders remained the pivotal actors in the redevelopment politics of the postwar era. In 1943 Richard King Mellon, the new head of the Mellon empire, gathered together the key business, government, and university leaders who created the Allegheny Conference on Community Development (ACCD), the main vehicle for spearheading the redevelopment effort. Through Mellon's leadership, the most powerful corporate executives in the region threw their political influence and corporate resources behind smoke control and renewal of the downtown district. To be sure, government officials played important roles. When Mayor David L. Lawrence made the strategic choice in 1945 to ally his Democratic political machine with the ACCD, a public-private partnership of Democratic politicians and Republican businessmen emerged, a partnership which proved essential for the redevelopment effort. Pollution control, new highway construction, and the building of Gateway Center in downtown Pittsburgh all required not just the cooperation of city government but enabling legislation from the state and the creation of new agencies and new forms of public authority which expanded the role of government in the region. Still, what made Pittsburgh's postwar redevelopment so distinctive, in addition to the sheer scale of the undertaking, was the extent to which business initiated and guided the process.[32]

Postwar redevelopment in Europe, in contrast, reflected and reinforced the predominance of the public sector. In Sheffield, for example, the Labour party, the municipal government, and national
planning policy all figured prominently in the effort to rebuild the
bombed-out sections of the central city and to shape the suburbanization of the city. Businessmen had relatively little power to influence
policy, and indeed, by the mid-1960s several of the important steel
companies in Sheffield had come under direct government control.
Some of the goals that Sheffield's municipal leaders pursued resembled the Pittsburgh Renaissance—smoke control, for example,
and new highways. But its postwar redevelopment was much more
oriented toward clearing slums and continuing the efforts which Labour officials had begun in the 1930s to resolve the long-term housing
problems of the working class. Moreover, city and national government alike went to considerable lengths to try to control the nature of
growth at the city's periphery, first through an attempt to build
"greenbelts" and discrete satellite suburbs, and then through a high-
density housing program in the inner city. The city government also
discouraged the growth of suburban shopping centers in an effort to
maintain the downtown district as the major commercial center.
This objective went hand-in-hand with a strategy of making bus transportation cheap and efficient.[33] Postwar redevelopment in Sheffield,
in short, had less to do with maintaining a corporate business sector
and more to do with steering the process of urban growth itself by
improving transport, services, and the housing stock. These goals
derived in large part from the rise of the Labour party and its postwar
agenda in urban planning at both the local and national level.

In France, state officials had an even greater impact on postwar
redevelopment. In the late 1940s, after having neglected urban planning and municipal reform during the early decades of the twentieth
century, French politicians and civil servants suddenly discovered
the city as a central problem in their society. Housing shortages were
acute, the pace of urbanization and city growth was accelerating, and
political leaders were becoming much more aware than before that
the economic revitalization of the country depended in large part on
rejuvenating regional cities like Toulouse, Grenoble, Strasbourg, and
Metz. Centuries of centralization in and around Paris had made provincial France, as the geographer Jean-François Gravier put it in 1947,
"le désert français." By the late 1940s, regional economic development became a major goal for Jean Monnet's planning council, and it
has remained a critical concern of the French state ever since. In 1963
de Gaulle created a new government agency, the Délégation à
l'Aménagement du Territoire et à l'Action Régionale (DATAR)

which channeled the considerable economic resources and bureau-cratic power of the state more effectively toward provincial urban growth. DATAR subsidized companies to locate factories in provincial cities and in the process checked the eastward drift of the industrial sector by coercing industry to develop the south and west. In addition, the government poured resources into provincial universities and relocated professional schools, government offices, and big new defense and research facilities in provincial cities. Although local citizens, interest groups, and business leaders became more involved in the process of city planning, the initiative came from Paris and the authority to chart new directions remained with the state.[34]

The sharp contrast between Pittsburgh's business-dominated re-development process, on the one hand, and the more state- and party-dominated processes of Britain and France, on the other, became a bit blurred in the 1980s. As Joel A. Tarr and Shelby Stewman have argued in their comparison of Renaissance I and Renaissance II, Pittsburgh's city officials became more influential partners in rede-velopment policymaking after the late 1970s, making the public-private relationship less one-sided than it was in the days of Richard King Mellon.[35] Moreover, when many of Pittsburgh's black residents rose in revolt during the 1960s against the adverse effects of down-town renewal—especially the physical destruction of residential property in the lower Hill district—planning politics had to change. Public and private officials were forced to become more responsive to the social consequences of redevelopment; citizens' groups de-manded a voice in city planning.[36] Though Pittsburgh's business leaders and politicians still remained more insulated from the kinds of grass-roots pressures that have long been part of local political life in Sheffield, black protest in the 1960s and early 1970s forced officials to subject their plans to greater public review.

Likewise, changes across the Atlantic weakened the role of state planning. Economic difficulties since the mid-1970s diminished the force with which French planners could channel urban growth. Under François Mitterrand, moreover, the French government tried to decentralize the state itself by abolishing the powerful prefectural corps—the key lieutenants of the centralized state—and by revitaliz-ing the role of the mayor and city council as holders of real adminis-trative authority. Although regional planning remained an impor-tant activity in France, there was more room for local initiative.[37] In Britain, a decade of Thatcherism shifted attention toward homeown-ership, rather than public housing, as the government's major prior-ity in urban policy. While the Sheffield city council, still dominated by the Labour party, remained committed to cheap public transport and controlled growth, it had to adjust to pressures to sell council housing and respond to the evolving concerns of a property-owning working class. As Anthony Sutcliffe has suggested, these changes gave Sheffield greater neighborhood stability and in this respect made it more like Pittsburgh than before.[38]

But if the events of the 1980s force us to modify the portrait of stark distinctions that characterized planning politics on the two sides of the Atlantic in the 1950s, the basic patterns in state structure, political culture, and class formation that have made for significant differences between these cities still have a powerful impact. Consider, for example, the way that people responded to deindustrialization from the mid-1970s onward. In Pittsburgh, the steelworkers' union made some effort to protest plant closings and massive layoffs, but for the most part workers and their families had to fall back on individualistic strategies of adjustment—moving elsewhere for work, accepting lower-paying jobs in Pittsburgh's growing service economy, retraining for something new, or just making do with casual labor, a spouse's income or an early retirement pension. In the end, families patched together their own life rafts.

In Europe, on the other hand, national governments and the European Economic Community were more directly involved in overseeing the contraction of the steel industry, and as a result, deindustrialization provoked a more political (and collective) response from workers. In France, for example, decisions to scale down the steel industry in 1978 sparked a veritable revolt in Lorraine. In the town of Longwy, young workers imprisoned their managers in factories, halted trains, assaulted a police station, and occupied local government offices. The CGT, the Communist-oriented trade union, gradually took control of the insurrection by organizing a national campaign to save jobs in Lorraine. The effort bore fruit. President Giscard d'Estaing, fearing massive disturbances like those of May 1968, pressured a number of companies outside the steel industry to set up plants in the steel towns, including the big automakers, Renault and Peugeot-Citroën. In addition, the government offered thousands of workers retraining programs with job guarantees at the end, while twelve thousand workers won generous retirement plans and another sixty-five hundred accepted lump sum benefits.[39] In a country where workers have traditionally been more reluctant to relocate for new jobs than has been the case in the United States, and where the government has a long-standing fear of wildcat strikes, state officials felt compelled to soften the blows of industrial contraction. And likewise, in an economy where the central government plays such a direct role in industrial planning and even day-to-day management, workers were more inclined to use tactics of militant protest and trade-union agitation to try to mitigate the effects of industrial decline.

Sheffield also faced industrial contraction, beginning in the early 1970s. And there, too, established habits and structures helped to shape the way people responded. Once again, the city's Labour-led municipal government took the lead; the public sector was the pivotal force in cushioning the impact of industrial decline. City officials pursued a strategy, on the one hand, of trying to minimize the contraction of the local steel industry by lobbying on its behalf in London, and, on the other hand, of seeking to expand the local service economy. City officials were able to convince the national government to shift some state bureaus from London to Sheffield, especially the Manpower Services Division of Employment.[40] Sheffield also attracted the International Division of the Midland Bank. With a growing tertiary sector, Sheffield, like Pittsburgh, developed a more diversified local economy by the mid-1980s. But unlike Pittsburgh, Sheffield was able to maintain a major portion of its manufacturing sector, since specialty steels were less vulnerable to overseas competition than bulk steel production.

None of the European cities examined here changed as dramatically in the 1980s as Pittsburgh did—at least in economic terms. The precipitous collapse of heavy industry, combined with the concomitant expansion of advanced technology and the service sector made the Pittsburgh region at once both depressed and resurgent. While to some extent this was true of Sheffield and Birmingham, the Ruhr and Lorraine, the contrasts between food kitchens in Homestead and high-priced eateries in Shadyside were particularly sharp in Pittsburgh. But even as Pittsburgh's economy changed, the legacy of the past remained. The business elite still had the decisive voice in debate over long-term planning, and policymaking in Pittsburgh remained a product of the decentralized structure of the American state. While ethnic distinctions lost much of their political significance, race and class remained important in shaping the identities and life choices of the young. Racial prejudice continued to divide the city in two, and Pittsburgh's less affluent citizens, both white and black, remained largely outside the city's real corridors of power. Postwar redevelopment, whether in its 1940s form of the Pittsburgh Renaissance or in its 1980s form of economic restructuring, did little to alter the governmental structures and political balance of power that had emerged in the city by 1940.

By emphasizing the ways in which basic features of American government, class relations, and immigration in the nineteenth century shaped the subsequent development of Pittsburgh, and by contrasting these American patterns with comparable ones in Europe, I do not mean to suggest that Pittsburgh was or is a quintessential American city. The pecularities of Pittsburgh's geography and economic development made it unique within the United States. Pittsburgh's upper class has probably been more homogeneous than that of most major northern cities, and its working class has been peopled to a greater extent by Eastern European immigrants than was commonly the case elsewhere. The Pittsburgh Renaissance, moreover, remains the most spectacular example anywhere of postwar redevelopment masterminded by a business elite. Nor should we mistake Sheffield or Birmingham for all of Britain, nor Essen and Longwy for Germany and France. Nonetheless, all these cities evolved within particular national landscapes, and it is only by making cross-national comparisons that we can see what a difference these peculiarities made. By the same token, the comparative themes I have emphasized here by no means exhaust the possibilities for comparison. Race relations, gender, religion, regional economic development, community-building, transportation, and city services—all topics explored in this volume—also deserve comparative analysis. As Europe's cities become increasingly multiracial, and not just multiethnic, race in particular should become a central theme in comparative urban history. Furthermore, by stressing the enduring power of social and political structures to shape the evolution of cities like Pittsburgh, Sheffield, and Essen, I do not mean to suggest that people who lived and fought over issues in these cities were condemned to reinforce the structures they inherited from the past. On the contrary, the story of each of these cities is as much a tale of social change, individual initiative, and human invention as it is of continuities in government, culture, and class. Still, even as people broke new ground—as entrepreneurs, immigrants, reformers, and labor organizers—they could not escape the gravitational pull of larger national developments. And in Pittsburgh, the consolidation of a powerful business elite in the relatively open social terrain of the nineteenth-century American city, the emergence of trans-class political parties and relatively weak bread-and-butter unions, the formation of an ethnically diverse and racially divided working class, and the presence of a decentralized government all had an enormous impact on the evolution of the city. Pittsburgh may not be a typical American city, but when we compare it transnationally, we have to acknowledge how American a city it really is.

NOTES

I wish to thank Liz Cohen, Samuel Hays, and Peter Stearns for suggestions and criticism.

1. John Bodnar, Roger Simon, and Michael P. Weber, *Lives of Their Own: Blacks, Italians, and Poles in Pittsburgh, 1900–1960* (Urbana: University of Illinois Press, 1982), pp. 113–51.

2. Shelby Stewman and Joel A. Tarr, "Public-Private Partnerships in Pittsburgh: An Approach to Governance," in *Pittsburgh—Sheffield, Sister Cities*, ed. Joel A. Tarr (Pittsburgh: Carnegie Mellon University Press, 1986), p. 141.

3. David B. Houston, "A Brief History of the Process of Capital Accumulation in Pittsburgh: A Marxist Interpretation," in *Pittsburgh—Sheffield*, p. 48.

4. John N. Ingham, *The Iron Barons: A Social Analysis of an American Urban Elite, 1874–1965* (Westport, Conn.: Greenwood Press, 1978), p. 18.

5. Ibid., pp. 228–29.

6. Paul M. Hohenberg and Lynn Hollen Lees, *The Making of Urban Europe, 1000–1950* (Cambridge, Mass.: Harvard University Press, 1985), pp. 287–88; Norman J. G. Pounds, *The Ruhr: A Study in Historical and Economic Geography* (Bloomington: Indiana University Press, 1952), pp. 114–17; David F. Crew, *Town in the Ruhr: A Social History of Bochum, 1860–1914* (New York: Columbia University Press, 1979), pp. 103–12.

7. On Britain's "municipal revolution" in the 1830s, see Derek Fraser, *Power and Authority in the Victorian City* (Oxford: Basil Blackwell, 1979).

8. On Birmingham's elite, see Asa Briggs, *Victorian Cities* (New York: Harper and Row, 1970), pp. 38, 184–240; and Dennis Smith, *Conflict and Compromise: Class Formation in English Society, 1830–1914: A Comparative Study of Birmingham and Sheffield* (London: Routledge & Kegan Paul, 1982).

9. Smith, *Conflict and Compromise*, pp. 87–88, 234–35.

10. Ira Katznelson, "Working-Class Formation and the State: Nineteenth-Century England in American Perspective," in *Bringing the State Back In*, ed. by Peter B. Evans, Dietrich Rueschemeyer, and Theda Skocpol (New York: Cambridge University Press, 1985); on the importance of urban political machines to American workers, see also Amy Bridges, "Becoming American: The Working Classes in the United States Before the Civil War," and Martin Shefter, "Trade Unions and Political Machines: The Organization and Disorganization of the American Working Class in the Late Nineteenth Century," in *Working-Class Formation: Nineteenth-Century Patterns in Western Europe and the United States*, ed. Ira Katznelson and Aristide R. Zolberg (Princeton, N.J.: Princeton University Press, 1986).

11. Aristide R. Zolberg, "How Many Exceptionalisms?" in Katznelson and Zolberg, eds., *Working-Class Formation*, pp. 450–51.

12. For comparisons of working-class politics across countries, see Katz-

nelson and Zolberg, eds., *Working-Class Formation;* Charles Tilly, Louise
Tilly, and Richard Tilly, *The Rebellious Century* (Cambridge, Mass.: Har-
vard University Press, 1975); Peter N. Stearns, *Lives of Labor: Work in a
Maturing Industrial Society* (New York: Holmes and Meier, 1975); Val R.
Lorwin, "Working-Class Politics and Economic Development in Western
Europe," *American Historical Review* 63 (Jan. 1958): 338–51; and Adolf
Sturmthal, *Unity and Diversity in European Labor* (Glencoe, Ill.: The Free
Press, 1953).

13. On working-class politics in Sheffield and Birmingham, see Smith,
Conflict and Compromise; Briggs, *Victorian Cities*, pp. 36–38, 186–90; and
Sidney Pollard, *A History of Labour in Sheffield* (Liverpool: Liverpool Uni-
versity Press, 1959).

14. Gianfausto Rosoli, "Italian Migration to European Countries from
Political Unification to World War I," in *Labor Migration in the Atlantic
Economies: The European and North American Working Classes During
the Period of Industrialization*, ed. Dirk Hoerder (Westport, Conn.: Greenwood Press, 1985), pp. 95–97.

15. Ibid., pp. 98–99, 108–09.

16. On the role of the Communist party and the Catholic church in
Italian communities in Lorraine, see Serge Bonnet, "Political Alignments
and Religious Attitudes Within the Italian Immigration to the Metallurgical
Districts of Lorraine," *Journal of Social History* 2 (Winter 1968): 123–55,
and *Sociologie politique et réligieuse de la Lorraine* (Paris: A. Colin, 1972).

17. Nancy L. Green, " 'Filling the Void': Immigration to France Before
World War I," in Hoerder, ed., *Labor Migration in the Atlantic Economies*,
p. 149.

18. For a comparison of how Polish immigrant workers responded to the
more authoritarian policies of the German government before 1914 and the
relatively more liberal policies of the French government in the 1920s and
1930s, see Christoph Klessmann, "Comparative Immigrant History: Polish
Workers in the Ruhr Area and the North of France," *Journal of Social History* (Winter 1986): 335–53.

19. Christoph Klessmann, "Polish Miners in the Ruhr District: Their
Social Situation and Trade Union Activity," in Hoerder, ed., *Labor Migration in the Atlantic Economies*, p. 259.

20. Ibid., pp. 263–64.

21. Ibid., p. 272.

22. Ewa Morawska, *For Bread with Butter: Life-Worlds of East Central
Europeans in Johnstown, Pennsylvania, 1890–1940* (New York: Cambridge
University Press, 1985), p. 171.

23. Lizabeth Cohen, "Learning to Live in the Welfare State: Industrial
Workers in Chicago Between the Wars, 1919–1939" (Ph.D. diss., University
of California at Berkeley, 1986), chap. 6.

24. Michael P. Hanagan, *The Logic of Solidarity: Artisans and Industrial Workers in Three French Towns, 1871–1914* (Urbana: University of
Illinois Press, 1980).

25. Samuel P. Hays, "The Politics of Reform in Municipal Government in the Progressive Era," *Pacific Northwest Quarterly* 55 (Oct. 1964): 157–69.

26. Roy Lubove, *Twentieth-Century Pittsburgh: Government, Business and Environmental Change* (New York: Knopf, 1969), pp. 20–23.

27. Ibid., pp. 20–105.

28. Andrew Lees, *Cities Perceived: Urban Society in European and American Thought, 1820–1940* (New York: Columbia University Press, 1985), pp. 239–47.

29. Ibid., p. 246.

30. On municipal reform in Birmingham, see Briggs, *Victorian Cities*, pp. 184–240; and Smith, *Conflict and Compromise*, pp. 225–47.

31. On housing reform in Sheffield before and after the First World War, see Anthony R. Sutcliffe, "Planning the British Steel Metropolis: Sheffield Since World War II," in Tarr, ed., *Pittsburgh—Sheffield*, pp. 185–89.

32. Lubove, *Twentieth-Century Pittsburgh*, pp. 106–41.

33. Sutcliffe, "Planning," pp. 188–98.

34. On urban development in postwar France, see John Ardagh, *France in the 1980s* (London: Penguin, 1982), pp. 123–205, 258–346; Brian J. L. Berry, *Comparative Urbanization: Divergent Paths in the Twentieth Century* (New York: St. Martin's Press, 1981), pp. 148–58; J.-F. Gravier, *Paris et le désert français en 1972* (Paris: Flammarion, 1972); Jerome Monod and Philippe Castelbajac, *L'Aménagement du territoire* (Paris: Presses Universitaires de France, 1972); and Peter Alexis Gourevitch, *Paris and the Provinces: The Politics of Local Government Reform in France* (Berkeley and Los Angeles: University of California Press, 1980).

35. Tarr and Stewman, "Public-Private Partnerships," pp. 176–79.

36. Ibid., pp. 163–66; Lubove, *Twentieth-Century Pittsburgh*, pp. 142–76.

37. On state decentralization under the Socialists in France, see Mark Kesselman, "The Tranquil Revolution at Clochemerle: Socialist Decentralization in France," and Irene B. Wilson, "Decentralizing or Recentralizing the State? Urban Planning and Centre-Periphery Relations," in *Socialism, the State and Public Policy in France*, ed. Philip G. Cerny and Martin A. Schain (New York: Methuen, 1985).

38. Sutcliffe, "Planning," pp. 195–96.

39. Ardagh, *France*, pp. 57–62.

40. Sutcliffe, "Planning," pp. 197–98.

▪ *Notes on Contributors*

HERRICK CHAPMAN is Assistant Professor of History at Carnegie Mellon University and the author of *State-Capitalism and Working-Class Radicalism in Twentieth-Century France: Industrial Politics in the Aircraft Industry, 1928–1950* (1990).

NORA FAIRES is Associate Professor of History at The University of Michigan-Flint. Her published articles include "The Great Flint Sit-Down Strike as Theater" and "Transition and Turmoil: Social and Political Development in Michigan, 1917–1945." She is currently writing about sectarianism and ideology among German immigrants in Pittsburgh and Allegheny City, 1782–1882.

LAURENCE GLASCO is Associate Professor of History at the University of Pittsburgh. A specialist in Afro-American history and urban demography, his articles include "Life Cycles and Household Structures of American Ethnic Groups" (*Journal of Urban History*). His current work concerns class and color among Afro-Americans.

MAURINE WEINER GREENWALD is Associate Professor of History and Women's Studies at the University of Pittsburgh and editor of the Pittsburgh Series in Social and Labor History. She is the author of *Women, War, and Work: The Impact of World War I on Women Workers in the United States*, and is writing about working-class attitudes toward gendered issues in the early twentieth century, sex segregation in the beauty culture industry, and popular cultural images of women in the late nineteenth century.

SAMUEL P. HAYS is Professor of History at the University of Pittsburgh, specializing in the history of the relationship between society and politics, urban history, and environmental history. He is the author of *Beauty, Health, and Permanence: Environmental Politics in the United States, 1955–1985*.

JOHN N. INGHAM is Professor of History at the University of Toronto. He is the author of *The Iron Barons; Biographical Encyclopedia of American Business Leaders;* and is completing a book on the iron and steel firms in nineteenth-century Pittsburgh.

PAUL KLEPPNER is University Research Professor of History and Political Science at Northern Illinois University and director of the Social Science Re-

439

search Institute there. His most recent book is *Continuity and Change in Electoral Politics, 1893–1928*. He is currently working on an eight-city comparative study of the political mobilization of ethnic and racial minorities.

ROY LUBOVE is Professor of Social Welfare and History, University of Pittsburgh. His many publications in the areas of urban and social welfare history include *Community Planning in the 1920s: The Contribution of the Regional Planning Association of America* and *The Struggle for Social Security, 1900–1935*.

EDWARD K. MULLER is Associate Professor of History at the University of Pittsburgh where he specializes in the historical geography of American cities. He has co-edited the *Atlas of Pennsylvania*, and is working on a study of low-income black families in post-World War II public housing.

RICHARD OESTREICHER is Associate Professor of History at the University of Pittsburgh. His publications include *Solidarity and Fragmentation: Working People and Class Consciousness in Detroit, 1875–1900* and "Urban Working-Class Political Behavior and Theories of American Electoral Politics, 1870–1940" (*Journal of American History*). He is writing a book on American working-class formation and development from 1800 to 1900.

LINDA K. PRITCHARD is Associate Professor of History at the University of Texas at San Antonio. She has completed *Evangelical Environments: The Social Contexts of Evangelical Religion in the Upper Ohio Valley, 1790–1860* and is working on a study of religious change in Texas from 1845 to 1900.

JOEL A. TARR is Professor of History and Public Policy and Academic Dean for the College of Humanities and Social Science at Carnegie Mellon University. He has written widely on the history of urban infrastructure, environmental pollution, and the process of city-building in Pittsburgh. His most recent book, edited with Gabriel Dupuy, is *Technology and the Rise of the Networked City in Europe and America*.

MICHAEL P. WEBER is Vice-President for Academic Affairs and Professor of Urban History at Duquesne University. He is the author of *Don't Call Me Boss: David L. Lawrence, Pittsburgh's Renaissance Mayor* and a co-author of *Lives of Their Own: Blacks, Italians, and Poles in Pittsburgh, 1900–1960*. He is currently working on a study of social change and deindustrialization in Pittsburgh's Turtle Creek Valley.

- *Bibliography*

This bibliography is not intended to be a comprehensive listing of sources about the history of Pittsburgh. It is confined to seminar papers, doctoral dissertations, printed articles and books written by students in the graduate history programs of Carnegie Mellon University and the University of Pittsburgh since 1960. References to the many valuable papers and dissertations produced by other schools and departments are included in the notes accompanying the individual essays.

Adams, John C. "Machine Politics in Action: Pittsburgh in the Early 1890s." Seminar paper, Carnegie Mellon University, 1972.

Agoratus, Steve, Phil Bateman, Shelly Giordano, Skip Stong, and Barry Whittemore. "Donora: An Historical Perspective." Seminar paper, Carnegie Mellon University, 1982.

Aitchison, Brian. "The Development of the Allegheny River from a Rural Backwater to an Industrial Artery, 1828–1930." Seminar paper, Carnegie Mellon University, 1979.

Allen, Marjorie. "The Negro Upper Class in Pittsburgh, 1910–1964." Seminar paper, University of Pittsburgh, 1964.

Angerman, Gerald. "McKeesport: A Preliminary Study." Seminar paper, University of Pittsburgh, 1970.

Bachand, Will. "Municipal Lighting in Pittsburgh, 1880–1920." Seminar paper, Carnegie Mellon University, 1973.

Barnes, John, Jr. "The Ku Klux Klan in Western Pennsylvania." Seminar paper, Carnegie Mellon University, n.d.

Bateman, Philip. "Early Reapportionment in Pittsburgh: 1845–1848." Seminar paper, Carnegie Mellon University, 1983.

Bauman, Geoffrey. "The Greek Immigrant and the Greek Orthodox Church in Pittsburgh." Seminar paper, University of Pittsburgh, 1974.

Bedeian, George. "Social Stratification Within a Metropolitan Upper Class: Early Twentieth-Century Pittsburgh as a Case Study." Seminar paper, University of Pittsburgh, 1974.

Bennett, John William. "The Iron Workers of Woods Run (1870–1880)." Seminar paper, University of Pittsburgh, 1973.

———. "Iron Workers in Woods Run and Johnstown: The Union Era, 1865–1895." Ph.D. diss., University of Pittsburgh, 1977.

Billings, Warren. "Steamboat Building in Pittsburgh, 1811–1840." Seminar paper, University of Pittsburgh, 1963.

Birney, Thomas C. "Railroad Acquisition of Rights of Way in Pittsburgh, 1830–1860." Seminar paper, Carnegie Mellon University, 1982.

Blais, Gerard. "Lawyers of Pittsburgh, 1860–1890." Seminar paper, University of Pittsburgh, 1966.

Bodnar, John, Roger Simon, and Michael P. Weber. *Lives of Their Own: Blacks, Italians, and Poles in Pittsburgh, 1900–1960.* Urbana: University of Illinois Press, 1983.

Bogin, Russell. "The Role of Education in Black Employment in the A. M. Byers & Co." Seminar paper, Carnegie Mellon University, 1981.

Brant, John W. " 'Those Damn Ignorant Somerset Dutch': An Analysis of Antimasonic and Whig Voting in Somerset County, Pa., from 1828 to 1840." Seminar paper, University of Pittsburgh, 1969.

Brewer, James. "Robert L. Vann and the Pittsburgh Courier." MA thesis, University of Pittsburgh, 1941.

Brown, Robert Kevin, Jr. "1830: The Select and Common Councils of Pittsburgh: The Birth of a City." Seminar paper, Carnegie Mellon University, 1980.

Bryan, Sam. "Fifth Avenue High School: An Inner City School as a Reflection of a Changing Neighborhood." Seminar paper, Carnegie Mellon University, 1969.

Bucki, Cecelia F. "The Evolution of Poor Relief Practice in nineteenth-Century Pittsburgh." Seminar paper, University of Pittsburgh, 1977.

Burkett, Ray. "The AntiMasonic Party in Pittsburgh." Seminar paper, University of Pittsburgh, 1971.

———. "Vandergrift: Model Worker's Community." Seminar paper, University of Pittsburgh, 1972.

Burstin, Barbara. "The Response of the Pittsburgh Jewish Community to Holocaust Survivors, 1946–1951." Seminar paper, University of Pittsburgh, 1980.

———. "From Poland to Pittsburgh: The Experience of Jews and Christians Who Migrated to Pittsburgh After World War II." Ph.D. diss., University of Pittsburgh, 1986.

———. *After the Holocaust: The Migration of Polish Jews and Christians to Pittsburgh.* Pittsburgh: University of Pittsburgh Press, 1989

Caffee, Anna-Mary. "The Socialization of Rural Children in Southwestern Pennsylvania, 1840–1900." Seminar paper, University of Pittsburgh, n.d.

Caine, Mark. "Student Life at Carnegie Technical Schools, 1905–1909." Seminar paper, Carnegie Mellon University, 1981.

Callister, Tom. "The Reaction of the Presbytery of Pittsburgh to the New Immigrants." Seminar paper, University of Pittsburgh, n.d.

Caloyer, Mark. "Minority Immigrant Groups of 1850 in Pittsburgh." Seminar paper, Carnegie Mellon University, n.d.

Campbell, Duane E. "Anti-Masonry in Pittsburgh." Seminar paper, Carnegie Mellon University, n.d.

Carroll, Ruth. "Income and Expenditure of 1917 Mill Workers in Pittsburgh." Seminar paper, Carnegie Mellon University, 1981.

Caye, James Francis, Jr. "Violence in the Nineteenth Century Community: The Roundhouse Riots in Pittsburgh, 1877." Seminar paper, University of Pittsburgh, 1969.

———. "Crime and Violence in the Heterogeneous Urban Community: Pittsburgh, 1870–1889." Ph.D. diss., University of Pittsburgh, 1977.

Chamovitz, Marcia. "The Persistence of Ethnic Identity in Two Nationality Groups in a Steel Mill Community." Seminar paper, University of Pittsburgh, 1976.

———. "The San Rocco Celebration of Aliquippa; An Italian Saint in an American Setting." Seminar paper, University of Pittsburgh, 1977.

Chasan, Joshua. "The Election of 1912 in Allegheny County: A Comparative Study of Progressive, Standpat, and Socialist Leadership." Seminar paper, University of Pittsburgh, 1967.

Claeren, Wayne H. "The Ministry in Pittsburgh During the Progressive Era." Seminar paper, University of Pittsburgh, 1964.

Clarke, Gerry. "Natural Gas in Pittsburgh." Seminar paper, Carnegie Mellon University, n.d.

Corn, Jacqueline Karnell. "Municipal Organization for Public Health in Pittsburgh, 1851–1895." D.A. diss., Carnegie Mellon University, 1972.

Covert, Patricia. "Pittsburgh Reaction to the Mexican War." Seminar paper, University of Pittsburgh, 1962.

Daly, Janet R. "The Political Context of Zoning in Pittsburgh, 1900–1923." Seminar paper, University of Pittsburgh, 1984.

———. "Zoning: Its Historical Context and Importance in the Development of Pittsburgh." *Western Pennsylvania Historical Magazine* 71 (Apr. 1988): 99–125.

Dankowski, John. "The Ante-Bellum Temperance Movement in Pennsylvania." Seminar paper, University of Pittsburgh, 1971.

———. "Pittsburgh City Government, 1816–1850." Seminar paper, University of Pittsburgh, 1971.

Davidson, J. E. " 'God Speed the Plough': A View of Agricultural Society." (Peters Township, Washington Co., Pa., 1800–1850.) Seminar paper, University of Pittsburgh, 1969.

Derringer, Dennis. "Impact of Industry on Harrison Township." Seminar paper, Carnegie Mellon University, 1975.

Deubel, Susan. "The Progressive Party in Pittsburgh." Seminar paper, University of Pittsburgh, 1961.

DeVault, Ileen. "The Students of the Pittsburgh High School Commercial Department: 'From Inclination or Necessity.' " Seminar paper, University of Pittsburgh, 1979.

DiCiccio, Carmen. "The 1890s Political Realignment and Its Impact on Pittsburgh's Political Structure." Seminar paper, University of Pittsburgh, 1983.

Dillenberg, Thomas. "Pittsburgh 'Child Saving': Early Twentieth Century." Seminar paper, Carnegie Mellon University, 1973.

Donahoe, Daniel. "The Catholic Community in Pittsburgh, 1865–1890." Seminar paper, Carnegie Mellon University, n.d.

Dornan, Jaquelin. "Gateway Teachers' Strike." Seminar paper, University of Pittsburgh, 1972.

Doyle, Kelly. "Elite Spatial Change Due to Transit, 1860–1870." Seminar paper, Carnegie Mellon University, 1982.

Dries, Tom. "Poor Relief in Pittsburgh, 1840–1860." Seminar paper, Carnegie Mellon Universtiy,, 1980.

Dykstra, Ann Marie. "Region, Economy, and Party: The Roots of Policy Formation in Pennsylvania, 1820 to 1860." Ph.D. diss., University of Pittsburgh, 1988.

Early, Trisha. "The Pittsburgh Survey." Seminar paper, University of Pittsburgh, 1972.

Edelman, A. "The Dilemma of Quack Medicine in the City of Pittsburgh." Seminar paper, Carnegie Mellon University, n.d.

Elecking, Alan. "Early Times in Pittsburgh Public Schools." Seminar paper, Carnegie Mellon University, 1982.

Elliot, Jeffrey M. "The Engineers' Society of Western Pennsylvania." Seminar paper, Carnegie Mellon University, 1975.

Ellison, Duane C. "The Development of the Public School System in Pittsburgh." Seminar paper, University of Pittsburgh, 1962.

Fadgen, Joseph P. "McKees Rocks: Study in Political Change." Seminar paper, University of Pittsburgh, 1971.

Faires, Nora. "The Germans in Allegheny, 1850–1860." Seminar paper, University of Pittsburgh, 1971.

———. "The Germans in Pittsburgh in 1860." Seminar paper, University of Pittsburgh, 1972.

———. "Ethnicity in Evolution: The German Community in Pittsburgh and Allegheny City, Pennsylvania, 1845–1885." Ph.D. diss., University of Pittsburgh, 1981.

Farrington, Peter M. "The Pittsburgh Subway." Research paper, Carnegie Mellon University, 1981.

Fennell, Dorothy E. "A Factory Study." Seminar paper, University of Pittsburgh, 1975.

———. "Uncommon People." Seminar paper, University of Pittsburgh, n.d.

———. "From Rebelliousness to Insurrection: A Social History of the Whiskey Rebellion, 1765–1802." Ph.D. diss., University of Pittsburgh, 1981.

Fleckenstein, John V. "The Reluctant Revolution: A View of Popular Attitudes in Pittsburgh to the Merger of United Steel as Reflected in the Press." Seminar paper, Carnegie Mellon University, n.d.

Floccos, Alice. "The Greek Immigrant and the Greek Orthodox Church." Seminar paper, University of Pittsburgh, 1974.

Foltz, Roger N. "The Story of the Finnish Community in Monessen, Pennsylvania." Seminar paper, University of Pittsburgh, 1964.

Frank, Charles. "Dramatic Art in Pittsburgh—the Old Drury Theater as Managed by Charles S. Porter Between 1845 and 1947." Seminar paper, Carnegie Mellon University, 1980.

Gabriel, Scott. "Historic Birmingham of Allegheny County, Pa." Seminar paper, Carnegie Mellon University, 1982.

Galloway, James S. "The Negro in Politics in Pittsburgh, 1928–1940." Seminar paper, University of Pittsburgh, n.d.

———. "The *Pittsburgh Courier* and FEPC Agitation." MA thesis, University of Pittsburgh, 1954.

Garland, William E.. "Police Reform and Patrolman Behavior: The Pittsburgh Police Bureau, 1887–1899." Seminar paper, Carnegie Mellon University, 1971.

Gelfand, Mitchell. "Vertical Integration in the Pittsburgh Iron and Steel Industry, 1870–1886." Seminar paper, Carnegie Mellon University, n.d.

Giltinan, John. "Lockout in Johnstown." Seminar paper, University of Pittsburgh, 1969.

Giovengo, Annette. "Pittsburgh's Music, 1865–1895." Seminar paper, Carnegie Mellon University, 1981.

Good, Patricia K. "Irish Adjustment to American Society: Integration or Separation? A Portrait of an Irish-Catholic Parish, 1863–1886." *Records of the American Catholic Historical Society of Philadelphia* 86 (Mar.–Dec. 1975): 7–23.

Gordon, Gerd. "Pittsburgh, USA, Conditions and Lifestyles of Some Immigrant Groups at the Turn of the Twentieth Century." Seminar paper, University of Pittsburgh, 1971.

Gottlieb, Peter. "Migration and Jobs: The New Black Workers in Pittsburgh, 1916–1930." Seminar paper, University of Pittsburgh, 1974.

———. "Making Their Own Way: Southern Blacks' Migration to Pittsburgh, 1916–1930." Ph.D. diss., University of Pittsburgh, 1977.

———. *Making Their Own Way: Southern Blacks' Migration to Pittsburgh, 1916–30.* Urbana: University of Illinois Press, 1987.

Greenawald, G. Dale. "Germans in Pittsburgh, 1850, 1880, 1930: Residency, Occupations, and Assimilation." Seminar paper, Carnegie Mellon University, n.d.

Greer, Clyde P. "The Site Selection of Pittsburgh Three Rivers Stadium." Seminar paper, Carnegie Mellon University, n.d.

Gregory, George. "Sewage Treatment in Pittsburgh." Seminar paper, Carnegie Mellon University, n.d.

———. "A Study in Local Decision Making: Pittsburgh and Sewage Treatment." *Western Pennsylvania Historical Magazine* 57 (Jan. 1974): 25–42.

Hahl, Andrew. "Popular Opinion in Pittsburgh Concerning the Issues of Specie and the United States Bank." Seminar paper, Carnegie Mellon University, n.d.

Hahn, Edward H. "Social Changes in a Small Community: 1860 to 1880." (Cross Creek Township, Washington Co., Pa.) Seminar paper, University of Pittsburgh, 1974.

Hall, Bradley W. "Elites and Spatial Change in Pittsburgh: Minersville as a Case Study." *Pennsylvania History* 48 (Oct. 1981): 311–34.

Hammar, Sven. "Wilkinsburg and Edgewood: Commuter Suburbs." Seminar paper, Carnegie Mellon University, 1972.

Haner, Mark. "The Growth of the Mackintosh-Hemphill Corporation, 1840–1860." Seminar paper, Carnegie Mellon University, 1980.

Harper, Robert Eugene. "Pittsburgh and the Election of 1894." Seminar paper, University of Pittsburgh, 1961.

———. "Fayette County, 1790–1793: A Study of the Economic Base and Local Government." Seminar paper, University of Pittsburgh, 1962.

———. "The Class Structure of Western Pennsylvania in the Late Eighteenth Century." Ph.D. diss., University of Pittsburgh, 1969.

———. "Town Development in Early Western Pennsylvania." *Western Pennsylvania Historical Magazine* 71 (Jan. 1988): 3–26.

Harris, Robert. "The Sculling Rivalry Between Joshua War and James Hamil." Seminar paper, Carnegie Mellon University, n.d.

Henry, Tom. "The Pittsburgh Architect—1910 and Municipal Improvement." Seminar paper, University of Pittsburgh, 1964.

Hiller, Jack L., Jarrell McCracken, and Ted C. Soens. "Morningside: An Urban Village." Seminar paper, Carnegie Mellon University, 1969.

Hobart, Buddy. "Nineteenth-Century Church Politics." Seminar Paper, Carnegie Mellon University, n.d.

Hoffman, Jerry H. "Andrew Carnegie: A Study of His Expressed Political Attitudes." Seminar paper, University of Pittsburgh, n.d.

Holmberg, James C. "The Industrializing Community: Pittsburgh, 1850–1880." Ph.D. diss., University of Pittsburgh, 1981.

Huey, Mary C. "Occupational and Nationality Structure of McKeesport, 1880." Seminar paper, University of Pittsburgh, n.d.

Hughes, Allen R. "Crime, Law Enforcement, and Penology in Pittsburgh, with Emphasis on the Period 1846–1848." Seminar paper, University of Pittsburgh, 1962.

Hutzler, Bette. "Pittsburgh in the Civil War Decade." Seminar paper, Carnegie Mellon University, 1984.

Ingham, John Norman. "Iron and Steel Families of Pittsburgh, 1875–1960." Seminar paper, University of Pittsburgh, 1964.

——. "A Strike in the Progressive Era: McKees Rocks, 1909." *Pennsylvania Magazine of History and Biography* 90 (July 1966): 353–77.

——. "Elite and Upper Class in the Iron and Steel Industry, 1874 to 1965." Ph.D. diss., University of Pittsburgh, 1973.

——. *The Iron Barons: A Social Analysis of an American Urban Elite, 1874–1965.* Westport, Conn.: Greenwood Press, 1978.

Jackson-Mann, Marvin. "Oakland, the Sixties and Beyond." Seminar paper, Carnegie Mellon University, n.d.

James, William A. "The History of Urban Transportation in Pittsburgh and Allegheny County with Emphasis on the Major Technological Developments." MA thesis, University of Pittsburgh, 1947.

Jirak, Ivan. "The Effects of the New Immigrants on Pittsburgh." Seminar paper, Carnegie Mellon University, n.d.

Johnston, Joseph. "National Origins and Ethnic Groups of the People of Allegheny, Pennsylvania, in 1880." Seminar paper, University of Pittsburgh, n.d.

Judd, Barbara. "The Search for Order in the City Beautiful and the City Efficient." Seminar paper, Carnegie Mellon University, 1970.

——. "Edward M. Bigelow: Creator of Pittsburgh's Arcadian Parks as Instruments of Social Control." Seminar Paper, Carnegie Mellon University, 1971.

——. "Edward M. Bigelow: Creator of Pittsburgh's Arcadian Parks." *Western Pennsylvania Historical Magazine* 58 (Jan. 1975): 53–67.

Kalloway, Lois J. "Sisters All: Polish, Irish, German and Lithuanian Religious Sisters in Pittsburgh, 1890–1940." Ph.D. diss., University of Pittsburgh, in progress.

Kammer, Richard C. "A Population Study of Wheeling Island in the Industrializing Era of 1859–1890." Seminar paper, Carnegie Mellon University, 1972.

Kanitra, Diane. "The Westinghouse Strike of 1916." Seminar paper, University of Pittsburgh, 1971.

Kaplan, Robert. "The Know-Nothings in Pittsburgh." Seminar paper, University of Pittsburgh, 1977.

Kelso, Thomas. "Allegheny Elites: 1850–1907." Seminar paper, University of Pittsburgh, 1964.

——. "Pittsburgh's Mayors and City Councils, 1794–1844: Who Governed?" Seminar paper, University of Pittsburgh, 1963.

Kelts, Gail. "An Analysis of the Election of the Mayor of Pittsburgh in 1906." Seminar paper, University of Pittsburgh, 1961.

Kern, Frank. "History of Pittsburgh Water Works, 1821–1842." Seminar paper, Carnegie Mellon University, 1982.

Kerr, Kathel A. "World War I and Pittsburgh: The Impact of Efficient Mobilization." Seminar paper, University of Pittsburgh, 1961.

Kimmins, Hal. "Joseph Barker, Mayor of Pittsburgh, 1850–51." Seminar paper, University of Pittsburgh, 1963.

————. "Westmoreland County, 1783–1790: A Study of the Economic Base and Local Government." Seminar paper, University of Pittsburgh, 1964.

Kleinberg, S. J. "A Study of a Women's Organization: The Women's Christian Temperance Union of Wilkinsburg, Pa." Seminar paper, University of Pittsburgh, 1970.

————. "Technology's Stepdaughters: The Impact of Industrialization Upon Working-Class Women, Pittsburgh, 1865–1890." Ph.D. diss., University of Pittsburgh, 1973.

————. "Technology and Women's Work: The Lives of Working-Class Women in Pittsburgh, 1870–1900." *Labor History* 17 (Winter 1976): 58–66.

————. *The Shadow of the Mills: Working-Class Families in Pittsburgh, 1870–1907.* Pittsburgh: University of Pittsburgh Press, 1989.

Kleppner, Paul. "Lincoln and the Immigrant Vote: A Case of Religious Polarization." *Mid-America* 48 (July 1966): 176–95.

Knapp, Mark. "Urban Changes Through Civil Litigation in Pittsburgh, 1865–1895." Seminar paper, Carnegie Mellon University, 1981.

Krause, Corinne Azen. "Italian, Jewish, and Slavic Grandmothers in Pittsburgh: Their Economic Roles." *Frontiers* 2 (Summer 1977): 15–23

————. "Urbanization Without Breakdown: Italian, Jewish, and Slavic Immigrant Women in Pittsburgh, 1900 to 1945." *Journal of Urban History* 4 (May 1978): 291–306.

————. *Grandmothers, Mothers, and Daughters: An Oral History Study of Ethnicity, Mental Health, and Continuity of Three Generations of Jewish, Italian, and Slavic-American Women.* New York: American Jewish Committee, n.d.

Kreidler, Robert T. "Managerial Displacement in the United States Steel Corporation and Its Relation to Reform, 1901–1920." Seminar paper, University of Pittsburgh, n.d.

Kreiling, Judith A.. "Universal: A Study of a Proposed Redevelopment Area in Penn Hills Township." Seminar paper, Carnegie Mellon University, 1969.

Kuo, Thomas. "Pittsburgh Attitudes Toward the Recharter of the Second Bank of the United States." Seminar paper, University of Pittsburgh, 1962.

Kurtz, Beth. "Women's Charity Work: Its Place and Value in America During the Second Half of the Nineteenth Century." Seminar paper, Carnegie Mellon University, 1980.

Lammie, Wayne. "Political Attitudes of Small Pittsburgh Merchants in the Progressive Era." Seminar paper, University of Pittsburgh, n.d.

Lamperes, Bill. "Smoke Control: Keystone of the Renaissance." Seminar paper, Carnegie Mellon University, 1978.

Larner, John W., Jr. "A Community in Transition: Pittsburgh's South Side, 1880–1920." Seminar paper, University of Pittsburgh, 1961.

Lasser, Ronald. "The Electrification of the City of Pittsburgh, 1880–1930." Seminar paper, Carnegie Mellon University, 1982.

LaVere, David B. "The Class Nature of Leisure in Pittsburgh, 1850–1914." Seminar paper, Carnegie Mellon University, 1979.

Lawther, Dennis E. "The Impact of the Monongahela Incline Plane on the Social, Economic, and Physical Development of Mt. Washington, 1870–1910." Seminar paper, Carnegie Mellon University, 1977.

Lazarus, Stuart. "The Functions of Boosters in Emerging Industrial Cities: Booster Activities in Pittsburgh, 1850–1910." Seminar paper, Carnegie Mellon University, 1973.

Lehman, D. "The Jewish Community in Pittsburgh During the Civil War." Seminar paper, Carnegie Mellon University, n.d.

Lettre, Michel, Jerry Kleinman, Bobby Bloom, and Kathy Sloss. "Relocation in the Urban Renewal Process." Seminar paper, Carnegie Mellon University, 1969.

Lettrich, Pam. "The Natural Gas Boom in Pittsburgh." Seminar paper, Carnegie Mellon University, 1982.

Lichtenberg, Mitchell P. "Persistence and Migration in a Nineteenth-Century Ward." Seminar paper, Carnegie Mellon University, 1967.

Linaberger, James. "The Rolling Mill Riots of 1850." Seminar paper, University of Pittsburgh, n.d.

Lloyd, Anne. "Pittsburgh's 1923 Zoning Ordinance." *Western Pennsylvania Historical Magazine* 57 (July 1974): 289–306.

Lloyd, Paul. "A Study of Social Control in Bethlehem Steel Company and the Westinghouse Companies, 1900–1916." Seminar paper, University of Pittsburgh, n.d.

Lonich, David. "A General Statement About Working-Class Housing in Pittsburgh, 1900–1910." Seminar paper, University of Pittsburgh, 1979.

Lukaszewicz, Frank. "Regional and Central Boards of Directors of Pittsburgh Banks in 1912." Seminar paper, University of Pittsburgh, 1966.

Lynch, Patrick M. "The Pittsburgh Stogy Industry in Transition and the IWW, 1906–1920." Seminar paper, University of Pittsburgh, 1976.

———. "Pittsburgh, the I.W.W., and the Stogie Workers." In *At the Point of Production: The Local History of the I.W.W.* Ed. Joseph R. Conlin. Westport, Conn.: Greenwood Press, 1981.

McCauley, Kevin R. "A History of Port Authority Transit, 1964–1980." Seminar paper, Carnegie Mellon University, 1985.

MeClain, Raymond. "The Immigrant Years: Irene Kaufmann Settlement, 1895–1915." Seminar paper, Carnegie Mellon University, 1969.

McDormott, Harriet. "The Steel Strike of 1919: Newspaper Coverage." Seminar paper, University of Pittsburgh, 1983.

McIlvaine, Josephine. "Twelve Blocks: A Study of One Segment of the South Side of Pittsburgh, 1880–1915." *Western Pennsylvania Historical Magazine* 60 (Oct. 1977): 35–70.

McMillan, Walter G. "Location of Steel Manufacturers in Pittsburgh–Allegheny County in 1881–1884 and 1915–1916." Seminar paper, Carnegie Mellon University, 1975.

McPherson, Donald Scott. "Mechanics' Institutes and the Pittsburgh Working Man, 1830–1840." Seminar paper, University of Pittsburgh, 1972.

———. "The Fight Against Free Schools in Pennsylvania: Popular Opposition to the Common School System, 1834–1874." Ph.D. diss., University of Pittsburgh, 1977.

Makarewicz, Joseph Thomas. "The Impact of World War I on Pennsylvania Politics with an Emphasis on the Election of 1920." Ph.D. diss., University of Pittsburgh, 1972.

Martin, Scott C. "Fathers Against Sons, Sons Against Fathers: Antimasonry in Pittsburgh." Seminar paper, University of Pittsburgh, n.d.

Matthews, Michael D.. "The Pittsburgh Brick Industry, 1865–1890." Seminar paper, Carnegie Mellon University, 1981.

Maxwell, Marianne. "Pittsburgh's Frick Park: A Unique Addition to the City's Park System." *Western Pennsylvania Historical Magazine* 68 (July 1985): 243–64.

Mayfield, Loomis. " 'The Figures Speak for Themselves': The Institution of Personal Registration for Voting in Pittsburgh, Pennsylvania, 1906." Seminar paper, University of Pittsburgh, 1984.

Medvin, Joshua. "The Development of District Heating in Pittsburgh: A Case Study." Seminar paper, Carnegie Mellon University, 1980.

Merges, Robert. "The Enemy Within: Civil Rights in Pittsburgh During the Civil War." Seminar paper, Carnegie Mellon University, 1979.

Messer, Ross. "The Medical Profession and Urban Reform in Pittsburgh, 1890–1920." Seminar paper, University of Pittsburgh, 1964.

Metzger, Elizabeth A. "A Study of Social Settlement Workers in Pittsburgh, 1893 to 1927." Seminar paper, University of Pittsburgh, 1974.

Miller, Joan. "The Early Historical Development of Hazelwood." Seminar paper, University of Pittsburgh, n.d.

Misko, Louise. "A Study of Political Activities and Attitudes of Pittsburgh Poles Relative to Achieving the Independence of Poland Through Preservation of Religious, Fraternal, and Cultural Institutions." Seminar paper, University of Pittsburgh, 1975.

Miskoff, Ronald. "The Building Trades and the Contractors." Seminar paper, University of Pittsburgh, 1970.

Mitchell, Anne. "Jane Grey Swisshelm: An Independent Commentator on Abolitionism." Seminar paper, Carnegie Mellon University, 1982.

Mizrahl, Laurie. "The History of the Jewish Community of Pittsburgh, 1847–1890." Seminar paper, Carnegie Mellon University, 1981.

Moser, Edwin M. "Jewish Labor in Pittsburgh: Its Ideology and Its Relation to the National Scene, 1905–1914." Seminar paper, University of Pittsburgh, 1961.

Mulcahy, Richard. "The Turner and Colodny Cases: A Comparative Study in Academic Freedom and Power at the University of Pittsburgh." Seminar paper, University of Pittsburgh, 1984.

Murray, Susan L. "Transportation and Recreation: Streetcar Systems in Pittsburgh." Seminar paper, Carnegie Mellon University, 1984.

Myers, Glenn E. "Reflections of the Revival of 1857–1858 in Greater Pittsburgh." Seminar paper, Carnegie Mellon University, 1980.

Myers, Miller. "An Analysis of Voting Behavior in Pittsburgh, 1848–1856." Seminar paper, University of Pittsburgh, 1963.

Nathan, Mitchell A. "The Jewish Community of Pittsburgh: A Beginning." Seminar paper, Carnegie Mellon University, 1982.

Nellis, Catherine M. "The Roots of Morewood Heights." Seminar paper, Carnegie Mellon University, 1980.

Netterville, George. "Pittsburgh Mayors." Seminar paper, Carnegie Mellon University, 1969.

Nyden, Linda. "Black Miners in Western Pennsylvania, 1925–1931: The National Miners Union and the United Mine Workers of America." Seminar paper, University of Pittsburgh, 1974.

———. "Women Electrical Workers at Westinghouse Electric Corporation's East Pittsburgh Plant, 1907–1945." Seminar paper, University of Pittsburgh, 1975.

———. "Black Miners in Western Pennsylvania, 1925–1931: The National Miners Union and the United Mine Workers of America." *Science and Society* 41 (Spring 1977): 69–101.

O'Connor, Richard. "Cinderheads and Iron Lungs: Window Glassworkers and Their Unions, 1865–1920." Ph.D. diss., University of Pittsburgh, in progress.

Osborn, Gina. "The W.C.T.U. of Pittsburgh, 1874–1875." Seminar paper, Carnegie Mellon University, n.d.

Owens, Barbara C. "The Consolidation of Allegheny." Seminar paper, Carnegie Mellon University, n.d.

Pailthorpe, Michelle. "The German-Jewish Elite of Pittsburgh: Its Beginings and Background." Seminar paper, University of Pittsburgh, 1967.

Pearce, George F., and Jim McCabe. "Two Census Tracts in Pittsburgh's Ward Seven: A Study in Contrast." Seminar paper, Carnegie Mellon University, 1969.

Peiffer, Layne. "The German Upper Class in Pittsburgh, 1850–1920." Seminar paper, University of Pittsburgh, 1964.

Peles, Robert. "Crisis in Johnstown: The 'Little Steel' Strike of 1937." Seminar paper, University of Pittsburgh, 1974.

———. "Labor Interlude—Johnstown 1919." Seminar paper, University of Pittsburgh, 1974.

Penna, Anthony N. "Changing Images of Twentieth-Century Pittsburgh." *Pennsylvania History* 43 (Jan. 1976): 49–63.

Perkins, Virginia. "A Political Analysis of Baldwin Township." Seminar paper, University of Pittsburgh, 1970.

Peterson, Roger. "The Reaction to a Heterogeneous Society: A Behavioral and Quantitative Analysis of Northern Voting Behavior, 1845–1870: Pennsylvania, A Test Case." Ph.D. diss., University of Pittsburgh, 1970.

Phipps, Robert. "The Building of the Boulevard: A Roadway to the Suburbs." Seminar paper, Carnegie Mellon University, 1972.

Price, Tom. "The Westinghouse Strikes of 1914 and 1916: Workers' Control in America?" Seminar paper, University of Pittsburgh, 1983.

Pritchard, Linda K. "Ministers: A Comparative Framework." Seminar paper, University of Pittsburgh, 1970.

———. "Presbyterian Clergy in the Nineteenth Century." Seminar paper, University of Pittsburgh, 1970.

———. "Religious Change in a Developing Region: The Social Contexts of Evangelicalism in Western New York and the Upper Ohio Valley During the mid-Nineteenth Century." Ph.D. diss., University of Pittsburgh, 1980.

Reitman, Renee. "The Elite Community in Shadyside, 1880–1920." Seminar paper, University of Pittsburgh, 1964.

Renner, Marguerite. "A Study of Women's Participation in Voluntary Organziations." Seminar paper, University of Pittsburgh, 1971.

———. "Who Will Teach? Changing Job Opportunity and Roles for Women in the Evolution of the Pittsburgh Public Schools, 1830–1900." Ph.D. diss., University of Pittsburgh, 1981.

Rishel, Joseph Francis. "The Founding Families of Allegheny County: An Examination of Nineteenth-Century Elite Continuity." Ph.D. diss., University of Pittsburgh, 1975.

Rizika, Adam. "Housing Conditions and Reforms During the Years 1865–1900." Seminar paper, Carnegie Mellon University, 1981.

Romig, Robert. "The Role of Women in Government and Politics of Pennsylvania from 1880 to 1920." Seminar paper, University of Pittsburgh, 1964.

Rosen, Philip. "Thirty Years at Kingsley House." Seminar paper, Carnegie Mellon University, 1969.

Rosenfeld, Renee. "Theatres and the Production of Shakespeare in Pittsburgh, 1865–1890." Seminar paper, Carnegie Mellon University, 1981.

Rosner, Leeann. "The Pittsburgh Temperance Movement, 1830–1850." Seminar paper, Carnegie Mellon University, 1982.

———. "Pittsburgh Strip District." Seminar paper, Carnegie Mellon University, 1984.

Rotblatt, Mark. "Pittsburgh's Black Community, 1850–1860, and How It Changed in Ten Years." Seminar paper, Carnegie Mellon University, 1982.

Ruck, Robert Lewis. "Origins of the Seniority System in Steel." Seminar paper, University of Pittsburgh, 1977.

———. "Soaring Above the Sandlots: The Garfield Eagles." *Pennsylvania Heritage* 8 (Summer 1982): 13–18.

———. "Sandlot Seasons: Sport in Black Pittsburgh." 2 vols. Ph.D. diss., University of Pittsburgh, 1983.

———. *Sandlot Seasons: Sport in Black Pittsburgh.* Urbana: University of Illinois Press, 1987.

Sabol, Richard A. "Public Works in Pittsburgh Prior to the Establishment of the Department of Public Works." Seminar paper, Carnegie Mellon University, 1980.

Salsgiver, Richard O. "Crime and the Nature of Policing in Allegheny, 1884–1893." Seminar paper, Carnegie Mellon University, 1971.

Santos, Michael. "Iron Workers in the Steel Era: The Case of A. M. Byers Company, 1900–1969." D.A. diss., Carnegie Mellon University, 1985.

Sauers, Bernard J. "A Political Process of Urban Growth: Consolidation of the South Side with the City of Pittsburgh, 1972." *Pennsylvania History* 41 (July 1974): 265–87.

Scheerer, Laura. "The Effect of the Skyscraper in Pittsburgh's Downtown, 1890–1910." Seminar paper, Carnegie Mellon University, 1984.

Schmidt, Salvin. "The Telephone Comes to Pittsburgh." MA thesis, University of Pittsburgh, 1948.

Schubert, Jane Gray. "A Man and His City." (Frederick Bigger.) Seminar paper, Carnegie Mellon University, 1971.

Schuchman, Mary J. "Medical Care in Pittsburgh, 1860–1861." Seminar paper, Carnegie Mellon University, 1979.

Schuchman, Stephen J. "The Elite at Sewickley Heights, 1900–1940." Seminar paper, University of Pittsburgh, 1964.

Schultz, Mindella. "Pittsburgh Confronts the Problem of Unemployment Relief." Seminar paper, Carnegie Mellon University, n.d.

Schumacher, Carolyn S. "Education and Social Mobility: Class and Occupation of Nineteenth-Century High School Students." Seminar paper, University of Pittsburgh, 1970.

———. "School Attendance in Nineteenth-Century Pittsburgh: Wealth, Ethnicity, and Occupational Mobility of School Age Children, 1855–1965." Ph.D. diss., University of Pittsburgh, 1977.

Schwartz, Rachel. "Father James Renshaw Cox." Seminar paper, Carnegie Mellon University, 1969.

Selavan, Ida Cohen. "Immigrant Education in Pittsburgh: An Analysis of Some Roll Books from Grant School 'Foreign' Classes." Seminar paper, University of Pittsburgh, n.d.

———. "The Jewish Labor Movement in Pittsburgh." Seminar paper, University of Pittsburgh, 1971.

———. "The Social Evil in an Industrial Society: Prostitution in Pittsburgh, 1900–1925." Seminar paper, University of Pittsburgh, 1971.

———. "The Founding of the Columbian Council." *American Jewish Archives* 30 (April 1978): 26–27.

Serene, Frank Huff. "Immigrant Steelworkers in the Monongahela Valley: Their Communities and the Development of a Labor Class Consciousness." Ph.D. diss., University of Pittsburgh, 1979.

Sherman, Ralph. "A Tale of One City: Pittsburgh, Several of Her Mayors, and Some of Their Times." Seminar paper, Carnegie Mellon University, 1981.

Siegel, Fred. "Selective Out-Migration of the Fourth Ward of Allegheny City, 1850–1860." Seminar paper, University of Pittsburgh, n.d.

Siemon, Jeffrey O. "Division Within the Sixth Presbyterian Church, Pittsburgh, Pa., 1850–1862." Seminar paper, Carnegie Mellon University, 1982.

Simpson, Patricia. "The Drunk and the Teetotaler: Two Phases in Temperance Reform Among the Irish Working Class of Pittsburgh." Seminar paper, University of Pittsburgh, 1977.

Sing, Nina H. "The Allegheny Cotton Mill Riot of 1848: The Beginning of a Trend." Seminar paper, Carnegie Mellon University, 1980.

Skud, Bruce. "Ethnicity and Residence Within the Jewish Immigrant Community." Seminar paper, University of Pittsburgh, 1975.

Sloan, Elaine F. "The Political Behavior of Mining Communities." Seminar paper, University of Pittsburgh, 1962.

Smith, Roland M. "The Politics of Pittsburgh Flood Control, 1936–1960." Seminar paper, Carnegie Mellon University, n.d.

———. "The Politics of Flood Control on the Allegheny and Monongahela Rivers at Pittsburgh, 1908–1938." Seminar paper, Carnegie Mellon University, 1971.

———. "The Politics of Pittsburgh Flood Control, 1908–1936." *Pennsylvania History* 42 (Jan. 1975): 5–24.

———. "The Politics of Flood Control, 1936–1960." *Pennsylvania History* 44 (Jan. 1977): 4–7.

Soens, Ted C. "Hazelwood in the 1890s." Seminar paper, Carnegie Mellon University, n.d.

Sporn, Nina. "Artistic Views of Pittsburgh, 1840–1860." Seminar paper, Carnegie Mellon University, n.d.

Spragg, Edwin Brownlee. "Antislavery Sentiment in Washington County in the 1830s." MA thesis, University of Pittsburgh, 1962.

Stave, Bruce M. "The New Deal and the Building of an Urban Political

Machine: Pittsburgh, a Case Study." Ph.D. diss., University of Pittsburgh, 1966.

———. *The New Deal and the Last Hurrah: Pittsburgh Machine Politics.* Pittsburgh: University of Pittsburgh Press, 1970.

Storch, Howard V., Jr. "Changing Functions of the Center-City: Pittsburgh, 1850–1912." Seminar paper, University of Pittsburgh, 1966.

Stromquist, Shelton. "Working-Class Organization and Industrial Change in Pittsburgh, 1860–1890: Some Themes." Seminar paper, University of Pittsburgh, 1973.

Stuart, Jim. "The History of the Pittsburgh Park System from 1867 to 1893." Seminar paper, Carnegie Mellon University, 1981.

Szwarc, Carole T. "Manchester: A Study in Contrast, 1930–1968." Seminar paper, Carnegie Mellon University, n.d.

Taylor, Donald L. "The Woods Run Settlement, 1895–1932." Seminar paper, Carnegie Mellon University, 1970.

Thompson, Marion. "Accident Reforms in the United States Steel Corporation." Seminar paper, Carnegie Mellon University, 1969.

Tierno, Mark J. "The Search for Pure Water in Pittsburgh: The Urban Response to Water Pollution, 1893–1914." *Western Pennsylvania Historical Magazine* 60 (Jan. 1977): 23–36.

Trusilio, Sharon. "William Martin, First Executive Secretary of the Amalgamated Association of Iron and Steel Workers." Ph.D. diss., Carnegie Mellon University, in process.

Turner, Barbara. "Organized Labor in Pittsburgh, 1826–1837." Seminar paper, University of Pittsburgh, 1973.

Twiss, Harold L. "The Development of Missionary Support by Baptist Churches and Associations in Western Pennsylvania, 1815–1845." Seminar paper, University of Pittsburgh, 1963.

———. "The Pittsburgh Business Elite, 1850–1890–1929." Seminar paper, University of Pittsburgh, 1964.

Vafis, John. "Socialist and Progressive Parties: Election of 1912 in Pittsburgh." Seminar paper, University of Pittsburgh, 1962.

Wallhausser, Fred. "The Upper-Class Society of Sewickley Valley, 1830–1910." Seminar paper, University of Pittsburgh, 1964.

Walsh, Victor Anthony. "Class, Culture, and Nationalism: The Irish Catholics of Pittsburgh, 1870–1883." Seminar paper, University of Pittsburgh, 1976.

———. " 'A Fanatic Heart': The Cause of Irish-American Nationalism in Pittsburgh During the Guilded Age." *Journal of Social History* 15 (Winter 1981): 187–204.

———. "Across the 'Big Wather' ": Irish Community Life in Pittsburgh and

Allegheny City, 1850–1885." Ph.D. diss., University of Pittsburgh, 1983.

———. "Across the 'Big Wather' ": The Irish-Catholic Community of Mid-Nineteenth-Century Pittsburgh." *Western Pennsylvania Historical Magazine* 66 (Jan. 1983): 1–23.

Warner, Marc. "Where the Pittsburgh Subway Hid: An Analysis of Seventy-five Years of Indecision." Seminar paper, Carnegie Mellon University, 1981.

Watson, Mary R. "The Abolitionist Movement in Pittsburgh, 1840–1860, with Particular Emphasis on the Western Pennsylvania Antislavery Society." Seminar paper, Carnegie Mellon University, 1980.

Weber, Michael P. "Patterns of Progress: Social Mobility in a Pennsylvania Oil Town, 1870–1910." Ph.D. diss., Carnegie Mellon University, 1972.

———. *Social Change in an Industrial Town: Patterns of Progress in Warren, Pennsylvania, from the Civil War to World War I.* University Park: Pennsylvania State University Press, 1976.

———. "Residential and Occupational Patterns of Ethnic Minorities in Nineteenth-Century Pittsburgh." *Pennsylvania History.* 44 (Oct. 1977): 316–34.

———. *Don't Call Me Boss: David L. Lawrence, Pittsburgh's Renaissance Mayor.* Pittsburgh: University of Pittsburgh Press, 1988.

Weingartner, Sharon. "Technology of Pleasure: The Beginnings of Amusement Parks and Moving Pictures." Seminar paper, Carnegie Mellon University, n.d.

Werner, Philip R. "The Transformation of Urban Land Use by Bridges: Pittsburgh, 1800–1860." Seminar paper, Carnegie Mellon University, 1982

Williams, Douglas M. "The Use of the Telegraph and Telephone by the Fire and Police Departments of Pittsburgh." Seminar paper, Carnegie Mellon University, 1984.

Williams, Jim. "The Control of Public Schools by a Business Elite: Pittsburgh, 1912–1929." Seminar paper, Carnegie Mellon University, 1968.

Williamson, David. "The Early History of Carnegie Borough." Seminar paper, Carnegie Mellon University, n.d.

Wolf, Noel R. "Period of Transition: The Entry of Railroads Into Pittsburgh." Seminar paper, Carnegie Mellon University, 1982.

Wood, Diana M. "A Case Study in Local Control of Schools: Pittsburgh, 1900–1906." *Urban Education* 10 (Apr. 1975): 7–26.

Yosie, Terry F. "Retrospective Analysis of Water Supply and Wastewater Policies in Pittsburgh, 1800–1959." D.A. diss., Carnegie Mellon University, 1981.

Young, Mary. "The Pittsburgh Chamber of Commerce and the Allied Boards of Trade in 1910." Seminar paper, University of Pittsburgh, 1966.

Zabrosky, Frank. "Some Aspects of Negro Civic Leadership in Pittsburgh, 1955–65." Seminar paper, University of Pittsburgh, n.d.

Zarychta, Ronald M. "Municipal Reorganization: The Pittsburgh Fire Department as a Case Study." *Western Pennsylvania Historical Magazine* 58 (Oct. 1975): 471–86.

▪ Index

459

Pittsburgh Series in Social and Labor History
Maurine Weiner Greenwald, Editor

And the Wolf Finally Came: The Decline of the American Steel Industry
John P. Hoerr

City at the Point: Essays on the Social History of Pittsburgh
Samuel P. Hays, Editor

The Correspondence of Mother Jones
Edward M. Steel, Editor

Distribution of Wealth and Income in the United States in 1798
Lee Soltow

Don't Call Me Boss: David L. Lawrence, Pittsburgh's Renaissance Mayor
Michael P. Weber

*The Shadow of the Mills: Working-Class Families in Pittsburgh,
1870–1907*
S. J. Kleinberg

The Speeches and Writings of Mother Jones
Edward M. Steel, Editor

The Steel Workers
John A. Fitch

Trade Unions and the New Industrialisation of the Third World
Roger Southall, Editor

What's a Coal Miner to Do? The Mechanization of Coal Mining
Keith Dix

Women and the Trades
Elizabeth Beardsley Butler

Other titles in the series

The Emergence of a UAW Local, 1936–1939: A Study in Class and Culture
Peter Friedlander

Homestead: The Households of a Mill Town
Margaret F. Byington

The Homestead Strike of 1892
Arthur G. Burgoyne

*Immigration and Industrialization: Ethnicity in an American Mill Town,
1870–1940*
John Bodnar

Out of This Furnace
Thomas Bell

Steelmasters and Labor Reform, 1886–1932
Gerald G. Eggert

Steve Nelson, American Radical
Steve Nelson, James R. Barrett, and Rob Ruck

*Working-Class Life: The "American Standard" in Comparative
Perspective, 1899–1913*
Peter R. Shergold